113

Advances in Biochemical Engineering/Biotechnology

Series Editor: T. Scheper

Advances in Biochemical Engineering/Biotechnology

Series Editor: T. Scheper

Recently Published and Forthcoming Volumes

Biotechnology in China I
From Bioreaction to Bioseparation and Bioremediation
Volume Editors: Zhong, J.-J., Bai, F.-W., Zhang, W.
Vol. 113, 2009

Bioreactor Systems for Tissue Engineering
Volume Editors: Kasper, C., van Griensven, M., Poertner, R.
Vol. 112, 2008

Food Biotechnology
Volume Editors: Stahl, U., Donalies, U. E. B., Nevoigt, E.
Vol. 111, 2008

Protein – Protein Interaction
Volume Editors: Seitz, H., Werther, M.
Vol. 110, 2008

Biosensing for the 21st Century
Volume Editors: Renneberg, R., Lisdat, F.
Vol. 109, 2007

Biofuels
Volume Editor: Olsson, L.
Vol. 108, 2007

Green Gene Technology
Research in an Area of Social Conflict
Volume Editors: Fiechter, A., Sautter, C.
Vol. 107, 2007

White Biotechnology
Volume Editors: Ulber, R., Sell, D.
Vol. 105, 2007

Analytics of Protein-DNA Interactions
Volume Editor: Seitz, H.
Vol. 104, 2007

Tissue Engineering II
Basics of Tissue Engineering and Tissue Applications
Volume Editors: Lee, K., Kaplan, D.
Vol. 103, 2007

Tissue Engineering I
Scaffold Systems for Tissue Engineering
Volume Editors: Lee, K., Kaplan, D.
Vol. 102, 2006

Cell Culture Engineering
Volume Editor: Hu,W.-S.
Vol. 101, 2006

Biotechnology for the Future
Volume Editor: Nielsen, J.
Vol. 100, 2005

Gene Therapy and Gene Delivery Systems
Volume Editors: Schaffer, D.V., Zhou,W.
Vol. 99, 2005

Sterile Filtration
Volume Editor: Jornitz, M.W.
Vol. 98, 2006

Marine Biotechnology II
Volume Editors: Le Gal, Y., Ulber, R.
Vol. 97, 2005

Marine Biotechnology I
Volume Editors: Le Gal, Y., Ulber, R.
Vol. 96, 2005

Microscopy Techniques
Volume Editor: Rietdorf, J.
Vol. 95, 2005

Regenerative Medicine II
Clinical and Preclinical Applications
Volume Editor: Yannas, I. V.
Vol. 94, 2005

Biotechnology in China I

From Bioreaction to Bioseparation and Bioremediation

Volume Editors:
Jian-Jiang Zhong · Feng-Wu Bai · Wei Zhang

With contributions by

J. Chen · J. Feng · E.-J. Hahn · F. Li · G.-Q. Lin · F.-F. Liu
X.-W. Liu · W.-Y. Lu · C. Ma · H.N. Murthy · B.-J. Ni
K.-Y. Paek · Y.-C. Shen · G.-P. Sheng · Q.-H. Shi · Y. Sun
J.-H. Xiao · J.-H. Xu · P. Xu · B. Yu · H.-L. Yu · H.-Q. Yu
Y. Zhang · R.-C. Zheng · Y.-G. Zheng · J.-J. Zhong

Springer

Editors
Prof. Dr. Jian-Jiang Zhong
Shanghai Jiao Tong University
School of Life Sciences & Biotechnology
800 Dongchuan Road
Minhang, 200240 Shanghai
China, People's Republic
jjzhong@sjtu.edu.cn

Wei Zhang
Chinese Academy of Sciences
Dalian Institute of Chemical Physics
116023 Dalian
China, People's Republic
weizhang@dicp.ac.cn

Feng-Wu Bai
Dalian University of Technology
Department of Bioscience & Bioengineering
116023 Dalian
China, People's Republic
fwbai@dlut.edu.cn

ISSN 0724-6145 e-ISSN 1616-8542
ISBN 978-3-540-88414-9 e-ISBN 978-3-540-88415-6
DOI 10.1007/978-3-540-88415-6
Springer Heidelberg Dordrecht London New York

Library of Congress Control Number: 2009933991

Cover design: WMXDesign GmbH, Heidelberg, Germany

Printed on acid-free paper

Springer is part of Springer Science+Business Media (www.springer.com)

Series Editor

Prof. Dr. T. Scheper

Institute of Technical Chemistry
University of Hannover
Callinstraße 3
30167 Hannover, Germany
scheper@iftc.uni-hannover.de

Volume Editors

Prof. Dr. Jian-Jiang Zhong

Shanghai Jiao Tong University
School of Life Sciences &
Biotechnology
800 Dongchuan Road
Minhang, 200240 Shanghai
China, People's Republic
jjzhong@sjtu.edu.cn

Prof. Dr. Wei Zhang

Chinese Academy of Sciences
Dalian Institute of Chemical Physics
116023 Dalian
China, People's Republic
weizhang@dicp.ac.cn

Prof. Dr. Feng-Wu Bai

Dalian University of Technology
Department of Bioscience & Bioengineering
116023 Dalian
China, People's Republic
fwbai@dlut.edu.cn

Honorary Editors

Advances in Biochemical Engineering/ Biotechnology Also Available Electronically

Advances in Biochemical Engineering/Biotechnology is included in Springer's eBook package *Chemistry and Materials Science*. If a library does not opt for the whole package the book series may be bought on a subscription basis. Also, all back volumes are available electronically.

For all customers who have a standing order to the print version of *Advances in Biochemical Engineering/Biotechnology*, we offer the electronic version via SpringerLink free of charge.

If you do not have access, you can still view the table of contents of each volume and the abstract of each article by going to the SpringerLink homepage, clicking on "Chemistry and Materials Science," under Subject Collection, then "Book Series," under Content Type and finally by selecting *Advances in Biochemical Bioengineering/Biotechnology*

You will find information about the

- Editorial Board
- Aims and Scope
- Instructions for Authors
- Sample Contribution

at springer.com using the search function by typing in *Advances in Biochemical Engineering/Biotechnology*.

Color figures are published in full color in the electronic version on SpringerLink.

Aims and Scope

Advances in Biochemical Engineering/Biotechnology reviews actual trends in modern biotechnology.

Its aim is to cover all aspects of this interdisciplinary technology where knowledge, methods and expertise are required for chemistry, biochemistry, microbiology, genetics, chemical engineering and computer science.

Special volumes are dedicated to selected topics which focus on new biotechnological products and new processes for their synthesis and purification. They give the state-of-the-art of a topic in a comprehensive way thus being a valuable source for the next 3-5 years. It also discusses new discoveries and applications.

In general, special volumes are edited by well known guest editors. The series editor and publisher will however always be pleased to receive suggestions and supplementary information. Manuscripts are accepted in English.

In references *Advances in Biochemical Engineering/Biotechnology* is abbreviated as *Adv. Biochem. Engin./Biotechnol.* and is cited as a journal.

Special volumes are edited by well known guest editors who invite reputed authors for the review articles in their volumes.

Impact Factor in 2008: 3,569; Section "Biotechnology and Applied Microbiology": Rank 48 of 138

Attention all Users of the "Springer Handbook of Enzymes"

Information on this handbook can be found on the internet at springeronline.com

A complete list of all enzyme entries either as an alphabetical Name Index or as the EC-Number Index is available at the above mentioned URL. You can download and print them free of charge.

A complete list of all synonyms (more than 25,000 entries) used for the enzymes is available in print form (ISBN 3-540-41830-X).

Save 15%

We recommend a standing order for the series to ensure you automatically receive all volumes and all supplements and save 15% on the list price.

Preface

In recent years, biotechnology research and development (R&D) in China has been receiving increasing attention from the world. With the open-door policy of the Chinese government, many international publications (for academia) and large market potential (for industry) constitute the two big reasons for the above phenomenon. Biotechnology has become one of the priorities in Mainland China for solving many important problems, such as food supply, health care, environment protection, and even energy. The central government has been implementing a couple of programs which cover a wide spectrum in basic research, high-tech development and industrialization, such as Basic Research Program (973 Plan), Hi-Tech R&D Program (863 Plan), Key Science & Technology Problem Solving Program (Gong-guan Plan), as well as the establishment of centers of excellence - Key Laboratories and Engineering Centers, etc. The funding from various local governments and industry for R&D has also been increasing continuously. Biotechnology centers in Shenzhen, Shanghai and Beijing have been established. There are more than 400 universities, research institutes and companies and a total of over 20,000 researchers involved in biotechnology in the Mainland. The number of research papers published internationally and patent applications is also increasing rapidly. In addition, the huge market potential with about 1.4 billion population, which is already open to the outside world, has provided numerous opportunities for international and domestic companies to invest in biotechnology, which pushes forward the biotechnology industrialization in China.

After organizing the 13th International Biotechnology Symposium & Exhibition (IBS2008, October 12–17, 2008, Dalian, China), under the auspices of the International Union of Pure and Applied Chemistry (IUPAC), recognized as the premier international biotechnology event, we are honored to be invited by the Series Editor Professor T. Scheper and Springer to edit a special volume on "Biotechnology in China," which is expected to provide a window for international colleagues to observe the current status of Chinese biotechnology. In this special volume, some examples of biotechnology R&D activities in Chinese universities and institutes are presented, covering biocatalysis and biotransformation, secondary metabolites from higher fungi and plant tissues, protein production by animal cell culture, HPLC purification of proteins, biodegradation/bioremediation and wastewater treatment technologies.

The importance of chiral issues in active pharmaceutical ingredients has been widely recognized not only by pharmacologists, but also by chemists, chemical engineers and administrators. Chapter 1 by Professor Jian-He Xu and his colleagues focuses on the biocatalytic synthesis of chiral compounds useful for the pharmaceutical industry. Biotransformation of nitriles mediated by nitrile-amide converting enzymes has attracted much attention and developed rapidly in China in recent years as it offers a valuable alternative to the traditional chemical reaction which requires harsh conditions. Professor Yu-Guo Zheng and his co-workers have overviewed the microbial transformation of nitriles to high-value acids or amides (Chap. 2).

Secondary metabolites from higher fungi and plant tissues include various unique bioactive compounds. Zhong and Xiao have reviewed the discovery, bioactivity and bioproduction of secondary metabolites from higher fungi including medicinal mushrooms (Chap. 3). In Chap. 4, Professor Kee-Yoeup Paek and his collaborators have demonstrated the large scale cultivation of ginseng adventitious roots for the production of ginsenosides, which is one of the most famous oriental medicinal plants used as crude drugs in Asian countries and is now being used worldwide for preventive and therapeutic purposes.

Some high value proteins and vaccines for medical and veterinary applications through animal cell culture have an increasing market in China. Chapter 5 by Professor Yuanxing Zhang summarizes the studies on optimization of animal cell culture processes in view of substrate metabolism regulation and protein expression improvement. Purification of proteins from cell cultures is generally a crucial step for a cost-effective process. Preparative liquid chromatography is widely used for the purification of chemical and biological substances. Professor Yan Sun and his colleagues demonstrates several approaches to high-performance preparative chromatography of proteins (Chap. 6).

Our society is facing serious environmental challenges such as curbing greenhouse gas emissions, finding renewable energy sources, managing waste and controlling pollution and diseases. In recent years, environmental pollution has become a big problem in the world and has been receiving great attention from people and governments. In this book, two chapters are dedicated to the recent development of environmental biotechnology research in China focusing on biodegradation and bioremediation. Professor Ping Xu and his co-workers have reviewed the recent progress in biodesulfurization of fossil fuels (Chap. 7). In Chap. 8, Professor Han-Qing Yu and his group members demonstrate the characterization, modeling and application of aerobic granular sludge for wastewater treatment.

I thank all the authors for their cooperation and significant contribution, Professor Scheper and Dr. Hertel for their strong support, Ms. Ingrid Samide and Mrs. Kreusel for their coordination and assistance, and my colleagues Profs. Wei Zhang and Fengwu Bai as co-editors. Also, I greatly appreciate the generosity of Shanghai Jiao Tong University, my colleagues and lab members at both SJTU and ECUST, and my friends and my family for their continuous support.

Shanghai, Summer 2009 Jian-Jiang Zhong

Contents

Adv Biochem Engin/Biotechnol (2009) 113: 1-31
DOI: 10.1007/10_2008_30

Published online: 21 July 2009

Discovery and Utilization of Biocatalysts for Chiral Synthesis: An Overview of Chinese Scientists Research and Development

Hui-Lei Yu, Jian-He Xu, Wen-Ya Lu, and Guo-Qiang Lin

Abstract The importance of chiral issues in active pharmaceutical ingredients has been widely recognized not only by pharmacologists, but also by chemists, chemical engineers and administrators. In fact, the worldwide sales of single-enantiomer drugs have exceeded US $150 billion. Among them the contribution rate of biocatalysis technology is ever increasing (up to 15–20%). This chapter will focus on the biocatalytic synthesis of chiral compounds useful for pharmaceutical industry. Diverse enzymes, such as oxidoreductases, epoxide hydrolases, nitrilases/nitrile hydratases and hydroxy nitrile lyases which were isolated from various sources including microorganisms and plants, and the methodology for utilizing these enzymes in enantioselective or asymmetric synthesis will be discussed briefly.

Keywords Biocatalysis, Biocatalysts screening, China, Chiral synthesis, Screening method

Contents

H.-L. Yu and J.-H. Xu (✉)
State Key Laboratory of Bioreactor Engineering, East China University of Science and Technology, 130 Meilong Road, Shanghai 200237, China
e-mail: jianhexu@ecust.edu.cn

W.-Y. Lu and G.-Q. Lin
Shanghai Institute of Organic Chemistry, Chinese Academy of Sciences, 354 Fenglin Road, Shanghai 200032, China

1 Introduction

The need for new chiral pharmaceutical reagents is believed to grow continuously since the human body usually functions via chiral catalysis of enzymes. In 2000, 35% of pharmaceutical intermediates were chiral and this number is expected to increase to 70% by 2010 [1]. Compounds with a chiral centre are usually manufactured in single isomeric form. Current USA Food and Drug Administration (FDA) regulations demand proof that the nontherapeutic isomer should be non-teratogenic. Moreover, the increasing size and complexity of these molecules frequently results in multiple chiral centres [2]. The target for a commercial process, however, is demanding and an enantiomeric excess (*ee*) of 98% is a minimum acceptable level.

Chemists have been performing organic chemistry for hundreds of years, while microbes have been doing it for at least millions of years. Enzymes from microbes and other sources are extremely chemo-, regio-, and enantio-selective across a diverse range of reactions under mild conditions of pH, temperature, and pressure. “When it comes to wanting selectivities of 98% or higher, you are probably bound to a bioprocess, because getting beyond 95% otherwise is really tough,” says Kurt Faber [3], professor of chemistry at the University of Graz and a member of the Research Center for Applied Biocatalysis (RCAB) in Austria. Biocatalysis can also open up new or even greener reaction routes [4]. As reported, the worldwide sales of single-enantiomer drugs have exceeded US $150 billion [5]. Among them the contribution rate of biocatalysis technology is ever increasing (up to 15–20%). Table 1 summarizes some chiral products in industrial biocatalysis.

Recently, based on the impact of biocatalysis on the synthesis of chiral small-molecule drug candidates, several pharmaceutical process R&D groups have placed a greater emphasis on biocatalysis for organic synthesis and have expanded or reinitiated activities in this area [3]. BMS's (Bristol–Myers Squibb) multidisciplinary effort was created about 20 years ago to find means to access chiral intermediates, Patel, who heads the BMS Group, explains. He can offer dozens of examples where biocatalysis has been successful [7]. Degussa's Service Center Biocatalysis (SCB) was founded in 2004 after the company had invested for 3 years in one of its Project House technology development initiatives. In February of 2006, SCB became a part of the company's Exclusive Synthesis and Catalysts business unit. Other fine chemicals companies, such as Dowpharma, Cambrex, and Archimica, have also added biocatalysis capabilities largely through acquisitions.

Some of the drug developers will pay more attention to biocatalysis soon, but by and large it still is a niche technology [3]. Although biocatalysis has many advantages in chiral synthesis, especially for its excellent selectivity, there are three main drawbacks of today's biocatalysts [8–10]: (1) biocatalysts stability is often not sufficient in the desired media; (2) too few biocatalysts exist for the desired reactions from available substrates to targeted products; and (3) too long a time is needed for development of new or improved biocatalysts. Researchers all over the world including Chinese researchers are making great a effort to solve these problems.

Table 1 Chiral products produced in quantities by industrial biocatalysis[a]

Product	Enzyme	Company
>1,000 tons per annum		
(*S*)-Aspartic acid	Aspartase	Tanabe
(*S*)-Methoxyisopropyl amine	Lipase	BASF
(*R*)-Pantothenic acid	Aldolactonase	Fuji Chem. Ind.
(*R*)-Phenyllglycine	Hydantoinase/carbamoylase	Several
(*S*)-Amino acids	Aminoacylase	Degussa, Tanabe
>10 tons per annum		
(*S*)-DOPA	β-Tyrosinase	Ajinomoto
Sterically pure alcohols and amines	Lipase	BASF
(*R*)-Mandelic acid	Nitrilase	BASF

[a]Cited with modification from [6]

2 Biocatalysts Source

Biocatalysts can be obtained from animals, plants or microorganisms. At present microorganisms are still the most prominent source of enzymes, while animal organs and plant materials contribute 8 and 4%, respectively, to the total amount of enzymes processed. Since the advent of recombinant DNA technology for over-expression of an enzyme in a host microorganism, novel and unique plant enzymes are also preponderant [11]. A very interesting research area for microorganism screening is extremophiles [12]. Extremophilic microorganisms are adapted to survive in ecological niches, and they produce unique biocatalysts that function under extreme conditions comparable to those prevailing in various industrial processes.

There are several techniques for getting a biocatalyst, including: (1) commercial enzyme companies and strain culture collection agencies; (2) screening for new activities in different environments (soil, polluted areas, deep vents); (3) screening for new activities from plant material and animal tissue; (4) discovery of unnatural activities of existing enzymes; (5) utilization of novel reaction conditions or altered reaction media resulting novel performance of the biocatalyst; (6) application of genetic engineering techniques, e.g., such as protein engineering or directed evolution; and (7) chemical modification and immobilization for improved biocatalyst.

2.1 Biocatalysts from Commercial Enzyme Companies or Strain Culture Collection Agencies

The most rapid and convenient way for biocatalyst screening is perhaps to search from the known commercial enzymes. The commercial enzyme companies such as Sigma, Novozyme, Amano and Genencor (now a division of Danisco) supply large quantities of enzymes as off-the-shelf and ready-to-use products. Generally speaking, the industrial strains can be obtained from the culture collections according to their related species. In addition, many countries have built up national collections administrated by government agencies, for instance, ATCC (American Type Culture Collection, USA), DSMZ (Deutsche Sammlung von Mikroorganismen und Zellkulturen, Germany), NCTC (National Collection of Type Cultures, UK), and CGMCC (China General Microbiological Culture Collection Center, China).

2.2 Screening Microorganisms from the Natural Environment for Target Biocatalysts

Most of the novel biocatalysts are discovered from a suitable environment in nature. It is very crucial to analyze the characteristics of the target microorganism and chose the suitable collection environment before the samples collection. For instance, when the biocatalytic reaction needs a condition of higher temperature, acidic or alkali environment, the comparable environment is better to be used for screening.

Although various microorganisms are present in a natural environment, they are often subjected to nutrient-poor conditions. Thus organisms in nature must be able to adapt quickly to a rapid change in nutrient availability. This requirement is perhaps one of the main reasons for the existence of highly sophisticated regulation mechanisms and also provides the diversity of diets upon which microorganisms can feed. Nutrient condition is one of the most important factors for screening the ideal microorganism from the diverse organism realm.

The criteria for a successful screening process with the goal include [8]: (1) the organism must produce the desired enzyme in good yield and within a reasonable short period if possible; (2) the organism must synthesize the enzyme with cheap and accessible nutrients; (3) the enzyme produced by the organism should have high selectivity with few by-product; (4) the organism should be nonpathogenic and should not produce undesired or toxic compounds; and (5) the organism should be genetically stable.

2.3 Screening Biocatalysts from Plant Materials

Plants are valuable sources of a variety of unique biocatalysts for performing a complex reaction. A wide variety of compounds, such as aromatics, steroids, alkaloids, coumarins and terpenoids, can be transformed by plant-originated biocatalysts [11].

The advantages of plant biocatalyst are as follows. First, being a different branch from microorganism in cladistics, the plant biocatalyst has its peculiarity, performing differently in bioconversion reaction, with respect to the stability, the substrate specificity and so on. Genetic manipulation approaches offer great potential to express heterologous genes for key enzymes [13]. Second, plant enzymes can be regarded as a more direct help than the microorganism biocatalyst for the organic chemist since most of them were unfamiliar with the microorganism fermentation process [14]. Third, the source of plant biocatalyst is very rich all over the world. Taking China as an example, the total number of plants can reach 43,000 and these plants grow diversely and cover almost any kind of plants in the Northern Hemisphere. Therefore, it is preponderant to develop the plant biocatalyst for areas rich of plant materials. Finally, most of the plant biocatalysts are absolutely 'green catalyst' ; sometimes they are even edible [15–18].

The possibility of success in exploring the biocatalytic capability of plant cells and enzymes is enormous. However, advancement in the biotransformation studies using plant biocatalysts, especially *in vitro*, is quite slow. More specifically spea-king, the characterization of secondary metabolic pathway is a multidisciplinary challenge. The limited and fragmented knowledge in this area is a real bottleneck for the exploitation of biotransformations *in vitro* as there are not many commercially available metabolites to examine these reactions. A coordinated approach of mole-cular studies on metabolic pathway engineering to understand the involved genes and enzymes may contribute toward utilizing biotransformations *in vitro* for practical use.

2.4 Metagenome Approach for Biocatalyst Screening

The development of direct techniques to extract, clone and heterogenously express genomic DNA from entire uncultivated microbial consortia, namely the so-called 'metagenome' approach, will result in revolutionary progress in the field of enzyme discovery by providing genetic access to the 'unseen' majority of microbial diversity and its enzymatic constituents [19–22]. The Diversa Company (San Diego, USA) specializes in improving the accessibility of enzymes from unculturable organisms [23]. The technique was successfully applied to samples from both sea water and soil.

3 Methods for Rapid Screening

3.1 Screening by Conventional and Novel Methods

Conventional screening methods are increasingly being replaced or supplemented by novel methods (Table 2). The system for HTS (high-throughput screening) has the ability to handle large amounts of samples. A robotic system, which is able to localize single colonies on a plate, to grow colonies on a plate, and to inoculate in liquid culture, allows screening of typically 1,000 strains per week. Given the

Table 2 Conventional and novel methods for biocatalyst screening [8]

Conventional methods	Novel methods
Isolation of microorganisms in the neighborhood of habits with enhanced concentration of substrate	Single cell selection by identification of interesting cells with the desired activity, and picking with robotic arms for further cultivation
Selection of strains based on taxonomic closeness to prior successful strains	Based on published nucleotide sequences of proteins with a similar activity profile, searching for similar enzymes in different organisms by using PCR and rRNA methods
Enrichment cultures	Cloning for high expression with highly transformable hosts
Use of 96-wells plate	Use of 384-wells plate and even 1,536-well plate

necessity to screen about 10,000–100,000 strains in order to find a useful biocatalyst, this means one could expect the process to take 10–100 weeks to a breakthrough.

3.2 *Screening by Detection*

One of the main challenges is to develop a simple assay sensitive enough to pick up enzyme activities that might be below their optimal level under the detection conditions. The ideal assay system for HTS would include automated manipulation of single wells and each should be tested for a variety of activities simultaneously. Nowadays novel colorimetric, luminescence and fluorescence methods have already been established in HTS with automated multiplex compound testing. Reyond et al. published several important reviews on the HTS screening methods [24–27]. Reymond et al. summarized both the off-line and on-line assays for biocatalyst screening (Table 3).

The most convenient reactions to test in high-throughput are those involving chromogenic or fluorogenic substrates. Fluorogenic and chromogenic assays can be employed for screening of most of the enzymes, such as lipases and esterases [28–31], proteases [32], aryl sulfatases [33], monooxygenases [34], amidase [35] and nitrilase [36]. Some fluorogenic substrates [24–27] were summarized in Fig. 1. Detection of genetic diversity via random mutagenesis can be achieved by using the Fluorescence Enzyme Bead Assay System (FEBAS), in which an evenly distributed substrate becomes a fluorescently localizable product after biocatalysis by the beads-attached enzymes [37].

3.3 *Screening for Stereoselectivity*

HTS for biocatalyst is often not only aimed at activity, but also at stereoselectivity, especially enantioselectivity for chiral synthesis. Compared to detect optical purity,

Table 3 Assays useful for biocatalyst screening[a]

Method type	Endpoint (off-line)	Real-time (on-line)
Chromometric/fluorimetric	TLC	Fluoro/chromogenic substrates
	Product staining	pH indicators
	Biosynthesis of color products	FRET substrates
Sophisticated reagents	QUEST	Homogeneous cat-ELISA
	cat-ELISA	
	Competitive cat-ELISA	
Sophisticated instruments	HPLC	IR-thermography
	MS	
	CE	

[a]Cited with modification from [24–27].

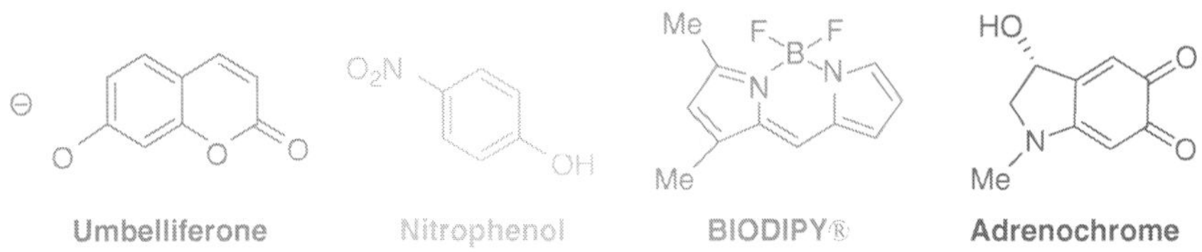

Fig. 1 Some fluorogenic substrates used for biocatalyst screening

the most direct method for analyzing enantioselectivity [38, 39], the HTS version used the novel approaches instead, such as fluorogenic assay [40–42], mass spectrum [43–46]. In addition, microarrays [47], fingerprinting [48], chips [49] and NMR [50] were also used for screening of enantioselective catalysts.

4 Examples of Biocatalysts for Chiral Synthesis

In the following examples, biocatalysts were discovered and utilized by researchers all over the world, especially in recent years by Chinese researchers.

4.1 Oxidoreductases (E.C. 1)

4.1.1 Dehydrogenases/Reductases (E.C. 1.1)

Asymmetric reduction of the carbonyl group is perhaps one of the most important, fundamental and practical reactions for producing chiral alcohols, which can be transformed into various functionalities for synthesis of industrially important chemicals such as pharmaceuticals, agrochemicals and natural products (Table 4) [51–53]. Dehydrogenases and reductases are enzymes that catalyze the reduction of carbonyl groups [54, 55]. In particular, the number of industrial processes using alcohol dehydrogenases (ADHs) is increasing [54–59]. These biocatalysts, used as isolated enzymes or whole cells, catalyze the stereoselective reduction of prochiral ketones with remarkable chemo-, regio-, and stereoselectivity.

In principle, virtually all organisms can serve as a source for ADHs. However, to date, commonly used, commercially available ADHs originate from horse liver (HLADH) and microorganisms, such as *Thermoanaerobium brockii* (TBADH), baker's yeast (*Saccaromyces cerevisiae*, YADH), *Candida boidinii*, *Candida parapsilosis*, *Rhodococcus erythropolis*, *Lactobacillus brevis*, and *Lactobacillus kefir*.

Table 4 Chiral alcohols produced by asymmetric reduction and their uses for synthesis of industrially important chemicals

Product	Yield	*ee* (%)	Application	Company
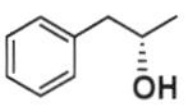	72%, 63.5 g (L.day)$^{-1}$	>99	Amphetamines	Forschungszentrum J¨lich GmbH
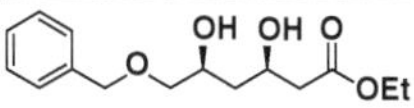	92%	99	Anticholesterol drugs	Bristol-Myers Squibb
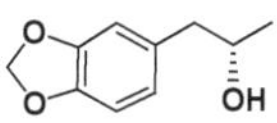	96%, 75 g (L.day)$^{-1}$	>99.9	LY 300164 (an orally active benzodi-azepine)	Eli Lilly
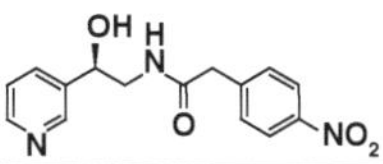	82.5%	>98	β-3-Agonist	Merck Research Laboratories
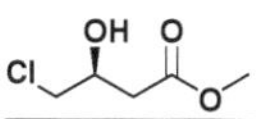	95%	99	Cholesterol antagonist	Bristol-Myers Squibb

Many ADHs obey the Prelog rule although exceptions (*Anti*-Prelog) have also been discovered (Table 5) [56, 60–63].

Due to the high cost of the pyridine cofactors needed by NAD(P)-dependent oxidoreductases, in situ cofactor regeneration is required for preparative applications. In recent years, existing regeneration methodologies have been improved and new approaches have been devised [64–66]. These include the use of newly discovered dehydrogenases that are stable in a high content of organic solvents such as 2-propanol [67, 68] and novel enzymes that can regenerate either the reduced or oxidized forms of the cofactor [69–71], such as phosphite dehydrogenase [69], pyridine nucleotide transhydrogenase [70] and NADH oxidase from *Lactobacillus brevis* [71]. The use of electrochemical methods has allowed cofactor regeneration [72, 73], and natural or engineered whole-cell systems provide alternatives to approaches relying on purified enzymes [57, 58, 74].

Our group employed whole-cells of newly isolated *Rhodotorula* sp. AS2.2241 for efficient reduction of a broad range of prochiral ketones (**a–k**) to the corresponding optically active secondary alcohols (Table 6). The microbial reduction system exhibited high activity and excellent enantioselectivity in the reduction of various aromatic ketones and acetylpyridines (>97% *ee*), but moderate to high enantioselectivity in the reduction of α- and β-keto esters [75, 76].

Xu Y and coworkers [77] reported an economical and convenient biocatalytic process for the preparation of (*S*)-1-phenyl-1,2-ethanediol (PED) by microbial deracemization from the corresponding racemate. The enantioselective conversion of racemic PED by *Candida parapsilosis* CCTCC M203011 was found to be the most efficient process to produce (*S*)-PED with high optical purity of 98% *ee* and yield of 92%. The (*S*)-enantiomer was produced from the intermediate identified as β-hydroxyacetophenone by asymmetric reduction, following the stereoselective

Table 5 Alcohol dehydrogenases from various sources and selectivities

ADH	Attack from	Configuration
Baker's yeast	*Re*	Prelog
Horse liver	*Re*	Prelog
Thermoanaerobacter ethanolicus	*Re*	Prelog
Thermoanaerobium brockii	*Re*	Prelog
Rhodotorula sp.	*Re*	Prelog
Pseudomonas sp.	*Si*	*Anti*-Prelog
Lactobacillus kefir	*Si*	*Anti*-Prelog
Mucor javanicus	*Si*	*Anti*-Prelog
Lactobacillus brevis	*Si*	*Anti*-Prelog
Geotrihum sp.	*Si*	*Anti*-Prelog

Table 6 Reduction of aromatic ketones by *Rhodotorula* sp. AS2.2241[a]

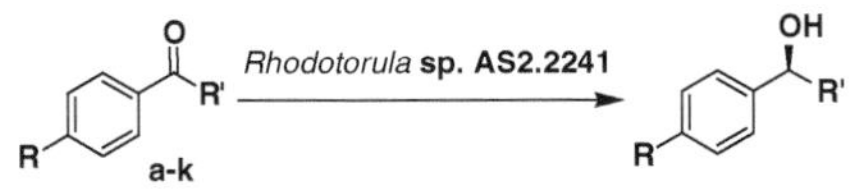

Substrate	R	R'	Time (h)	Yield (%)	*ee* (%)	Config.
a	H	CH_3	7	100 (65)	99	*S*
b	NO_2	CH_3	3	100 (84)	>99	*S*
c	Br	CH_3	3	100 (63)	>99	*S*
d	Cl	CH_3	4	100 (72)	99	*S*
e	MeO	CH_3	24	50 (35)	>99	*S*
f	NH_2	CH_3	24	ND (6)	ND	ND
g	H	CH_2Br	3	100 (69)	>99	*R*
h	H	CH_2Cl	3	100 (52)	>99	*R*
i	H	C_2H_5	9	100 (62)	>99	*S*
j	H	C_2H_4Cl	20	100 (50)	>99	*S*
k	NO_2	CH_2Br	5	100 (51)	97	*R*

[a] Cited with modification from Reference [75–76]

oxidation of (*R*)-PED to β-hydroxyacetophenone. The stereoselective conversion involves the oxidation of (*R*)-PED to the intermediate with $NADP^+$ as the cofactor and the reduction that forms the product using NADH as the cofactor, which was catalyzed by a novel cofactor-dependent oxidoreduction system (Fig. 2). The $NADP^+$-dependent (*R*)-specific alcohol dehydrogenase involved in the stereoinversion has been purified from *C. parapsilosis* CCTCC M203011, which has a relative molecular mass of 45 kDa.

Sun ZH et al. [78] utilized whole-cells of a fungus, *Aureobasidium pullulans* CGMCC1244, for highly stereoselective reduction of ethyl 4-chloro-3-oxobutanoate to ethyl (*S*)-4-chloro-3-hydroxybutanoate in a biphasic system composed of potassium phosphate buffer and dibutylphthalate.

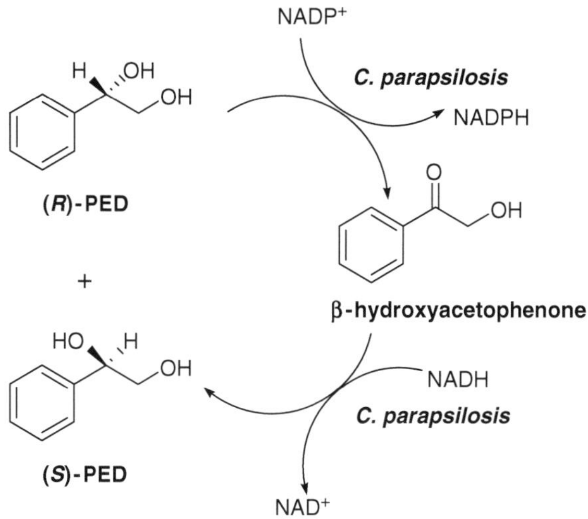

Fig. 2 Stereoinversion of (*R*)-PED to (*S*)-PED catalyzed by C. parapsilosis CCTCC M203011

Our group [61–63] reported the work on bioreduction of aryl ketones, aryl α-hydroxy ketones, 2-hydroxy and 2-acetoxy ketones and sulfur containing ketones by *Geotrichum* sp. 38, affording mostly the *anti-Prelog* alcohols.

4.1.2 Oxygenases (E.C. 1.14)

Oxygenases are enzymes that introduce one or two atoms of molecular oxygen into an organic molecule. They are attractive for chemical synthesis and bioremediation as they react with a wide range of organic substances [79]. However, their practical applications are still limited because most oxygenases are membrane-associated, not very stable, and display rather low activity. For their catalytic function, most oxygenases require reduction equivalents usually supplied by NADPH or NADH. Moreover, they require electron-transfer partners such as flavin reductases and iron–sulfur proteins, which might be also membrane-associated. Some of these limitations can be overcome by using whole-cell systems [74]. However, physiological effects such as low expression rates, limited substrate uptake, toxicity of substrate or product, and product degradation must be taken into account. The mainstream of current studies is thus to either engineer isolated oxygenases for altered selectivity or enhanced stability, or to engineer the metabolism of whole cells towards an improved yield of the desired oxidation product or to combine both approaches [80–82].

Chiral sulfoxides are formed as metabolites of many sulfur-containing drugs, and exhibit differential stereochemically-dependent metabolism and enzyme inhibition [83]. They are valuable asymmetric starting materials and chiral auxiliaries in organic synthesis, and have been the target of a number of chemical syntheses [84]. The enantioselective oxidation of a prochiral sulfide is no doubt the most direct and economical method for the synthesis of enantiopure sulfoxides. The promising results obtained in biological sulfoxidation in the past decade suggest that this approach will be of synthetic interest in the future [85]. Both isolated enzymes, such as cyclohexanone monooxygenase, haloperoxidases, monooxygenases, dioxygenases and so on, and whole cells including bacteria, fungi and yeasts have been used in the asymmetric oxidation of prochiral sulfides [85].

Recently, our group has isolated a new bacterium *Rhodococcus* sp. ECU0066 for biocatalytic oxidation of a broad spectrum of sulfides, affording quite high activity and enantioselectivity for most of the sulfides [86]. The optimization of the whole-cell reaction system, cofactor regeneration and substrate toxicity are still in progress.

4.2 Hydrolases (E.C. 3)

4.2.1 Lipase (E.C. 3.1.1.3)

Lipases are currently one of the best biocatalysts for the preparation of a wide range of optically active alcohols [87]. Applications of lipases in asymmetric synthesis include kinetic resolution of racemic alcohols, acids, esters or amines [88], as well as the desymmetrization of prochiral compounds [89, 90]. Ghanem summarized the recent developments in the rapidly growing field of lipase-catalyzed kinetic resolution of racemates, including kinetic resolution of primary alcohols, secondary alcohols, tertiary alcohols, chiral carboxylic acids and diols [91]. Novel biotechnological applications have been successfully established using lipases for the synthesis of biopolymers and biodiesel, the production of enantiopure pharmaceuticals, agrochemicals, and flavor compounds [92].

Lipases from a large number of bacterial, fungal and plant and animal sources have been purified to homogeneity [93]. Hasan reviewed numerous lipases from species of bacteria, yeasts and molds [94]. Many plant lipases have been isolated now which can be used as important biocatalysts. Many companies market digestive enzymes prepared from plant and fungal lipases. Doctor's Holistic Market manufactures Chiro-Zyme, the digestive enzymes formula containing lipase from *Aspergillus niger* and *Rhizopus oryzae*. Lipase discovered from extremophilic microorganisms in recent years resulted in several novel applications in industrial processes [95–97].

Discovery of Novel Lipase

Our group [98] succeeded in isolating a racemic glycidyl butyrate resolving strain, *Rhizopus* sp. Bc0-09. Its extracellular lipase was found to hydrolyze enantioselectively the (*S*)-enantiomer of the chiral ester (Fig. 3), with optimal activities at pH 5.3 and 42 °C. Higher enantioselectivity of the enzyme was observed at lower temperatures, while the best enantioselectivity was obtained at pH 5.5–6.0, with an *E* value (enantiomeric ratio) of 57.

Our group [99–101] had isolated a lipase-producing *Serratia marcescens* ECU1010, its extracellular lipase being used to resolve successfully *trans*-3-(4'-methoxyphenyl) glycidic acid methyl ester [(±)-MPGM] (Fig. 4). The product (2*R*,3*S*)-(−)-MPGM with optical purity of >99% *ee* was obtained in a yield of 37.2%, which is an important precursor for production of Diltiazem (a coronary vasodilator).

Wei et al. [102] screened a lipase-producing strain *Pseudomonas fluorescens* B68, with GenomeWalker. The open reading frame of lipase gene *lipB68*, encoding 476 amino acids, was cloned and expressed in *Escherichia coli* BL21 (DE3). By affinity chromatography, the recombinant LipB68 protein was purified to 95% purity. In chiral resolution, LipB68 effectively catalyzed the transesterification of both α-phenylethanol and α-phenylpropanol (Fig. 5) at 20 °C, achieving *E* values

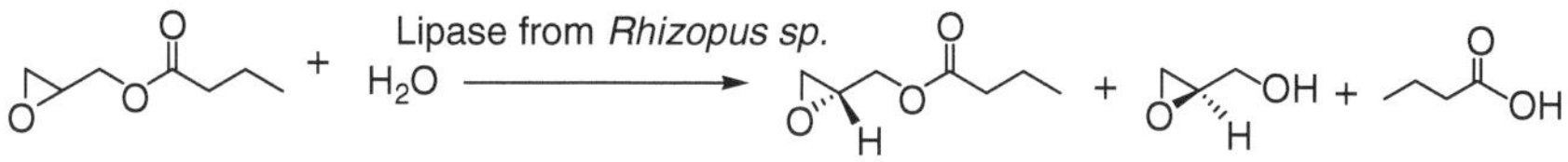

Fig. 3 Bioresolution of glycidyl butyrate by lipase from *Rhizopus* sp. (cited from [98])

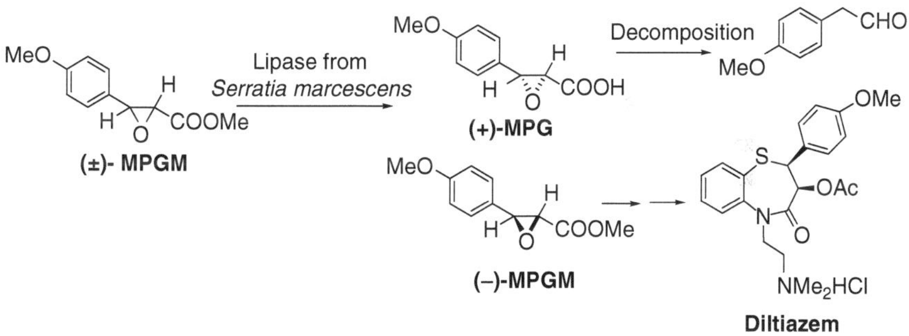

Fig. 4 Stereoselective hydrolysis of (±) MPGM by lipase from *Serratia marcescens* ECU1010 (cited from [101])

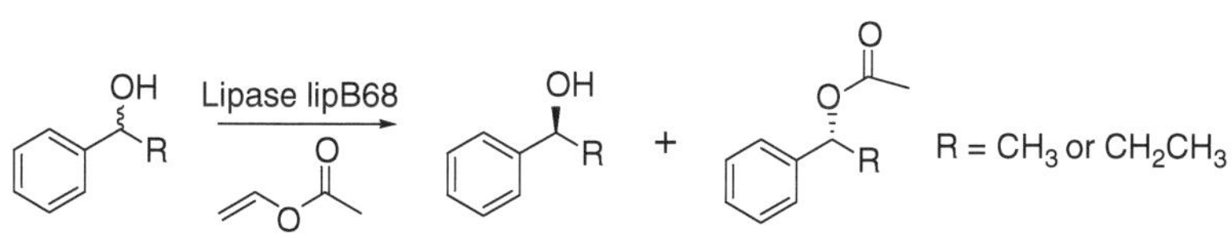

Fig. 5 Transesterification of α-phenylethanol and α-phenylpropanol by a lipase from *Pseudomonas fluorescens* (cited from [102])

greater than 100 and 60 after 120 h, respectively. Among the known biocatalysts used in biodiesel production, LipB68 gave a yield of 92% after 12 h.

Lipase for Kinetic Resolution

Our group [103–108] studied the kinetic resolution of 2-chloroethyl ester of *rac*-ketoprofen catalyzed by Lipase OF (lipase from *Candida rugosa*) systematically (Fig. 6).

Surfactants and pH of the reaction medium were found to be two critical factors affecting both the activity and enantioselectivity of Lipase OF. With the addition of either of the two emulsifiers at their optimal concentrations, i.e., 2% (w/v) Tween-80 or 3% (w/v) OP-10, the crude Lipase OF activity was greatly enhanced up to 13 and 15 times, respectively. On the other hand, the *E* value increased from 1.2 to 6.7 for the crude lipase and from 8 to 100 for the purified lipase in the presence of 8 or 2% (w/v) of Tween-80 [103]. The optimal activity of Lipase OF for hydrolysis was at pH 4.0, while the best enantioselectivity was observed at pH 2.2 where the enzyme was still 60% active and stable [104]. By spectroscopic studies [105], the origin of the activity and enantioselectivity enhancement induced by pH variation could be attributed to the changes in the conformation and/or the flexibility of the lipase. By combination of the two effectors, the *E* value of the lipase was increased up to 60 for the crude lipase (42% yield, >95% *ee*) [106].

Various materials have been used as supports for the immobilization of Lipase OF. The enzyme adsorbed on silica gel showed the maximum activity recovery (37%), and was thus used to construct a packed bed reactor for the continuous production of (*S*)-ketoprofen. Using a packed bed bioreactor, optically pure (*S*)-ketoprofen (*ee* >99%) was produced with a conversion of 30% and productivity of 1.5 mg (g biocatalyst)$^{-1}$ h^{-1} [107].

The two processes respectively used for the partial purification and for the immobilization of Lipase OF have been successfully integrated into one by simple adsorption of the enzyme onto a cation ion exchanger resin (SP-Sephadex C-50) at pH 3.5 [108]. Due to selective removal of the unfavorable lipase isoenzyme (L1), the enzyme components (mainly L2 and L3) that are tightly fixed on the resin displayed a significantly improved enantioselectivity. By using such an immobilized enzyme as biocatalyst, the process for preparing enantiopure (*S*)-ketoprofen becomes simpler and more practical as compared with the previously reported procedures and the product was obtained with >94% *ee* at 22.3% conversion in the

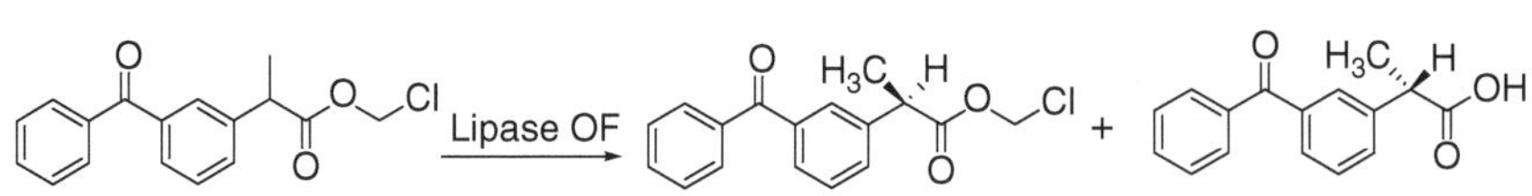

Fig. 6 Enantioselective hydrolysis of 2-chloroethyl ester of ketoprofen

presence of an optimal concentration (0.5 g l^{-1}) of Tween-80 at pH 3.5. Furthermore, the operational stability of the immobilized biocatalyst was examined in different types of reactors. Under optimal conditions, the air-bubbled column reactor could be operated smoothly for at least 350 h, remaining nearly 50% activity.

Yuan et al. [109] reported the preparation of optically active β-hydroxy-β-aryl-propionates, δ-hydroxy-δ-aryl-β-oxo-pentanoates and their butyryl derivatives via *Candida rugosa* lipase (CRL) catalyzed hydrolysis (Fig. 7). The optically active products are potential precursors of some chiral pharmaceuticals and natural products.

Cao et al. [110] reported a highly active lipase expressed from a newly constructed strain of *Bacillus subtilis* (BSL2) for enantioselective lipase-catalyzed kinetic resolution of *N*-(2-ethyl-6-methylphenyl)alanine (NEMPA) (Fig. 8). A high enantiomeric ratio ($E = 60.7$) is reached in diisopropyl ether/water (10 vol.%) and the enantioselectivity is about 22-fold higher than that in pure buffered aqueous solution. The results show that the reaction medium greatly influences BSL2 reaction and its enantioselectivity in the hydrolysis of racemic methyl ester.

Immobilization of Lipase

A synthetic surfactant, didodecyl *N*-D-glucono-l-glutamate (DGG), was used for modification of *Pseudomonas* sp. lipase (Lipase PS) to prepare optically pure (*S*)-4-hydroxy-3-methyl-2-(2-propynyl)-cyclopent-2-enone [(*S*)-HMPC] in a solvent-free system (Fig. 9) [111]. The V_{max} of the modified enzyme was improved by as

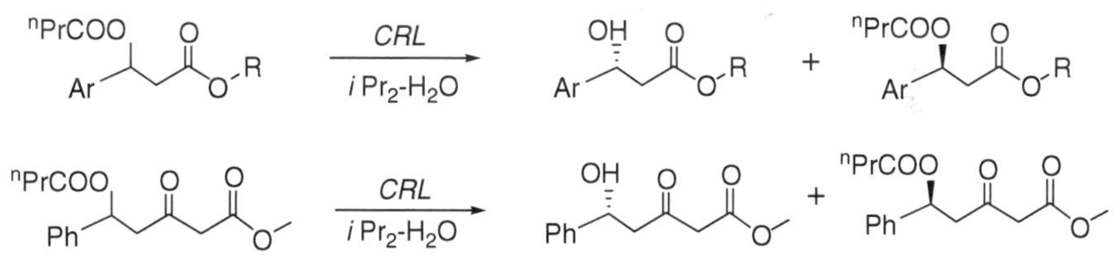

Fig. 7 *Candida rugosa* lipase-catalyzed kinetic resolution of β-hydroxy-β-aryl propionates and δ-hydroxy-δ-aryl-β-oxo-pentanoates (cited from [109])

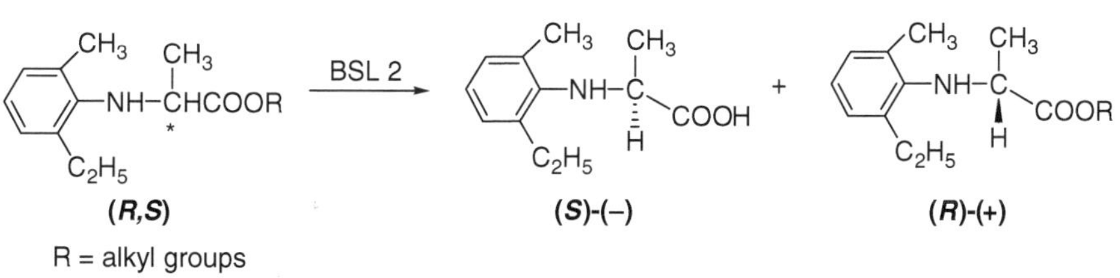

Fig. 8 Enantioselective hydrolysis of racemic alkyl esters of NEMPA using BSL2 (cited from [110])

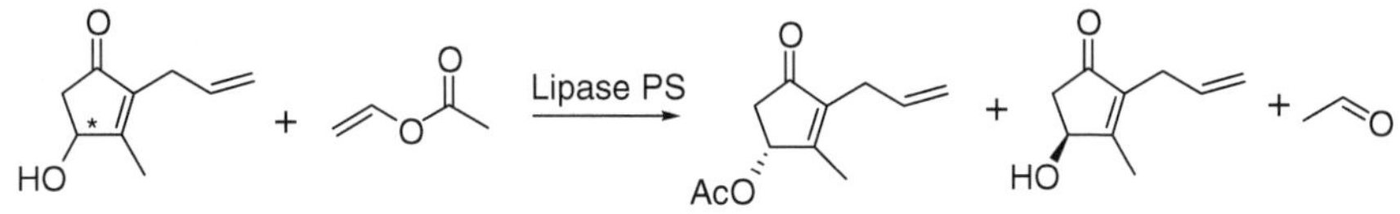

Fig. 9 Enantioselective transesterification of HMPC with vinyl acetate by Lipase PS (cited from [111])

much as 160 times over the native lipase, in spite of a similar K_M value to that of the native lipase. When the substrate concentration of the racemic alocohol (HMPC) was as high as 1 M, the reaction reached 40% conversion in 20 h with merely 1.0 g L^{-1} of the DGG-lipase, producing (*R*)-HMPC acetate in nearly 100% *ee*. The stearic acid-coated Lipase PS showed the highest catalytic activity, with a specific activity improved by 54 times over the native lipase [112]. The microcrystal salt-supported lipase and celite-adsorbed Lipase PS also displayed a much better performance as compared with the native lipase.

A lipase from *Penicillium expansum* PED-03 lipase (PEL) was immobilized on a modified ultrastable-Y (USY) molecular sieve. The immobilized lipase catalyzed the kinetic resolution of (*R,S*)-2-octanol in microaqueous media. The conversion of the reaction catalyzed by PEL immobilized on modified USY molecular sieve reached 48.8% with excellent enantioselectivity (average *E* of eight batches >460) in *n*-hexane with 0.8 vol.% water, which was much higher than that of the reaction catalyzed by free PEL and PEL immobilized on other supports [113].

Lauric acid-stabilized magnetic particles were prepared by coprecipitation in the presence of lauric acid and used for the covalent immobilization of *Candida rugosa* lipase via carbodiimide activation [114]. Biocatalytic resolution of (±)-menthol was performed by the immobilized lipase-catalyzed enantioselective esterification with propionic anhydride. As a result, (−)-menthyl propionate with a yield higher than 96% and over 88% *ee* of products was obtained. Good durability of the immobilized lipase to catalyze the resolution of (±)-menthol was also observed.

Lipase in Ionic Liquids

Enantioselective acylation of (*R,S*)-1-trimethylsilylethanol [(*R,S*)-1-TMSE] with vinyl acetate (Fig. 10) catalyzed by immobilized lipase from *Candida antarctica* B (i.e., Novozym 435) was efficiently conducted in ionic liquids (ILs) [115]. Of the six ILs examined, the IL $C_4MIm.PF_6$ gave the fastest initial rate and the highest enantioselectivity, and was consequently chosen as the favorable medium for the reaction. After a reaction time of 6 h, the *ee* of the remaining (*S*)-1-TMSE reached 97.1% at the substrate conversion of 50.7%. Additionally, Novozym 435 was effectively recycled and reused in $C_4MIm.PF_6$ for five consecutive runs without substantial loss in activity and enantioselectivity. The preparative-scale kinetic

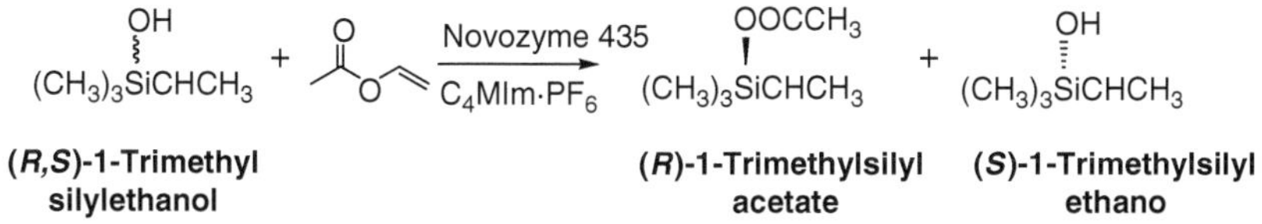

Fig. 10 Enantioselective acylation of (*R,S*)-1-TMSE in ionic liquids (cited from [115])

resolution of (*R,S)*-1-TMSE in C_4MIm. PF_6 is shown promising and useful for the practical production of enantiopure (*S*)-1-TMSE.

The lipase-catalyzed enantioselective transesterification of racemic secondary alcohols was studied in two imidazolium-based ionic liquids ([bmim][PF_6] and [bmim][BF_4]) *vs* hexane [116]. The system 1-butyl-3-methyl-1*H*-imidazolium hexafluorophosphate/*Candida antarctica* lipase B ([bmim][PF_6]/CALB) could be readily recycled four times without significant loss in activity or enantioselectivity.

4.2.2 Esterase (E.C. 3.1.1.1)

Esterases, as well as lipases, have industrially important properties such as chain length specificity, region-specificity, or chiral selectivity. A number of thermostable esterases have already been described from various thermophilic microorganisms within the three domains of life [117, 118].

A newly isolated esterase-producing bacterium *Acinetobacter* sp. CGMCC 0789 was used to hydrolyze stereoselectively the acetate of (*S*)-4-hydroxy-3-methyl-2-(2-propynyl)-cyclopent-2-enone (HMPC) (Fig. 11). The hydrolysis of 52 mM (*R,S*)-HMPC acetate to (*R*)-HMPC was completed within 8 h, with optical purity of 91.4% ee_p and conversion of 49% [119]. Using calcium alginate gel to entrap cells of *Acinetobacter* sp. CGMCC 0789, it was found that isopropanol (10 vol.%) as a cosolvent could markedly enhance the activity and enantioselectivity of the immobilized cells. After ten cycles of reaction, no significant decrease in the enzyme activity was observed [120].

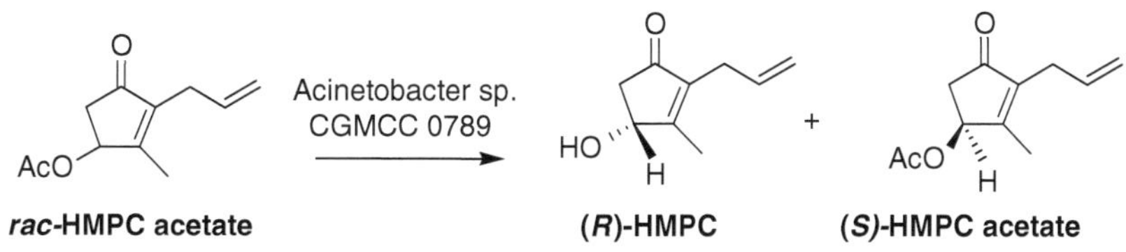

Fig. 11 Enantioselective hydrolysis of *rac*-HMPC acetate by cells of *Acinetobacter* sp. CGMCC 0789 (cited from [120])

The gene encoding pyrethroid-hydrolyzing esterase (EstP) from *Klebsiella* sp. strain ZD112 was cloned into *E. coli* and sequenced [121]. No similarities were found by a database homology search using the nucleotide and deduced amino acid sequences of the esterases and lipases. EstP was heterologously expressed in *E. coli* and purified. The molecular mass of the native enzyme was approximately 73 kDa as determined by gel filtration. The purified EstP not only degraded many pyrethroid pesticides and the organophosphorus insecticide malathion, but also hydrolyzed *p*-nitrophenyl esters of various fatty acids, indicating that EstP is an esterase with broad substrates.

Lactonases, which catalyze the reversible or irreversible hydrolysis of lactone compounds, belong to the esterase family of enzymes. A novel D-pantothenate synthesis process, which involves the enzymatic resolution of dl-pantoyl lactone with *Fusarium* lactonase, has been in practical use since 1999 [122]. It has been shown that the new process is highly satisfactory not only from an economic aspect but also from an environmental one [123]. Nowadays, about 30% of the world production of calcium D-pantothenate occurs through this chemo-enzymatic process.

A fungus strain ECU2002, capable of enantioselectively hydrolyzing chiral lactones to optically pure hydroxyl acids (Fig. 12), was newly isolated from soil in our laboratory and identified as *Fusarium proliferatum* [124]. From the crude extract of *F. proliferatum* ECU2002, a novel l-lactonase was purified to homogeneity, with a molecular mass of *ca.* 68 kDa and a p*I* of 5.7. The *F. proliferatum* lactonase preferentially hydrolyzed the l-enantiomer of butyrolactones, affording (+)-hydroxy acids with high enantioselectivity (94.8–98.2% *ee*) and good conversions (38.2–44.2%). The immobilized lactonase with simple glutaraldehyde cross-linking performed quite well in repeated batch resolution of dl-pantolactone at a concentration of 35% (w/v), retaining 67% of initial activity after ten cycles of reaction (corresponding to a half life of 20 cycles). The cells from other strains of *Fusarium* sp., which also produces lactonase, were immobilized with glutaraldehyde [125], *k*-carrageenan [126] and calcium alginate [127], and the stability of these biocatalysts was also increased significantly.

Sun ZH et al. successfully cloned the gene of D-pantonohydrolase from *Fusarium moniliforme* and expressed it in *Saccharomyces cerevisiae*. The recombinant yeast showed an excellent stereoselective activity toward D-pantolactone [128].

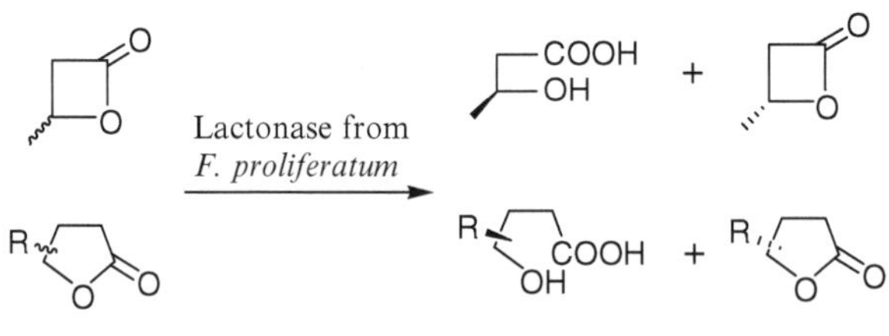

Fig. 12 Enantioselective hydrolysis of racemic lactones by l-lactonase from *Fusarium proliferatum* ECU2002 (cited from [124])

This D-pantonohydrolase was also selected for directed evolution through error-prone polymerase chain reaction (PCR) combined with DNA shuffling for improving the activity and pH stability via a convenient two-step high-throughput screening method based on the product formation and pH indicator [129].

4.2.3 Epoxide Hydrolases (E.C. 3.3.2.3)

Chiral epoxides and vicinal diols are versatile intermediates for the synthesis of enantiopure organic chemicals, especially in the case of pharmaceuticals. Epoxide hydrolases catalyze the stereoselective hydrolysis of racemic epoxides, yielding a mixture containing high optical purity of remaining epoxides and the formed diols [130–133]. Occasionally the single enantiomer of diols can be obtained with the throughout hydrolysis of epoxide substrates through so-called enantioconvergent hydrolysis if the enzymes have complementary enantioselectivities and opposite regioselectivities. Microbial epoxide hydrolases show great promise for the preparative scale hydrolysis of epoxides [134].

Our group [135–139] discovered two microbial strains with epoxide hydrolase activity toward phenyl glycidic ether (PGE) and its derivations (Fig. 13), the important precursors of β-adrenergic blockers. One strain is *Bacillus megaterium* ECU1001, which catalyzed the hydrolysis of the (*R*)-PGE preferentially, remaining the enantiopure (*S*)-PGE; the other is *Trichosporon loubierii* ECU1040, whose enantioselectivity was complement with ECU1001, *i.e.*, it catalyzed the hydrolysis of (*S*)-PGE preferentially.

Moreover, we interestingly found two kinds of novel epoxide hydrolases (mbEH-A and mbEH-B) from mung beans. These enzymes could catalyze the enantioconvergent hydrolysis of styrene oxide derivatives (Fig. 14) [140]. Using mung bean powder as the biocatalyst and *p*-nitrostyrene oxide as the substrate, the *ee* of the (*R*)-product was 82.4% when the substrate was hydrolyzed completely.

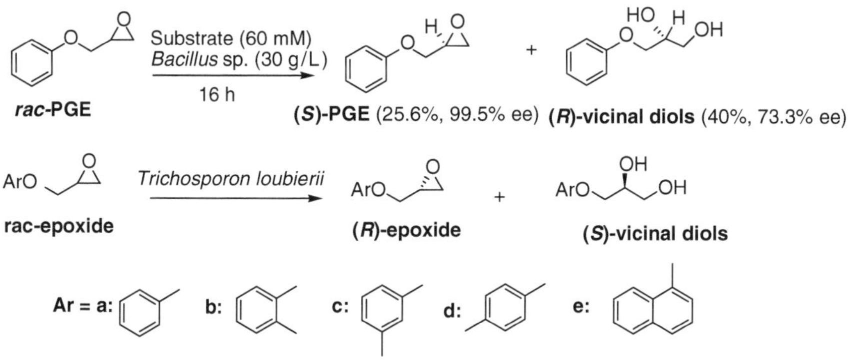

Fig. 13 Enantioselective hydrolysis of (*R*,*S*)-glycidyl aryl ethers by epoxide hydrolases

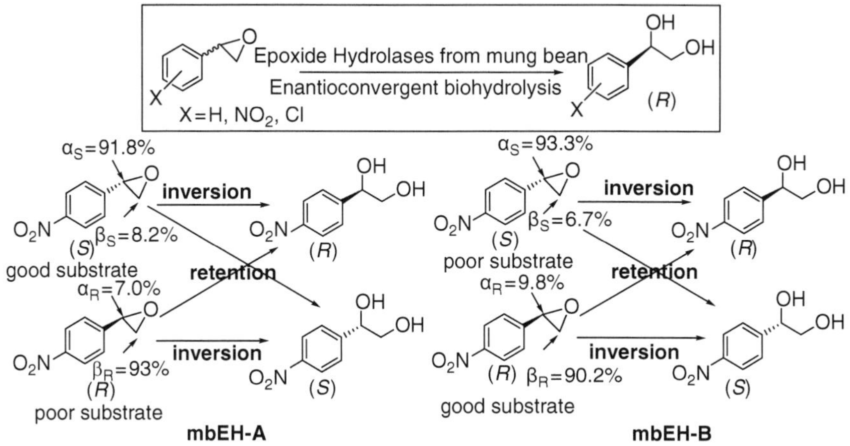

Fig. 14 Enantioconvergent biohydrolysis of racemic styrene epoxides catalyzed by mung bean epoxide hydrolases (mbEHs) A and B (cited with modification from [140])

The optical purity of (*R*)-diol could be improved to 99% *ee* via recrystalization, with a total yield of 68.7%.

Li et al. [140, 141] isolated a strain of *A. niger* CGMCC 0496, which has high stereoselectivity toward ortho- or para-substituted styrene-oxides. Zheng et al. [142] cloned, sequenced and expressed an epoxide hydrolase gene from *Rhodococcus opacus* in *E. coli*. This epoxide hydrolase showed a high stereospecificity for L(+)-tartaric acid, but not for D (–) tartaric acid.

4.2.4 Nitrilases (E.C. 3.5.5.1)/Nitrile Hydratases (E.C. 4.2.1.84, a Lyase)

Nitriles are an important type of synthetic intermediates in organic synthesis, because of their easy preparations and versatile transformations. Chemical hydrolysis of nitriles requires strong acid or base at high temperatures. In contrast, biotransformations of nitriles into the corresponding carboxamides and carboxylic acids proceed with excellent chemo-, regio- and stereoselectivities under very mild conditions [143–147]. Biocatalytic hydrolysis of nitriles is known to proceed through two distinct pathways: in the presence of a nitrilase, nitriles undergo direct bioconversion to corresponding carboxylic acids and ammonia; whereas a nitrile hydratase (NHase) catalyzes the hydration of nitriles followed by the biotransformation into the corresponding acids with the aid of an amidase (E.C. 3.5.1.4).

So far a large number of nitrile-hydrolyzing microorganisms have been isolated. Most of the strains have been shown to contain either a nitrilase or nitrile hydratase/amidase [148, 149]. Some nitrilases and nitrile hydratases have been purified [150], in which the three-dimensional structures of three enzymes [151–153] have been obtained. A few mechanisms have been proposed for the Fe–NHase and Co–NHase catalyzed hydration reaction of a nitrile [151].

Wang MX's group systematically studied enantioselective biotransformations of nitriles in organic synthesis utilizing nitrile-hydrolyzing microorganisms and enzymes. He has published an excellent review [147] on their work with emphasis on enantioselective biotransformations of various structures of nitriles and amides in organic synthesis, including racemic nitriles bearing one α-positioned chiral center [154–164], racemic nitriles bearing one chiral center β- or remote to cyano functional group [165–168], racemic nitriles containing a three-membered ring [169–180] and prochiral dinitriles [181, 182].

Other groups in China also discovered and utilized biocatalysts for production of carboxamides and carboxylic acids. For example, our group [183] isolated a thermostable nitrilase bacterium, *Alcaligenes* sp. ECU0401, from soil. The microbial nitrilase system exhibited high activity in the hydrolysis various nitriles. Significantly higher activity and enantioselectivity in the hydrolysis of mandelonitrile (50% yield, >99.9% *ee*) and higher degrees of glycolonitrile conversions were also obtained and glycolic acid was obtained in a yield of 96.5% from a total of 200 mM of glycolonitrile in the mode of sequential addition during the cultivation of *Alcaligenes* sp. ECU0401. Li ZY et al. [184] used the cells of *Rhodococcus* sp. CGMCC 0497 for synthesis of optically active α,α-disubstituted-α-cyanoacetamides from α,α-disubstituted -malononitriles (53% yield, >99.9% *ee*). Zheng YG et al. [185] used *Nocardia* sp. 108 to hydrate indole-3-acetonitrile to indole-3-acetamide; the maximal indole-3-acetonitrile conversion ratio and indole-3-acetamide yield were 34.3 and 32.7% (w/w), respectively.

4.3 *Lyases (E.C. 4)*

4.3.1 Hydroxynitrile Lyases (E.C. 4.1.2.–)

Hydroxynitrile lyases (Hnls) are important biocatalysts for the synthesis of optically pure cyanohydrins, which are used as precursors and building blocks for a wide range of high value-added fine chemicals [186]. Although two Hnl enzymes, from the tropical rubber tree *Hevea brasiliensis* and from the almond tree *Prunus amygdalus*, have already been used for large scale applications [187–191], the enzymes still need to be improved and adapted to the special demands of industrial processes. In many cases, directed evolution has been the method of choice to improve enzymes, and rapid screening procedure was generated in metagenome or directed evolution approaches [192, 193].

Our group [194] employed almond (*Prunus armeniaca L.*), peach (*Prunus persica L.*), Loquat (*Eriobottya L*) and Badamu (*Prunus communis* L. var. *dulcis* Borkh, almond from *Xinjiang*, China) kernels as (*R*)-oxynitrilase sources for the synthesis of (*R*)-cyanohydrins from aliphatic and aromatic aldehyde as well as methyl ketone under microaqueous conditions. Under the microaqueous conditions, cyanohydrins from aliphatic and aromatic aldehyde, methyl ketone, fluoro-substituted benzaldehydes, furanyl carboxaldehydes as well as thien-2-yl carboxaldehydes were synthesized with high enantioselectivities (Fig. 15) [194–197].

We also reported a practical high throughput continuous process for the synthesis of chiral cyanohydrins [198], in which pretreated almond meal (or other solid raw enzyme sources) was loaded in a column to form a reactor, a supplying system was attached to deliver a solution of substrate and HCN in isopropyl ether (IPE) on one end and a collecting–separating system on the other (Fig. 16). By controlling the flow rate, optimal conditions were achieved for the hydrocyanation of various aromatic carboxaldehydes in a micro-aqueous medium to produce chiral cyanohydrins in high yields and high enantiomeric excess (*ee*) with high substrate/catalyst ratio.

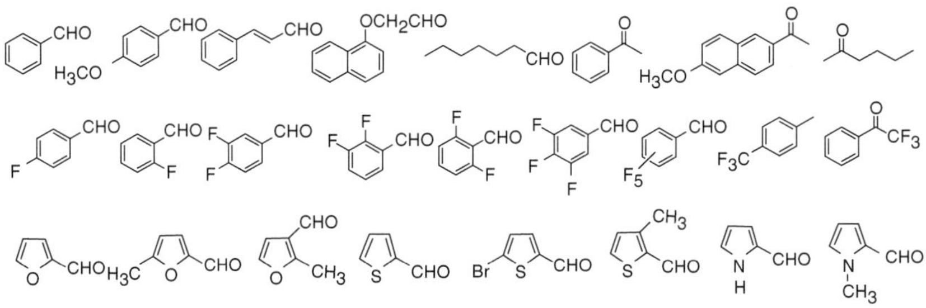

Fig. 15 Enantioselective synthesis of cyanohydrins from various aldehyde as well as methyl ketone

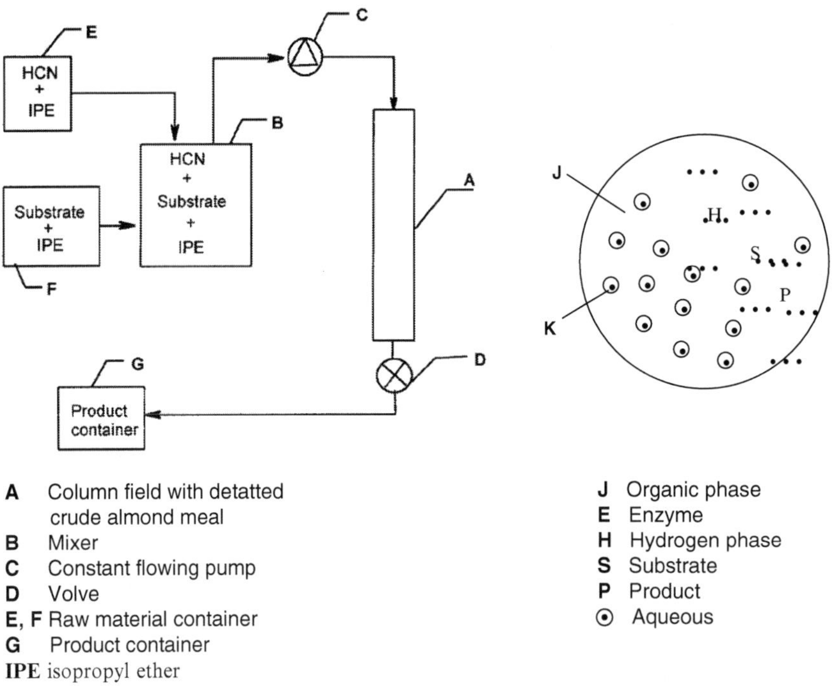

Fig. 16 A flow sheet of continuous process for the synthesis of chiral cyanohydrins under micro-aqueous conditions (cited with modification from [119, 205, 206, 120])

4.3.2 Aldolase (E.C. 4.1.2.-; E.C. 4.1.3.-)

Aldolases are a specific group of lyases catalyzing the reversible stereoselective addition of a nucleophilic donor onto an electrophilic acceptor. Aldolases are the most important group of asymmetric C–C bonding enzymes, which are currently used in bioorganic chemistry. The application of aldolases for asymmetric aldol reactions in chemoenzymatic syntheses has been reviewed in-depth over the years [199–205], mostly from the viewpoint of bioorganic chemists.

In a systematic study by researchers of Austra, various ring-substituted benzaldehydes (Fig. 17) were reacted with glycine under catalysis with a L-threonine aldolase (LTA) from *Pseudomonas putida* to form the corresponding β-hydroxy-α-amino acids [206].

Xu P et al. [207] efficiently synthesized *N*-acetyl-d-neuraminic acid (Neu5Ac) from lactate and a mixture of *N*-acetyl-d-glucosamine (GlcNAc) and *N*-acetyl-d-mannosamine (ManNAc) by whole cells (Fig. 18). The biotransformation utilized *E. coli* cells (Neu5Ac aldolase) and *Pseudomonas stutzeri* cells (lactate oxidase components, LOX). The results demonstrate that the reported Neu5Ac biosynthetic process can compare favorably with natural product extraction or chemical synthesis processes.

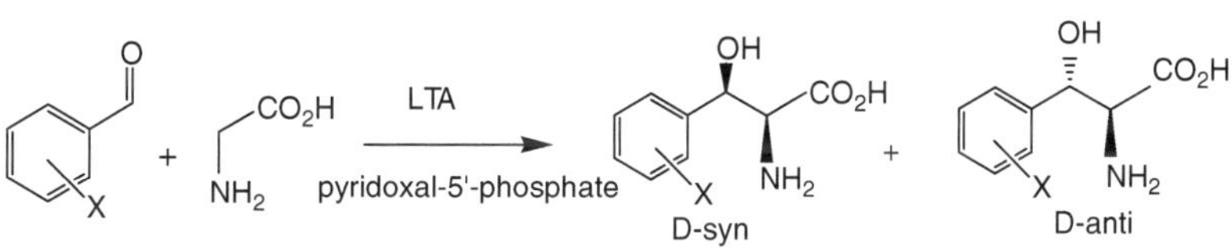

X = H, 2-F, 3-F, 4-F, 2-Cl, 3-Cl, 4-Cl, 2-Br, 3-Br, 4-Br, 2-NO_2, 3-NO_2, 4-NO_2, 4-$MeSO_2$, 2-NH_2SO_2, 3-OH, 4-OH,3, 4-OH,3, 4(O(Me), 3-MeOH-4-OH, 3,5-OH

Fig. 17 LTA-catalyzed synthesis of l-phenylserine derivatives

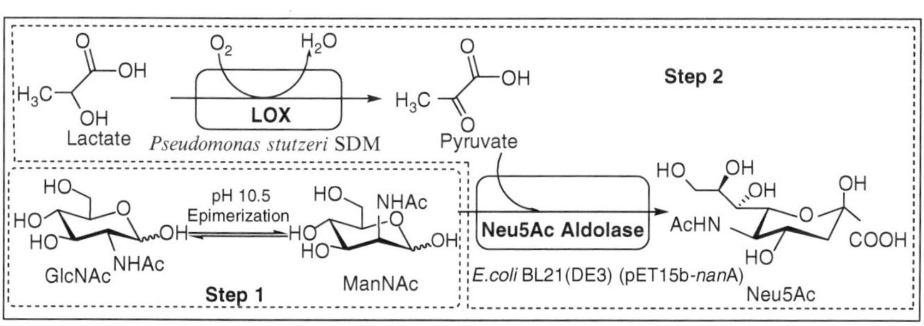

Fig. 18 A procedure for the production of Neu5Ac from GlcNAc/ManNAc and lactate (cited from [207]). Step 1: preparation of ManNAc/GlcNAc by the alkaline-catalyzed epimerization. Step 2: production of Neu5Ac by coupling whole-cell conversion of lactate and ManNAc

4.3.3 Other Lyases

Some novel lyases were used for enantioselective formation of carbon–carbon bonds in synthetic organic chemistry, such as benzoylformate decarboxylase (BFD) and benzaldehyde lyase (BAL) [208]. BFD catalyzes the acyloin or benzoin reaction between benzoylformate in the presence of acetaldehyde with concomitant decarboxylation of the former. Favorably, the enzyme also accepts various aldehydes instead of the expensive α-keto acids. As depicted in Fig. 19, a wide range of acyloin reactions (acceptor = acetaldehyde) and benzoin reactions (acceptor = aromatic aldehyde) were catalyzed by BFD, leading to the corresponding (*S*)-acyloins and (*R*)-benzoins, respectively. The limited tolerance of this enzyme for 2-substituted benzaldehydes was overcome by using L476Q or M365LL461S mutants. A biochemically related ThDP-dependent enzyme BAL also catalyzes the same carboligation reaction, but with opposite stereoselectivity, i.e., (*R*)-configuration, when acetaldehyde is used as acceptor.

4.4 *Isomerases (E.C. 5)*

Racemases and epimerases (E.C. 5.1.-.-) catalyze the inversion of stereocenters in organic synthesis. Racemases convert an enantiomer in a racemate, while epimerases

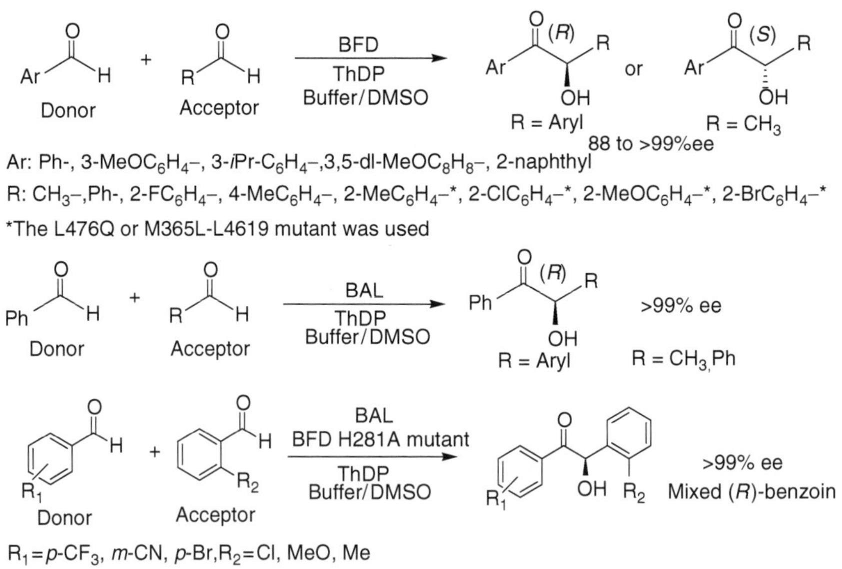

Fig. 19 Biocatalytic asymmetric C–C bond formation via acyloin and benzoin reactions by benzoylformate decarboxylase (BFD) and benzaldehyde lyase (BAL) (cited from [208])

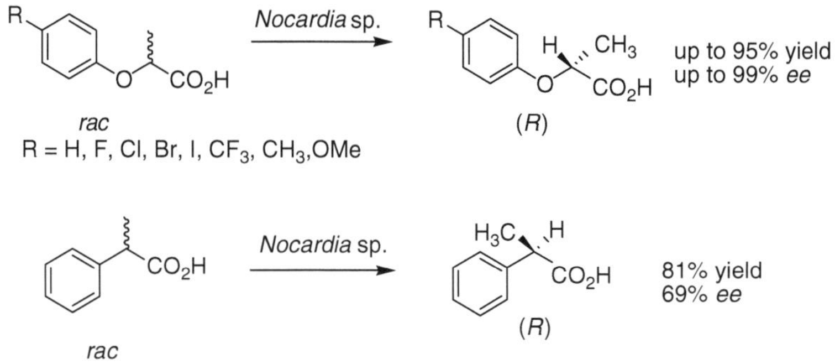

Fig. 20 Deracemization of α-phenyl- and ρ-substituted α-phenoxypropanoic acids by enzymes of *Nocardia diaphanozonaria* JCM3208 (cited from [214–216])

convert one diastereomer selectively to another diastereomer. Both enzymes are useful in the deracemization process, offering a potential of 100% yield [209, 210].

Faber's group has employed a few racemases for obtaining high yield in dynamic kinetic resolution or in auxiliary biocatalytic recycling processes. Faber and coworkers employed a mandelate racemase whose broad substrate tolerance for β,γ-unsaturated α-hydroxycarboxylic acids makes it a good candidate [211]. His group and collaborators at BASF have also found a lactate racemase that works with pharmaceutically important saturated aliphatic, arylaliphatic, and aromatic α-hydroxycarboxylic acids [212]. Employing a whole-cell system under mild conditions, they have achieved 'clean' racemization without unwanted side reactions. The same system also works with α-hydroxyketones [213].

Kato et al. [214–216] employed enzymes of *Nocardia diaphanozonaria* JCM3208 for deracemization of α-phenyl- and ρ-substituted α-phenoxypropanoic acids (Fig. 20), affording high yields and excellent enantiomeric excesses from the corresponding racemates.

5 Outlook

Biocatalysis is now becoming a key component in the toolbox of synthetic chemists. Biocatalytic technologies will ultimately gain universal acceptance when enzymes are perceived to be robust, specific and inexpensive. Classic cultivation-based methods and modern direct cloning strategies both have their advantages and limitations for biocatalyst screening. The access to novel natural sequence space, via direct cloning of metagenomic DNA, could significantly contribute to the identification of valuable biocatalysts.

The field of enzyme assays has undergone great advancement in recent years. Just a few years ago, classic methods such as nitrophenyl esters or ethers and alcohol dehydrogenase-coupled assays represented almost the entire known chemistry

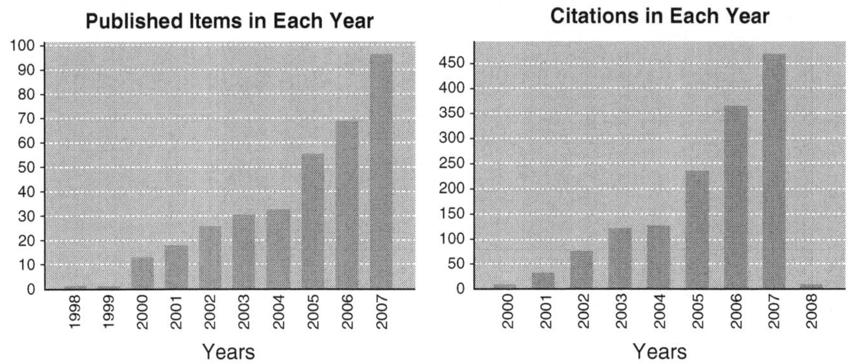

Fig. 21 Search results with the Web of Science® for publications on biotransformation by researchers in China and the citations thereof during 2000–2007

of enzyme assays. Now they have made room for a broad variety of labels, triggering chemistries, and signaling systems. There is no doubt that the field will continue to evolve as newer and better methods appear for screening enzymes. Challenges abound because many reaction types are still difficult to assay in high-throughput or with a sufficient level of accuracy.

In the worldwide and rapid progress of the industrial biocatalysis, China is making its increasing contributions. In addition, many research centers comprising biocatalysis units were being built in China recently. Chinese researchers have made their own contributions in most of this area (Fig. 21), although more attention should be paid to some new enzymes such as alkyl sulfatases, dehalogenases, decarboxylases, racemases and so on.

Acknowledgment Thanks are due for financial support from National Natural Science Foundation of China (Grant Nos. 20506037, 20672037 and 20672126) and Ministry of Science and Technology (Grant Nos. 2006AA02Z205, 2009CB724706 and 2007AA02Z225). We are indebted to the other members of our laboratories for their kind helps with the collection of related literature to this review.

References

1. Rouchi MA (2002) Chem Eng News 80:43
2. Buckland BC, Drew SW, Connors NC, Chartrain MM, Lee C, Salmon PM, Gbewonyo K, Zhou WC, Gailliot P, Singhvi R, Olewinski RCJr, Sun WJ, Reddy J, Zhang JY, Jackey BA, Taylor C, Goklen KE, Junker B, Greasham RL (1999) Metab Eng 1:63
3. Thayer AM (2006) Chem Eng News 84:15
4. Edegger K, Mang H, Faber K, Gross J, Kroutil W (2006) J Mol Catal A Chem 251:66
5. Rouhi AM (2003) Chem Eng News 81:45
6. Bornscheuer UT, Buchholz K (2005) Eng Life Sci 5:309
7. Patel RN (2006) Curr Org Chem 10:1289

8. Bommarius AS, Riebel BR (2004) Biocatalysis – fundamentals and applications. Wiley, Weinheim, pp 5, 45–47, 56
9. Chica RA, Doucet N, Pelletier JN (2005) Curr Opin Biotechnol 16:378
10. Otten LG, Quax WJ (2005) Biomol Eng 22:1
11. Giri A, Dhingra V, Giri CC, Singh A, Ward OP, Narasu ML (2001) Biotechnol Adv 19:175
12. Rothschild LJ, Mancinelli RL (2001) Nature 409:1092
13. Hong Y, Wang TW, Hudak KA, Schade F, Froese CD, Thompson JE (2000) Plant Biol 97:8717
14. Kutney JP (1993) Acc Chem Res 26:559
15. Andrade LH, Utsunomiya RS, Omori AT, Porto ALM, Comasseto JV (2006) J Mol Catal B Enzym 38:84
16. Bruni R, Fantin G, Medici A, Pedrini P, Sacchetti G (2002) Tetrahedron Lett 43:3377
17. Pras N, Woerdenbag JH, van Uden W (1995) Plant Cell, Tissue Organ Cult 43:117
18. Akakabe Y, Naoshima Y (1994) Phytochemistry 35:661
19. Lorenz P, Eck J (2004) Eng Life Sci 6:501
20. Wackett LP (2004) Curr Opin Biotechnol 15:280
21. Lorenz P, Liebeton K, Niehaus F, Eck J (2002) Curr Opin Biotechnol 13:572
22. Otten LG, Quax WJ (2005) Biomol Eng 22:1
23. Robertson DE, Steer BA (2004) Curr Opin Chem Biol 8:141
24. Wahler D, Reymond JL (2001) Curr Opin Chem Biol 5:152
25. Wahler D, Reymond JL (2001) Curr Opin Biotechnol 12:535
26. Goddard JP, Reymond JL (2004) Trends Biotechnol 22:363
27. Goddard JP, Reymond JL (2004) Curr Opin Biotechnol 15:314
28. Grognux J, Wahler D, Nyfeler E, Reymond JL (2004) Tetrahedron: Asymmetry 15:2981
29. Leroy E, Bensel N, Reymond JL (2003) Bioorg Med Chem Lett 13:2105
30. Leroy E, Bensel N, Reymond JL (2003) Adv Synth Catal 345:859
31. Yang YZ, Babiak P, Reymond JL (2006) Helv Chim Acta 89:404
32. Dean KES, Klein G, Renaudet O, Reymond JL (2003) Bioorg Med Chem Lett 13:1653
33. Ahmed V, Ispahany M, Ruttgaizer S, Guillemette G, Taylor SD (2005) Anal Biochem 340:80
34. Sicard R, Chen LS, Marsaioli AJ, Reymond JL (2005) Adv Synth Catal 347:1041
35. Zheng RC, Zheng YG, Shen YC (2007) Appl Microbiol Biotechnol 74:256
36. Zhu Q, Fan A, Wang YS (2007) Appl Environ Microbiol 73:6053
37. Winson MK, Bell DB (1997) Trends Biotechnol 15:120
38. Horeau A (1961) Tetrahedron Lett 2:506
39. Horeau A (1962) Tetrahedron Lett 3:965
40. Pérez CR, Jourdain N, Reymond JL (2000) Chem Eur J 6:4154
41. Bornscheuer UT (2004) Eng Life Sci 4:539
42. Huang HZ, Nishi K, Gee SJ, Hammock BD (2006) J Agric Food Chem 54:694
43. Reetz MT, Becker MH, Klein HW, Stockigt D (1999) Angew Chem Int Ed 38:1758
44. Guo J, Wu J, Siuzdak G, Finn MG (1999) Angew Chem Int Ed 38:1755
45. Finn MG (2002) Chirality 14:534
46. Zhang QS, Curran DP (2005) Chem Eur J 11:4866
47. Korbel GA, Lalic G, Shair MD (2001) J Am Chem Soc 123:361
48. Wahler D, Badalassi F, Crotti P, Reymond JL (2002) Chem Eur J 8:3211
49. Belder D, Ludwig M, Wang LW, Reetz MT (2006) Angew Chem Int Ed 45:2463
50. Reetz MT, Tielmann P, Eipper A, Ross A, Schlotterbeck G (2004) Chem Commun 1366
51. Liese A, Seelbach K, Wandrey C (2000) Industrial biotransformations. Wiley, Weinheim, p 99
52. Breuer M, Ditrich K, Habicher T, Hauer B, Keßeler M, Stürmer R, Zelinski T (2004) Angew Chem Int Ed 43:788
53. Liese A, Seelbach C, Wandrey C (2006) Industrial biotransformations, 2nd edn. Wiley, Weinheim
54. Kroutil W, Mang H, Edegger K, Faber K (2004) Curr Opin Chem Biol 8:120
55. Nakamural K, Matsuda T (2006) Curr Org Chem 10:1217

56. Kroutil W, Mang H, Edegger K, Faber K (2004) Adv Synth Catal 346:125
57. Goldberg K, Schroer K, Lütz S, Liese A (2007) Appl Microbiol Biotechnol 76:237
58. Goldberg K, Schroer K, Lütz S, Liese A (2007) Appl Microbiol Biotechnol 76:249
59. Nakamura K, Yamanaka R, Matsuda T, Harada T (2003) Tetrahedron: Asymmetry 14:2659
60. Buchholz K, Kasche V, Bornscheuer UT (2005) Biocatalysts and enzyme technology. Wiley, Weinheim, p 112
61. Wei ZL, Li ZY, Lin GQ (1998) Tetrahedron 54:13059
62. Wei ZL, Lin GQ, Li ZY (2000) Bioorg Med Chem 8:11297
63. Wei ZL, Li ZY, Lin GQ (2000) Chin J Chem 18:249
64. van der Donk WA, Zhao HM (2003) Curr Opin Biotechnol 14:421
65. Wichmann R, Vasic-Racki D (2005) Adv Biochem Eng Biotechnol 92:225
66. Leonida MD (2001) Curr Med Chem 8:345
67. Schubert T, Hummel W, Muller M (2002) Angew Chem Int Ed 41:634
68. St Clair N, Wang YF, Margolin AL (2000) Angew Chem Int Ed 39:380
69. Vrtis JM, White A, Metcalf WW, van der Donk WA (2002) Angew Chem Int Ed 41:3257
70. Boonstra B, Rathbone DA, French CE, Walker EH, Bruce NC (2000) Appl Environ Microbiol 66:5161
71. Geueke B, Riebel B, Hummel W (2003) Enzyme Microb Technol 32:205
72. Leonida MD, Fry AJ, Sobolov SB, Bartoszko-Malik A (2001) Int J Biochromat 6:207
73. Hollmann F, Schmid A, Steckhan E (2001) Angew Chem Int Ed 40:169
74. Duetz WA, van Beilen JB, Witholt B (2001) Curr Opin Biotechnol 12:419
75. Ni Y, Xu JH (2002) J Mol Catal B Enzym 18:233
76. Yang W, Xu JH, Xie Y (2006) Tetrahedron: Asymmetry 17:1769
77. Nie Y, Xu Y, Mu XQ (2004) Org Process Res Dev 8:246
78. He JY, Sun ZH, Ruan WQ, Xu Y (2006) Process Biochem 41:244
79. van Beilen JB, Duetz WA, Schmid A, Witholt B (2003) Trends Biotechnol 21:170
80. Urlacher VB, Schmid RD (2006) Curr Opin Chem Biol 10:156
81. Burton SG (2003) Trends Biotechnol 21:543
82. Bühler B, Schmid A (2004) J Biotechnol 113:183
83. Bentley R (2005) Chem Soc Rev 34:609
84. Fernández I, Khiar N (2003) Chem Rev 103:3651
85. Holland HL (2001) Nat Prod Rep 18:171
86. Li AT, Zhang JD, Xu JH, Lu NT, Lin GQ (2009) Appl Environ Microbiol 75:551
87. Gotor-Fernéndez V, Brieva R, Gotor V (2006) J Mol Catal B Enzym 40:111
88. Ghanem A, Aboul-Enein HY (2004) Tetrahedron: Asymmetry 15:3331
89. García-Urdiales E, Alfonso I, Gotor V (2005) Chem Rev 105:313
90. Tian P, Xu MH, Wang ZQ, Li ZY, Lin GQ (2006) Synlett 8:1201
91. Ghanem A, Aboul-Enein HY (2005) Chirality 17:1
92. Jaeger KE, Eggert T (2002) Curr Opin Biotechnol 13:390
93. Saxena RK, Sheoran A, Giri B, Davidson WS (2003) J Microbiol Methods 52:1
94. Hasan F, Shah AA, Hameed A (2006) Enzyme Microb Technol 39:235
95 Demirjian DC, Morís-Varas F, Cassidy CS (2001) Curr Opin Chem Biol 5:144
96. van den Burg B (2003) Curr Opin Microbiol 6:213
97. Ferrer M, Golyshina O, Beloqui A, Golyshin PN (2007) Curr Opin Microbiol 10:207
98. Jia SY, Xu JH, Li QS, Yu JT (2003) Appl Biochem Biotechnol 104:69
99. Long ZD, Xu JH, Zhao LL, Pan J, Yang S, Hua L (2007) J Mol Catal B Enzym 47:105
100. Long ZD, Pan J, Xu JH (2007) Appl Biochem Biotechnol 42:148
101. Gao L, Xu JH, Li XJ, Liu ZZ (2004) J Ind Microbiol Biotechnol 31:525
102. Luo Y, Zheng YT, Jiang ZB, Ma YS, Wei DZ (2006) Appl Microbiol Biotechnol 73:349
103. Liu YY, Xu JH, Hu Y (2000) J Mol Catal B Enzym 10:523
104. Liu YY, Xu JH, Xu QG, Hu Y (1999) Biotechnol Lett 21:143
105. Xu TW, Xu JH, Yu W, Zhong JH (2006) Biotechnol J 1:1293
106. Wu HY, Xu JH, Liu YY (2001) Synth Commun 31:3491
107. Xi WW, Xu JH (2005) Process Biochem 40:2161
108. Liu YY, Xu JH, Wu HY, Shen D (2004) J Biotechnol 110:209

109. Xu CF, Yuan CY (2005) Tetrahedron 61:2169
110. Zheng LY, Zhang SQ, Feng Y, Cao SG, Ma JS, Zhao LF, Gao G (2004) J Mol Catal B Enzym 31:117
111. Wu HY, Xu JH, Tsang SF (2004) Enzyme Microb Technol 34:523
112. Xu JH, Zhou R, Bornscheuer UT (2005) Biocatal Biotransf 23:415
113. Dai DZ, Xia LM (2006) Process Biochem 41:1455
114. Bai S, Guo Z, Liu W, Sun Y (2006) Food Chem 96:1
115. Lou WY, Zong MH (2006) Chirality 18:814
116. Wu XM, Xin JY, Sun W, Xia CG (2007) Chem Biodivers 4:183
117. Woese CR, Kandler O, Wheelis M (1990) Proc Natl Acad Sci U S A 87:4576
118. Ravot G, Buteux D, Favre-Bulle O, Wahler D, Veit T, Lefevre F (2004) Eng Life Sci 4:533
119. Qian JH, Xu JH (2004) J Mol Catal B Enzym 27:227
120. Chen Y, Xu JH, Pan J, Xu Y, Shi JB (2004) J Mol Catal B Enzym 30:203
121. Wu PC, Liu YH, Wang ZY, Zhang XY, Li H, Liang WQ, Luo N, Hu JM, Lu JQ, Luan TG, Cao LX (2006) J Agric Food Chem 54:836
122. Kataoka M, Honda K, Sakamoto K, Shimizu S (2007) Appl Microbiol Biotechnol 75:257
123. Honda K, Kataoka M, Shimizu S (2002) Biotechnol Bioprocess Eng 7:130
124. Zhang X, Xu JH, Xu Y, Pan J (2007) Appl Microbiol Biotechnol 75:1087
125. Hua L, Sun ZH, Zheng P, Xu Y (2004) Enzyme Microb Technol 35:161
126. Tang YX, Sun ZH, Hua L, Lv CF, Guo XF, Wang J (2002) Process Biochem 38:545
127. Yu MR, Tan TW (2005) Process Biochem 40:2609
128. Liu ZQ, Sun ZH ,(2004) Biotechnol Lett 26:1861
129. Liu ZQ, Sun ZH, Leng Y J Agric Food Chem (2006) 54:5823
130. Smit MS, Labuschagné M (2006) Curr Org Chem 10:1145
131. de Vries EJ, Janssen DB (2003) Curr Opin Biotechnol 14:414
132. Archelas A, Furstoss R (2001) Curr Opin Chem Biol 5:112
133. Fretland AJ, Omiecinski CJ (2000) Chem Biol Interact 129:41
134. Steinreiber A, Faber K (2001) Curr Opin Biotechnol 12:552
135. Tang YF, Xu JH, Ye Q, Schulze B (2001) J Mol Catal B Enzym 13:61
136. Gong PF, Xu JH, Tang YF, Wu HY (2003) Biotechnol Prog 19:652
137. Pan J, Xu JH (2003) Enzyme Microb Technol 33:527
138. Xu Y, Xu JH, Pan J, Zhao L, Zhang SL (2004) J Mol Catal B Enzym 27:155
139. Xu Y, Xu JH, Pan J, Tang YF (2004) Biotechnol Lett 26:1217
140. Xu W, Xu JH, Pan J, Gu Q, Wu XY (2006) Org Lett 8:1737
141. Jin H, Li ZY (2002) Biosci Biotechnol Biochem 66:1123
142. Jin H, Li ZY, Dong XW (2004) Org Biomol Chem 2:408
143. Liu ZQ, Li Y, Xu YY, Ping LF, Zheng YG (2007) Appl Microbiol Biotechnol 74:99
144. Kobayashi M, Shimizu S (2000) Curr Opin Chem Biol 4:95
145. Martínkové L, Mylerové V (2003) Curr Org Chem 7:1279
146. Singh R, Sharma R, Tewari N, Geetanjali, Rawat DS (2006) Chem Biodivers 3:1279
147. Banerjee A, Sharma R, Banerjee UC (2002) Appl Microbiol Biotechnol 60:33
148. Wang MX (2005) Top Catal 35:117
149. Martinkova L, Kren V (2002) Biocatal Biotrans 20:73
150. Cowan DA, Cameron RA, Tsekoa TL (2003) Adv Appl Microbiol 52:123
151. Huang W, Jia J, Cummings J, Nelson M, Schneider G, Lindvist Y (1997) Structure 5:691
152. Nagashima S, Nakasako M, Dohmae N, Tsujimura M, Takio K, Odeka M, Yohda M, Kamiya N, Endo I (1998) Nat Struct Biol 5:347
153. Miyanaga A, Fushinobu S, Ito K, Wakagi T (2001) Biochem Biophy Res Commun 288:1169
154. Wang MX, Lu G, Ji GJ, Huang ZT, Meth Cohn O, Colby J (2000) Tetrahedron: Asymmetry 11:1123
155. Gao M, Wang DX, Zhneg QY, Huang ZT, Wang MX (2007) J Org Chem 72:6060
156. Wang MX, Li JJ, Ji GJ, Li JS (2001) J Mol Catal B Enzym 14:77
157. Wang MX, Zhao SM (2002) Tetrahedron: Asymmetry 13:1695

158. Wang MX, Zhao SM (2002) Tetrahedron Lett 43:6617
159. Ma DY, Zheng QY, Wang DX, Wang MX (2006) Tetrahedron: Asymmetry 17:2366
160. Wang MX, Lin SJ (2001) Tetrahedron Lett 42:6925
161. Wang MX, Lin SJ (2002) J Org Chem 67:6542
162. Wang MX, Lin SJ, Liu J, Zheng QY (2004) Adv Synth Catal 346:439
163. Wang MX, Liu J, Wang DX, Zheng QY (2005) Tetrahedron: Asymmetry 16:2409
164. Liu J, Wang DX, Zheng QY, Wang MX (2006) Chin J Chem 24:1665
165. Wang MX, Wu Y (2003) Org Biomol Chem 1:535
166. Zhao SM, Wang MX (2002) Chin J Chem 20:1291
167. Ma DY, Wang DX, Zheng QY, Wang MX (2006) Org Lett 8:3231
168. Gao M, Wang DX, Zheng QY, Wang MX (2006) J Org Chem 71:9532
169. Wang MX, Feng GQ (2000) Tetrahedron Lett 41:6501
170. Wang MX, Feng GQ (2002) New J Chem 26:1575
171. Wang MX, Feng GQ (2003) J Org Chem 68:621
172. Feng GQ, Wang MX (2001) Chin J Chem 19:113
173. Wang MX, Feng GQ, Zheng QY (2003) Adv Synth Catal 345:695
174. Wang MX, Feng GQ, Zheng QY (2004) Tetrahedron: Asymmetry 15:347
175. Wang MX, Wang GQ (2002) J Mol Catal B: Enzym 18:267
176. Wang MX, Lin SJ, Liu CS, Sheng QS, Li JS (2003) J Org Chem 68:4570
177. Wang MX, Deng G, Wang DX, Zheng QY (2005) J Org Chem 70:2439
178. Yang L, Deng G, Wang DX, Huang ZT, Zhu J, Wang MX (2007) Org Lett 7:1387
179. Guo XL, Deng G, Xu J, Wang MX (2006) Enzyme Microb Technol 39:1
180. Wang JY, Wang DX, Zheng QY, Huang ZT, Wang MX (2007) J Org Chem 72:2040
181. Wang MX, Liu CS, Li JS, Meth Cohn O (2000) Tetrahedron Lett 41:8549
182. Wang MX, Liu CS, Li JS (2001) Tetrahedron: Asymmetry 12:3367
183. He YC, Xu JH, Xu Y, Ouyang LM, Pan J (2007) Chin Chem Lett 18:677
184. Wu ZL, Li ZY(2003) Tetrahedron: Asymmetry 14:2133
185. Wang YJ, Zheng YG, Xue JP, Shen YC (2006) Process Biochem 41:1746
186. Sharma M, Sharma NN, Bhalla TC (2005) Enzyme Microb Technol 37:279
187. Glieder A, Weis R, Skranc W, Pöchlauer P, Dreveny I, Majer S, Wubbolts M, Schwab H, Gruber K (2003) Angew Chem Int Ed 42:4815
188. Hasslacber M, Schall M, Hayn M, Griengl H, Kohlwein SD, Schwab H (1996) J Biol Chem 271:5884
189. Purkarthofer T, Pabst T, Broek CVD, Griengl H, Maurer O, Skranc W (2006) Org Process Res Dev 10:618
190. van Langen LM, Selassa RP, van Rantwijk F, Sheldon RA (2005) Org Lett 7:327
191. van Langen LM, van Rantwijk F, Sheldon RA (2003) Org Process Res Dev 7:828
192. Krammer B, Rumbold K, Tschemmernegg M, Pöchlauer P, Schwab H (2007) J Biotechnol 129:151
193. Andexer J, Guterl JK, Pohl M, Eggert T (2006) Chem Commun 4201
194. Lin GQ, Han SQ, Li ZY (1999) Tetrahedron 55:3531
195. Han SQ, Lin GQ, Li ZY (1998) Tetrahedron: Asymmetry 9:1835
196. Han SQ, Chen PR, Lin GQ, Huang H, Li ZY (2001) Tetrahedron: Asymmetry 12:843
197. Chen PR, Han SQ, Lin GQ, Huang H, Li ZY (2001) Tetrahedron: Asymmetry 12:3273
198. Chen PR, Han SQ, Lin GQ, Li ZY (2002) J Org Chem 67:8251
199 Samland AK, Sprenger GA (2006) Appl Microbiol Biotechnol 71:253
200. Sukumaran J, Hanefeld U (2005) Chem Soc Rev 34:530
201. Silvestri MG, Desantis G, Mitchell M, Wong CH (2003) Top Stereochem 23:267
202. Breuer M, Hauer B (2003) Curr Opin Biotechnol 14:570
203. Fessner WD, Helaine V (2001) Curr Opin Biotechnol 12:574
204. Machajewski TD, Wong CH (2000) Angew Chem Int Ed 39:1352
205. Takayama S, McGarvey GJ, Wong CH (1997) Annu Rev Microbiol 51:285
206. Steinreiber J, Fesko K, Reisinger C, Sch¨rmann M, van Assema F, Wolberg M, Mink D, Griengl H (2007) Tetrahedron 63:918

207. Xu P, Qiu JH, Zhang YN, Chen J, Wang PG, Yan B, Song J, Xi RM, Deng ZX, Ma CQ (2007) Adv Synth Catal 349:1614
208. Faber K, Kroutil W (2005) Curr Opin Chem Biol 9:181
209. Thayer AM (2006) Chem Eng News 84:29
210. Schnell B, Faber K, Kroutil W (2003) Adv Synth Catal 345:653
211. Felfer U, Goriup M, Koegl MF, Wagner U, Larissegger-Schnell B, Faber K, Kroutil W (2005) Adv Synth Catal 347:951
212. Glueck SM, Pirker M, Nestl BM, Ueberbacher BT, Larissegger-Schnell B, Csar K, Hauer B, Stuermer R, Kroutil W, Faber K (2005) J Org Chem 70:4028
213. Nestl BM, Kroutil W, Faber K (2006) Adv Syn Catal 348:873
214. Kato D, Mitsuda S, Ohta H (2002) Org Lett 4:371
215. Kato D, Mitsuda S, Ohta H (2003) J Org Chem 68:7234
216. Kato D, Miyamaoto K, Ohta H (2004) Tetrahedron Asymmetry 15:2965

Adv Biochem Engin/Biotechnol (2009) 113: 33-77
DOI: 10.1007/10_2008_25

Published online: 26 May 2009

Microbial Transformation of Nitriles to High-Value Acids or Amides

Jing Chen, Ren-Chao Zheng, Yu-Guo Zheng, and Yin-Chu Shen

Abstract Biotransformation of nitriles mediated by nitrile-amide converting enzymes has attracted considerable attention and developed tremendously in the recent years in China since it offers a valuable alternative to traditional chemical reaction which requires harsh conditions. As a result, an upsurge of these promising enzymes (including nitrile hydratase, nitrilase and amidase) has been taking place. This review aims at describing these enzymes in detail. A variety of microorganisms harboring nitrile-amide converting activities have been isolated and identified in China, some of which have already applied with moderate success. Currently, a wide range of high-value compounds such as aliphatic, alicyclic, aromatic and heterocyclic amides and their corresponding acids were provided by these nitrile-amide degrading organisms. Simultaneously, with the increasing demand of chiral substances, the enantioselectivity of the nitrilase superfamily is widely investigated and exploited in China, especially the bioconversion of optically active α-substituted phenylacetamides, acids and 2,2-dimethylcyclopropanecarboxamide and 2,2-dimethylcyclopropanecarboxylic acid by means of the catalysts exhibiting excellent stereoselectivity. Besides their synthetic value, the nitrile-amide converting enzymes also play an important role in environmental protection. In this context, cloning of the genes and expression of these enzymes are presented. In the near future in China, an increasing number of novel nitrile-amide converting organisms will be screened and their potential in the synthesis of useful acids and amides will be further exploited.

Keywords Amidase, Application, Nitrilase, Nitrile, Nitrile hydratase

J. Chen, R.-C. Zheng, Y.-G. Zheng (✉), and Y.-C. Shen
Institute of Bioengineering, Zhejiang University of Technology, Hangzhou 310014, People's Republic of China
e-mail: zhengyg@zjut.edu.cn

Contents

1 Introduction

Biotransformation and biocatalysis have gained increasing interest in recent years due to their mild conditions (physiological pH and ambient temperature), environmentally attractive catalysts, high activities and inherent excellent selectivities including chemo-, regio- and enantio- selectivities [1, 2]. Biocatalysis is now an established method in the synthesis of organic compounds and is especially useful for the production of chiral substances. By virtue of these obvious advantages, biotransformation is becoming a promising alternative to the traditional acid- or base-catalysed reactions, so is the case of nitrile hydrolysis. Nitriles, as the substrates, are widespread in the environment and they are produced by plants in various forms, such as cyanoglycosides, cyanolipids, ricinine, phenylacetonitrile, etc. [3]. Despite the fact that a majority of nitriles are highly toxic, mutagenic and carcinogenic in nature [4, 5], they are an important class of compounds for their ability to afford significant intermediates in the synthesis of acids, amides, amines, amidine, esters, aldehydes, ketones and so on by chemical and enzymatic hydrolysis. Chemical hydrolysis of nitriles was extensively applied to synthesize amides and acids previously; however, these applications may not be suitable for the hydrolysis of nitriles in the presence of sensitive groups. In sharp contrast, enzymatic hydrolysis of nitriles could alleviate this problem ascribed to the mild reaction conditions. Besides, three enzymes, namely nitrile hydratase (EC 4.2.1.84), nitrilase (EC 3.5.5.1) and amidase (EC 3.5.1.4) involved in the transformation of nitriles or amides exhibit great potential of chemo-, enantio- and regioselective synthesis [6].

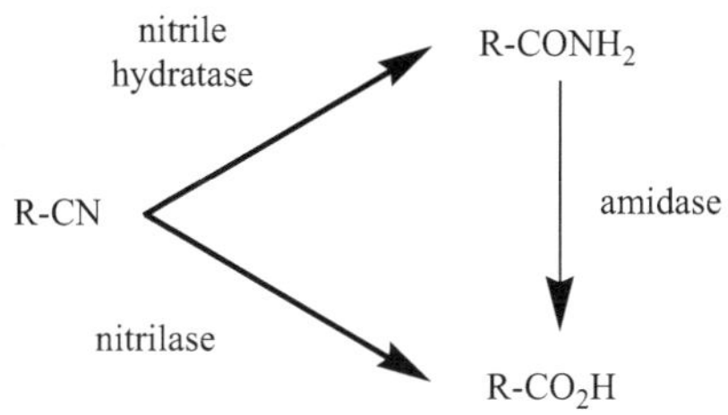

Scheme 1 The pathways of nitrile compounds by nitrile-amide converting enzymes

As a result, these enzymes have evoked substantial attention and they are becoming more and more demanding. These enzymes operate either by direct hydrolysis of nitrile to the corresponding acid (by a nitrilase enzyme) or by sequential action of an enzyme that hydrates the nitrile to the amide and the latter is transformed to the acid (by an amidase enzyme) (Scheme 1) [7, 8]. To date, various nitrile-amide converting organisms isolated from bacteria, fungi and plant have been described [9, 10]. Among them, most of them have been derived from bacterial species by enrichment strategies with nitriles as the sole nitrogen source [11]. Some reactions mediated by nitrile-converting enzymes have been applied on a large scale in industry. Productions of acrylamide [12] and nicotinic acid [13] on an industrial scale have proved the commercial value of these enzymes. With the fast development of these enzymes, an upsurge of biotransformation of nitriles has been taking place in China as well. According to the statistical data, an increasing number of reports have appeared and several institutes and universities have taken part in this research in recent years. As a result, a variety of microorganisms harboring nitrile-amide converting activities have been isolated, identified and characterized in various places, some of which have already applied with moderate success. Moreover, much work has focused on the organization and regulation of the genes encoding for nitrile metabolism. For example, research on the expressions of nitrile-degradation enzymatic system in recombinant strains has been carried out in China in recent years.

2 Description of Three Classes of Nitrile-Amide Converting Enzymes

2.1 Nitrile Hydratase

Nitrile hydratase, known as metalloenzyme, is a vital enzyme in the bienzymatic hydrolysis of nitriles to acids, which transforms nitriles to the corresponding amides. Asano et al. first reported the occurrence of nitrile hydratase from *Rhodocococcus rhodochrous* J-1 (formerly identified as *Arthrobacter* sp. J-1) to degrade acetonitrile, which was later applied with excellent success to the production of acrylamide from acrylonitrile on an industrial scale [14]. These findings promoted the intensive investigation of nitrile hydratase including physiochemical properties, substrate specificities and the reaction mechanism. According to the presence of metal co-factor, nitrile hydratase can be classified into two kinds: ferric nitrile hydratase and cobalt nitrile hydratase. The existence of metal ions in the active site of the enzyme is presumably effective in enhancing the folding or stabilization of the subunit that is dominantly consisted of α and β subunits. NHase can also be classified into high and low molecular weight (H- and L-NHases) on the basis of molecular weight of the enzyme. So far, a considerable number of microorganisms were successfully screened (Table 1).

Additionally, two new bacterial strains, *Pseudomonas marginales* MA32 and *Pseudomonas putida* MA113, containing nitrile hydratase resistant to cyanide were isolated from soil samples by an enrichment procedure [40]. In contrast to known nitrile hydratases, which rapidly lose activity at low to moderate cyanide concentrations, the enzymes tolerated up to 50 mM cyanide. Cyanide-resistant nitrile hydratase will find great application in the hydration of α-hydroxynitriles for the production of α-hydroxyamide because cyanide is always present in aqueous solutions of α-hydroxynitriles due to their tendency to decompose to the respective carbonyl compound and prussic acid.

Table 1 Some previously reported microorganisms with nitrile hydratase activity

Microorganisms	Substrates specificity
Agrobacterium tumefaciens d3 [15]	Arylnitriles, arylalkylnitriles, acrylonitrile
Arthrobacter sp. J-1 [16]	Alipatic nitriles
Bacillus cereus [17]	Acrylonitrile
Bacillus sp. BR 449 [18]	Acrylonitrile
Bacillus smithii SC-J05-1 [19]	Arylnitriles
Brevibacterium imperialis CBS 489-74 [20]	Acrylonitrile
Pseudomonas chlororaphis B23 [21]	Alkylnitrile
Pseudomonas putida [22]	Acetonitrile
Pseudonocardia thermophila JCM 3095 [23]	Acrylonitrile
Rhodococcus rhodochrous J-1 [24, 25]	Alkylnitrile, heterocyclic nitriles, arylnitriles
Rhodococcus rhodochrous R312 [26]	Alkylnitrile, benzonitrile
Rhodococcus rhodochrous LL 100-21 [27, 28]	Alkylnitriles, acrylonitrile, arylalkylnitriles, 3-cyanopyridine
Rhodococcus erythropolis BL1 [29]	Alkylnitriles, arylalkylnitriles
Rhodococcus rhodochrous A4 [30, 31]	Alkylnitriles, arylnitriles, cycloalkylnitriles, arylnitriles, heterocyclic nitriles, arylalkylnitriles
Rhodococcus sp. AJ270 [32]	Wide spectrum nitrile hydratase
Rhodococcus sp. SHZ-1 [33]	Acrylonitrile
Nocardia sp. 108 [34]	Acrylonitrile
Rhodococcus sp. ZJUT-N595 [35]	Acrylonitrile, glycolonitrile, 2,2-dimethylcyclopropanecarbonitrile
Rhodococcus sp. N 774 [36]	Aliphatic nitriles
Candida guilliermondii CCT 7207 [37]	Cycloalkylnitriles, arylnitriles, heterocyclic nitriles
Candida famata [38]	Alkylnitriles
Cryptococcus flavus UFMG-Y61 [39]	Isobutyronitrile

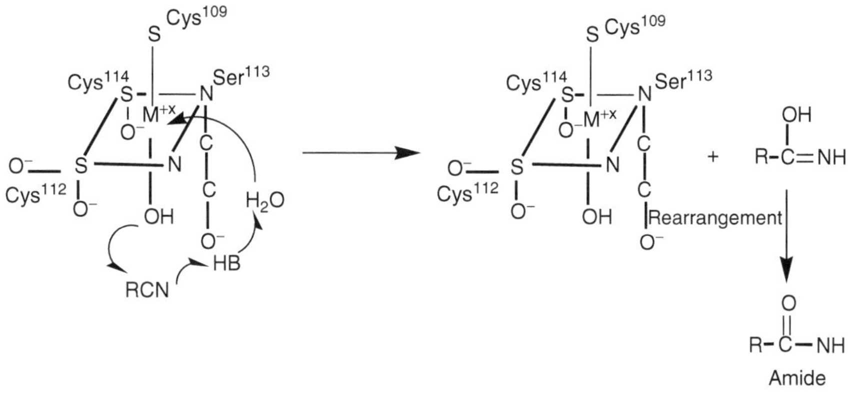

Fig. 1 Catalytic reaction mechanism of nitrile hydratase

To date, many studies focused on the mechanism of NHase mediated catalysis. Kobayashi et al. put forward a possible mechanism as follows. A water molecule was activated by the metal-bound hydroxide ion after the combination of the metal ion and water molecule. Imidate as an intermediate was initially formed via attacking on nitrile carbon by the activated water molecule. The imidate were then tautomerized to form the amide form (Fig. 1) [41].

2.2 *Nitrilase*

Nitrilase, the first nitrile-converting enzyme, was discovered in barley approximately 40 years ago and famous for the ability to convert indole-3-acetonitrile to the auxin indole-3-acetic acid [42]. From then on, several microorganisms harboring nitrilase activity have been screened, purified and characterized. Bacteria, fungi as well as plants provide excellent source of nitrilase, with bacterial species as the main source (Table 2), which are generally derived using enrichments from environmental samples. In the case of plants, numerous studies were carried out with *Arabidopsis thaliana*, from which four kinds of nitrilase were separated numbered NIT1, NIT2, NIT3 and NIT4. Although it was found that these enzymes were capable of transforming indole-3-acetonitrile to indole-3-acetic acid, recent results have indicated that NIT1, NIT2, NIT3 showed significant preference for 3-phenylpropionitrile, whose product, phenylacetic acid, is found in nasturtium. NIT4, however, was effective in hydrolyzing β-cyano-l-alanine [45].

Depending on the substrate specificity, nitrilase is differentiated into three subclasses: aliphatic nitrilase, aromatic nitrilase that shows preference for aromatic and heterocyclic nitriles and arylacetonitrilase which is highly specific for arylacetonitriles [61].

Table 2 Some previously reported microorganisms with nitrilase activity

Bacteria	Fungi	Plants
Acidovorax facilis 72W [43]	*Fusarium solani* IMI196840 [44]	*Arabidopsis thaliana* [45]
Bacillus pallidus Dac521 [46]	*Fusarium oxysporum* [47]	Barley [42]
Alcaligenes faecalis JM3 [48]	*Cryptococcus* sp. UFMG-Y28 [49]	Chinese cabbage [50]
Alcaligenes faecalis ATCC8750 [51]	*Aspergillus niger* [52]	Brassica rapa [53]
Rhodococcus rhodochrous J-1 [54]	*Penicillium multicolor* [55]	
Rhodococcus rhodochrous NCIMB 11216 [56]	*Exophiala oligosperma* R1 [57]	
Rhodococcus rhodochrous PA-34 [58]		
Rhodococcus rhodochrous K22 [59]		
Comamonas testosterone [60]		
Pseudomonas fluorescens DSM 7155 [61]		
Rhodococcus rubber [62]		
Acinetobacter sp. AK 226 [63]		
Klebsiella ozaenae [64]		
Arthrobacter sp. J1[65]		
Streptomyces sp. MTCC 7546 [66]		
Bacillus subtilis ZJB-063 [67]		

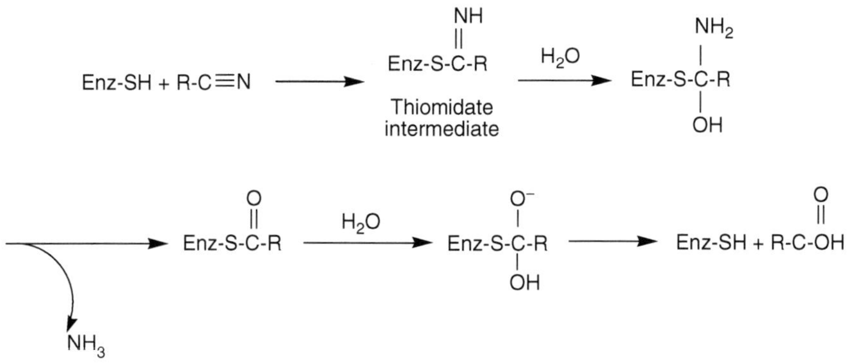

Fig. 2 Mechanism of nitrilase-catalysed reaction

Numerous investigations have provided insights into the mechanism of nitrilase catalyzed reaction. All nitrilases studied contain a cysteine residue in their catalytical center. The mechanism involves a nucleophilic attack by a thiol group in cysteine residue on the nitrile C-atom, forming an enzyme-linked tetrahedral thiomidate intermediate which is then attacked by H_2O and nitrogen atom is released as NH_3. Further addition of H_2O results in the production of acid and a regenerated enzyme (Fig. 2) [68].

2.3 Amidase

Amidase, amide bond-cleaving enzymes, exists ubiquitously in nature in both prokaryotic and eukaryotic forms. To the best of our knowledge, amidase-mediated processes have been extensively investigated, especially, the hydrolysis of amides to the corresponding carboxylic acid and ammonia. Additionally, hydroxamic acids were formed owing to the acyl transfer activity of amidase in the presence of hydroxylamine. Two reactions involved are shown in Scheme 2.

Besides, amidase is also capable of catalyzing diverse reactions such as ester hydrolysis, hydroxamic acid hydrolysis, acid hydrazide hydrolysis, amide transfer on hydrazine, ester transfer on hydroxylamine, ester transfer on hydrazine and so on [69].

Therefore, amidase turns out to be efficient and promising tools for the synthesis of various compounds. In addition, with regard to nitrile hydratase in Gram-positive bacterium, its enantioselectivity was always combined with the amidase's. Namely, their cooperation gave rise to the excellently pure products. However, the former usually displayed almost no stereoselectivity with the amidase being a major contributor. Consequently, significant attention has been paid to the isolation and discovery of amidase-producing organisms including bacteria, yeasts, fungi, plants and animals (Table 3).

Although amidase catalyzes many reactions, some of them proceed at a comparative low rate employing esters or carboxylic acids as acyl donors. In sharp contrast, high amidase activity is achieved in the presence of water (H_2O) and hydroxylamine (NH_2OH) as the cosubstrates when amide is used as the substrate, indicating that these two compounds functions as efficient acyl acceptors [69].

Both the amidase-catalyzed hydrolytic reaction and the acyl transfer reaction share the same reaction mechanism. In view of this, study on the mechanism of the acyl transfer reaction shed light on that of hydrolytic reaction, in which case, there is difficulty in investigating the mechanism where water is the cosubstrate. A possible mechanism suggested the reaction belonged to Ping Pong Bi Bi type: The carbonyl group of amide undergoes a nucleophilic attack by the enzyme, leading to the formation of a tetrahedral intermediate, which is consequently converted to an acyl–enzyme intermediate with the release of

Amide hydrolysis:

$$RCONH_2 + H_2O \longrightarrow RCOOH + NH_3$$

Amide acyl transfer reaction:

$$RCONH_2 + NH_2OH \longrightarrow RCONHOH + NH_3$$

Scheme 2 The pathways of hydrolysis and transfer of amide Amide hydrolysis: $RCONH_2 + H_2O \rightarrow RCOOH + NH_3$ Amide acyl transfer reaction: $RCONH_2 + NH_2OH \rightarrow RCONHOH + NH_3$

Table 3 Some previously reported microorganisms with amidase activity

Microorganisms	Substrate specificity
Rhodococcus erythropolis MP50 [70]	Aromatic amide
Geobacillus pallidus RAPc8 [71]	Aliphatic amide
Sulfolobus tokodaii strain 7 [72]	Aromatic amide
Delftia acidovorans [73]	d-Amino acid amide
Arthrobacter sp. J-1 [74]	Aliphatic amide
Rhodococcus rhodochrous M8 [75]	Aliphatic amide
Xanthobacter flavus NR303 [76]	l-Amino acid amide
Brevibacterium sp Strain R312 [77]	Aryloxypropionamides
Variovorax paradoxus [78]	d-Amino acid amide
Pseudonocardia thermophila [79]	Aliphatic, aromatic and amino acid amide
Pseudomonas sp. MCI3434 [80]	Heterocyclic amide
Klebsiella oxytoca [81]	Aliphatic amide
Brevibacillus borstelensis BCS-1 [82]	Aromatic and aliphatic amide
Ochrobactrum anthropi SV3 [83]	Amino acid amide
Stenotrophomonas maltophilia [84]	Peptide amide
Agrobacterium tumefaciens strain d3 [85]	Aromatic amide
Sulfolobus solfataricus MT4 [86]	Aliphatic and aromatic amide
Klebsiella pneumoniae NCTR1 [87]	Aliphatic amide
Bacillus stearothermophilus BR388 [88]	Wide spectrum amidase
Delftia tsuruhatensis ZJB-05174 [89]	2,2-Dimethylcyclopropanecarboxamide

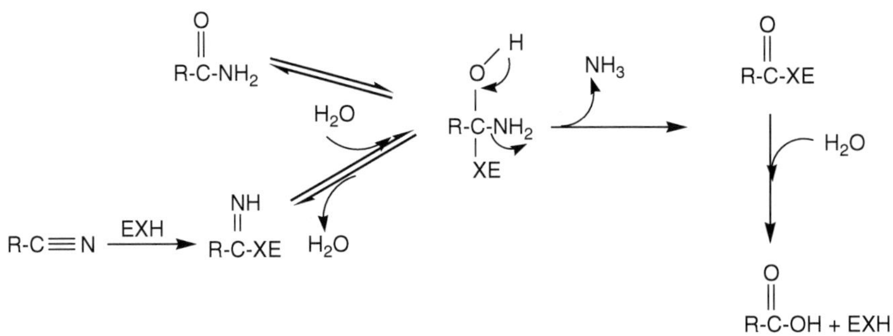

Fig. 3 Mechanism of amidase-catalysed reaction

ammonia. The acyl–enzyme complex in turn is subjected to attack by water or hydroxylamine (Fig. 3) [9, 90].

Importantly, a majority of amidases bear enantioselectivity, which contributed to the synthesis of chiral carboxylic acid via nitrile hydratase and amidase. However, these nitrile-hydratase-associated amidases are, surprisingly, mostly *S*-stereospecific. *R*-Enantioselective amidases are gaining more and more interest because of their potential application in the production of d-amino acids and other optically active compounds. Hydroxamic acids, products of the acyl transfer reactions, can be detected easily and fairly specifically by the addition of acidic ferric chloride solution, which results in the production of a deep magenta color [91]. Therefore, this acyl transfer reaction can be employed in a colorimetric screening procedure for active and enantioselective amidases. By employing the amidase-catalyzed acyl transfer reaction, *Delftia tsuruhatensis* producing *R*-enantioselective amidase was screened in our lab [89, 92].

3 The Isolation and Identification of the Nitrile-Amide Converting Organisms in China

Biotransformation of nitriles is of great potential in organic synthesis and it provides green access to various carboxylic acids and amides; thus nitrile-amide converting enzymes are of broad use and commercial interest. Recently, biotransformation of carboxylic acids and amides via these enzymes has been a hot issue in China. The scarcity of appropriate nitrile-amide converting biocatalysts and the difficulty in commercial availability of these enzymes promoted the screening and discovery of the novel nitrile-amide converting organisms in China. So far, a series of organisms producing nitrile-amide degrading enzymes were isolated, identified and characterized, some of which were purified. Among the obtained organisms, it was observed that some harbored nitrile hydratase, some produced nitrilase and others could form amidase. It was also found that the existences of nitrile hydratases were always accompanied by amidases, so amides and acids were formed in different proportions in such kinds of microbes mediated reactions. A prominent example is *Rhodococcus* sp. AJ270, which is a powerful and robust nitrile hydratase/amidase-containing microorganism isolated by Wang et al. Later, its broad applications in transforming various nitriles were substantially explored [32]. *Rhodococcus* sp. AJ270 as well as other nitrile-amide converting organisms was dominantly obtained by enrichment strategies where nitriles were employed as the sole nitrogen source ascribed to the highly toxic nature. Owing to the fact that screening a desired organism for a particular biocatalytic process is always a time-consuming and tedious job, some direct and sensitive readouts of the nitrile-amide converting enzyme activity have to be considered and developed. Conventional routes employ high-performance liquid chromatography, liquid chromatography-mass spectrometry, capillary electrophoresis, or gas chromatography to determine the enzyme activity, where determinations have to carry out one by one. Furthermore, those traditional enrichment strategies usually result in the isolation of a rather restricted group of microorganisms. A successful instance of application of high throughput screening method in our research was the isolation of *Delftia tsuruhatensis*, a *R*-stereospecific amidase producing bacterium. *R*-Enantioselective amidases are of considerable industrial interest due to potential applications in the production of optically active compounds [92]. More recently, Zhu et al., driven by the attempt to find a fast, convenient and sensitive method, reported a more accurate and innovative high throughput route. In their paper, a novel time-resolved luminescent probe: *o*-hydroxybenzonitrile derivatives could be applied to detect the activity of the nitrilases. By the action of nitrilases, *o*-hydroxybenzonitrile derivatives could be transformed to the corresponding salicylic acid derivatives, which, upon binding Tb^{3+}, served as a photon antenna and sensitized Tb^{3+} luminescence. Because of the time-resolved property of the luminescence, the background from the other proteins (especially in the fermentation system) in the assay could be reduced and, therefore, the sensitivity was increased. Moreover, because the detection was performed on a 96- or 384-well plate, the activity of the nitrilases from microorganisms could be determined quickly [93].

Moreover, some other high throughput methods have been reported as alternatives to conventional screening methods. A critical review on selection and screening strategy for enzymes of nitrile metabolism based on spectrophotometric and fluorimetric methods has been published [94]. Recently, convient screening methods have been developed on the basis of the color variation of indicators which are added to the mixture in advance. Once the acid formed, the color would have a dramatic change [95]. Additionally, a new method for nitrilase screening has been developed to detect nitrilase activity. The ammonia product of nitrilase mediated conversion of nitriles forms a complex with the cobalt ion results in a color change, which can readily be quantified using a spectrophotometer at 375 nm. The assay has the potential to be used for the real-time monitoring of nitrilase-catalyzed reactons [96]. More noticeably, Hu et al. introduced a simple and rapid high-throughput screening method based on a colorimetric reaction of glycolic acid with β-naphthol in sulfuric acid solution to isolate glycolonitrile-hydrolyzing microorganisms. Four strains able to convert glycolonitrile to glycolic acid were isolated from soil samples using this screening method, among which *Rhodococcus* sp. ZJUT-N595 displayed the highest hydrolytic activity [35].

These soil-derived nitrile-amide hydrolyzing organisms have been currently under active development and some have even achieved with small to moderate success. The advantage of applying whole cell biocatalysts lies in that they can be relatively easily and cheaply prepared and the whole cell catalyzed reactions can be operated much more easily. Nevertheless, some small aliphatic nitriles, hydroxyl- and amino-substituted nitriles give lower yields and appear to be alternatively metabolized when whole cell biocatalysts were applied [32]. Hence, on one hand, careful monitoring of the reaction is strongly recommended to achieve the maximal desired product. On the other hand, the use of purified enzymes is of substantial significance and benefit in case that substrate or product utilization by whole cells exists. Due to this, the purification and characterization of nitrile-amide converting are under progress in China.

4 Factors Affecting the Activity and Enantioselectivity of Nitrile-Amide Converting Enzyme

Numerous factors, such as some culture conditions like carbon source, nitrogen source, inducer and conversion conditions like temperature, pH, reaction time, cosolvent and so on, turn out to affect the activity and enantioselectivity, and consequently the biomass production.

4.1 Inducer

Nitrile hydratase and nitrilase are generally inducible, with a paucity of them being constitutive. Namely, the activity could be detected only in the presence of suitable inducers. Substrate, product, or their analogs are usually functioned as inducers, with the exception of some extremely toxic nitriles, such as mandelonitrile, in which case, the growth of the microorganism was completely inhibited. Urea and ε-caprolactam, potential nitrile hydratase inducers, play important roles in the induction of nitrile hydratase activity. Moreover, with the addition of different inducers, microorganism harboring versatile nitrile-converting enzyme activities exhibited various activities. A distinguished instance was *R. rhodochrous* J-1, a currently utilized organism in commercial synthesis of acrylamide in Japan, which was found to contain two inducible nitrile hydratases, one of which was specific for aliphatic nitriles induced by urea and the other for aromatic nitriles with cyclohexanecarboxamide and crotonamide as the inducers [24, 97]. Similar phenomena were found in *Nocardia Rhodochrous* LL100–21 [98], *R. rhodochrous* NCIMB 11216 [99, 100], *Bacillus pallidus* DAC521 [46, 101] and *Nocardia globerula* NHB-2 [102]. *Bacillus subtilis* ZJB-063, a newly isolated strain in our research, exhibited nitrilase activity without addition of inducers, indicating that the nitrilase in *B. subtilis* ZJB-063 is constitutive. Interestingly, the strain exhibited nitrile hydratase and amidase activity with the addition of ε-caprolactam. The versatility of this strain in the hydrolysis of various nitriles and amides makes it a potential biocatalyst in organic synthesis [67]. In a word, selecting a suitable inducer is of great significance in the formation of nitrile-amide converting enzymes, and in enhancing the activity as well.

4.2 Metal Ions

As far as nitrile hydratases are concerned, metal ions predominantly including Fe^{3+} and Co^{2+} are essential in the exhibition of its activity. In case of Fe^{3+} type nitrile hydratase, only with the addition of Fe^{3+} in the nutrient broth acted as co-factor could nitrile hydratase activity be observed, so is the Co^{2+} type nitrile hydratase [24]. Nitrile hydratase from the fungus *Myrrothecium verrucaria* even has Zn^{2+} in the active site and in such case Zn^{2+} is necessary for the formation of the enzyme [103].

Unlike nitrile hydratases, nitrilases show no requirement of any metal co-factor. Instead, they are proved to have cataltically essential cysteine residues. Effect of metal ions on the nitrilase activity in many studies indicated that thiol binding reagents like $CuCl_2$ and $AgNO_3$ were strong inhibitors of the nitrilase activity [61, 63, 67]. The high sensitivity to these metal ions suggested that one or more thiol residues were necessary for this enzyme and these ions should not be involved in the nutrient broth or reaction mixture.

4.3 Effect of Light on Nitrile Hydratase

As mentioned earlier, the activity of nitrile hydratase displayed unique features with the exposure to light. A nitrile hydratase producing strain *Rhodococcus* sp. N-771 exhibited extremely low activities when the cultivation was carried out in the dark. However, recovery of activity occurred with the irradiation of light [104, 105].

4.4 Amidase Inhibitor

As for amidase, it is a generally accepted fact that urea would have a negative effect on the activity, which is an important feature, especially in the production of some valuable amides via organisms with nitrile hydratase and amidase activity. Undesired acid would form by the cleavage on the amide by accompanying amidase. However, addition of urea could not only protect the amide from serious acid contamination, but also keep the amount of amide constant. Previous research also demonstrated the inhibitory effect derived from the competitive inhibition for active site of amidase between urea and the reactive amide [74, 106]. With that exception, no significant inhibition effect of urea on amidase catalyzed acyl transfer activity and hydrolytic activity from *D. tsuruhatensis* ZJB-05174 was observed, indicating that the inhibition effect of urea was not exclusive [89]. Bauer et al. indicated that diethyl phosphoramidate was also an excellent inhibitor of the amidase from *Agrobacterium tumefaciens* strain d3 [15]. In the presence of urea or chloroacetone, amidase activity in *Bacillus* spp. was inhibited and the amide intermediate was accumulated [106]. Bearing this in mind, we can procure amides in high yields and with excellent purity.

4.5 Temperature and pH

Temperature and pH significantly affect the enzyme activity, and sometimes enantioselectivity. Nitrile-converting biocatalysts mediated reactions are often operated in a narrow pH range, neutral or slightly alkaline. These enzymes mostly exhibit comparatively low activity in too acid or alkaline environments. Therefore, addition of HCl is employed to stop the reaction. With the exception, a nitrile hydrolyzing acidotolerant black yeast *Exophiala oligosperma* R1 was isolated in order to convert α-amino- and α-hydroxynitriles whose enzymatic conversion was hampered by their low stability under neutral conditions [107].

Besides, pH could sometimes affect enantioselectivity to a certain extent. Attempts were made by Wang et al. to improve the enantioselectivity of nitrile of the biotransformation of phenylglycine nitrile by altering the buffer pH from 7.62 to 7.13.

Results indicated that the enantioselectivity of the transformation was enhanced and both the amide and acid form were obtained with ee values over 99% at a pH of 7.13 within 8 h though slower rate was observed [108].

The effect of temperature on the enzyme activity is duple. On one hand, according to Arrhenius's equation ($k = Ae^{-\frac{Ea}{RT}}$), enhancement of the activity occurs with an increase of temperature. On the other hand, enzyme inactivation accompanies an increase of temperature. A large portion of nitrile-amide converting enzymes are not thermal stable and they are usually inactive above 50 °C. A small alteration in temperature may lead to substantial change in enzyme activity. *B. subtilis* ZJB-063, for example, displayed an abrupt decrease at 40 °C and only 13.36% of the activity at 32 °C was inspected, indicating that enzyme activity are sensitive to temperature [109]. Enzymes of good thermostability are of significant importance in pharmaceutical industry. The discovery of thermophilic nitrile-metabolizing microorganisms enables nitrile conversion at high temperature above 50 °C. Zheng et al. succeeded in screening an amidase producing bacterium with excellent thermostability [89].

Along with enzyme activity, enantioselectivity is also influenced by temperature. Wang and coworker observed that lowering the reaction temperature from 30 to 20 °C would lead to increased enantioselectivity of *Rhodococcus* sp AJ270 mediated synthesis of optically active 2, 2-dimethylcyclopropanecarboxylic acid [110].

More interestingly, temperature played an unexpected role in the stereoselectivity of amidase in *D. tsuruhatensis* ZJB-05174. It was detected that enzyme underwent an unusual temperature-dependent reversal of stereospecificity. With the increase in the reaction temperature, the E-value dropped from 27 (30 °C) to 5 (46 °C). When the reaction temperature reached 56 °C, it exhibited reversal stereospecificity, though with a comparatively low E value of 0.03 [89]. This phenomenon can be interpreted by temperature variation caused change on the difference in activation free energy $\Delta\Delta G^{\dagger}$, which can be divided into its enthalpic $\Delta\Delta H^{\dagger}$and entropic $\Delta\Delta S^{\dagger}$ components [111].

4.6 Organic Solvents

As is generally accepted, organic solvents are widely applied in the lipase or esterase mediated reactions because of the hydrophobic nature of esters. Most nitriles also exhibit low solubility in aqueous solutions. Supplementing organic cosolvents to the reaction mixture containing a nitrile-amide converting biocatalyst is considered to be a useful way to increase the availability of the substrate. However, addition of organic solvents might lead to the inactivation of enzyme. Hence, the amount and suitable kind of the solvents should be carefully evaluated in terms of activity and the availability of soluble substrates. There is only a limited portion of *p*-methoxyphenylacetonitrile available to the enzyme in aqueous solution when fed to *B. subtilis* ZJB-063 due to its low solubility, which in turn resulted in low activity.

DMSO and methanol of 5 vol.% in the reaction mixture, as the cosolvents, resulted in approximately 62.7% and 15.2% enhancement of activity, respectively, compared to the control where there was no cosolvent. Further increase in concentration of the two solvents, on the contrary, caused a loss of activity, implying that protein denaturation occurred at concentrations above 5 vol.% [109].

Besides, some foreign researches indicated that high percentages of organic solvents were also acceptable by some nitrile-amide converting enzymes. A purified nitrile hydratase from *Rhodococcus equi* A4 showed high resistance to organic solvents and it could tolerate up to 90 vol.% isooctane or pristine [112].

Organic cosolvent showed an effect not only on the activity but also the enantioselectivity of nitrile-amide converting enzymes. Hydrocarbons and methanol (5 vol.%) increased the enantioselectivity of the nitrile hydratase from *Rhodococcus equi* for the conversion of 2-(6-methoxynaphthyl)propionitrile from moderate to good (E = 14.8–41) [112].

4.7 Steric and Electronic Factors

Both steric and electronic factors dramatically affected the reactivity and more importantly, the enantioselectivity. Considerable investigations concerning the effect of substituents have been carried out. Both the kind and position of substituents have much to do with enzyme activity. Wu and Li conducted a successful asymmetric hydrolysis of α,α-disubstituted malonamides to afford enantiopure (*R*) -α,α-disubstituted malonamic acids employing the strain *Rodococcus* sp. CGMCC 0497 (Scheme 3). In their study, it seemed that the results were not only influenced by steric hindrance but by an electronic effect of the substrates as well. Among the substrates bearing aromatic ring substituents in the *ortho-*, *meta-*, and *para-*positions, all para-substituted ones gave products with excellent ee values, slightly higher

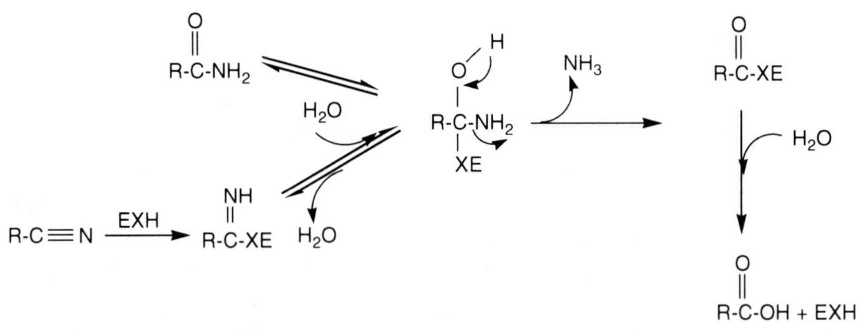

Scheme 3 Asymmetric hydrolysis of α,α-disubstituted malonamides by *Rhodococcus* sp. CGMCC 0497

than *ortho*- and *meta*-substituted ones(Table 4) [113]. More noticeably, when α-substituted phenylacetonitriles and phenylacetamides were subjected to the *Rhodococcus* sp. AJ270 catalyzed hydrolysis, the reaction outcome was remarkably affected by the nature of α-substituent (Scheme 4). As shown in Table 5, small groups appeared to have no obvious effect on enzyme activity, whereas introduction

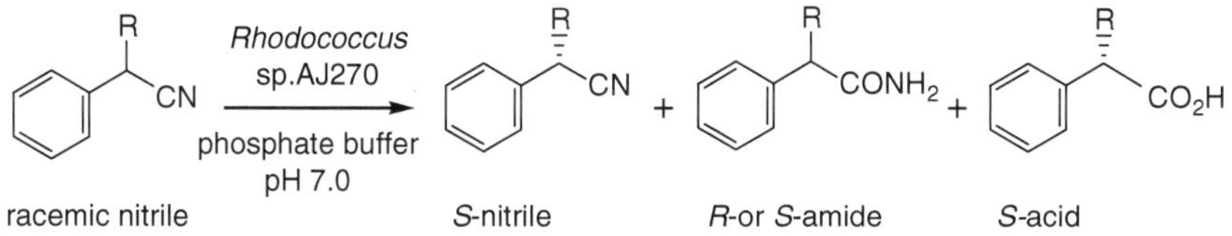

(R=Me, Et, *n*-Pr, *i*-Pr, *n*-Bu, MeO, MeS)

Scheme 4 Enantioselective biotransformations of racemic α-substituted phenylacetonitriles and phenylacetamides by *Rhodococcus* sp. AJ270

Table 4 Enantioselective hydrolysis of various α-substituted-α-methylmalonamides by *Rhodococcus* sp. CGMCC 0497

Substrate (R)	Yield (%)	*e.e.* (%)
C_6H_5	94	97
o-ClC_6H_4	94	97
m-ClC_6H_4	94	95
p-ClC_6H_4	92	>99
p-$CH_3C_6H_4$	97	>99
p-MeOC_6H_4	95	>99
p-FC_6H_4	95	>99
p-BrC_6H_4	97	>99
$C_6H_5CH_2$	94	>99
C_3H_7	94	91

Table 5 Enantioselective hydrolysis of α-substituted phenylacetonitriles by *Rhodococcus* sp. AJ270

Substrate (R)	Time (h)	nNtrile (%)	Configuration *e.e.*(%)	Amide 2 (%)	Configuration *e.e.* (%)	Acid 3 (%)	Configuration *e.e.* (%)
Me	10	-	-	42	*R*, >99	48	*S*, 90
Me	13.5	-	-	36	*R*, >99	58	*S*, 67
Et	70	-	-	58	*R*, 35	39	*S*, >99
Et	96	-	-	34	*R*, 96	40	*S*, >99
n-Pr	150	55	*S*, 24	27	*S*, 41	8	*S*, >99
n-Pr	214	33	*S*, 28	40	*S*, 13	13	*S*, >99
i-Pr	120	-	-	47	*R*, >99	46	*S*, >99
n-Bu	300	36	*S*, 36	34	*S*, 20	23	*S*, 98
MeO	46	-	-	78	0	-	
MeO	72	-	-	56	0	-	-
MeS	120	-	-	64	*R*, 15	10	*S*, 96

of bulky and conformationally flexible groups such as *n*-propyl and *n*-butyl resulted in a much slower rate. Surprisingly, polar group like methoxy and methylthio exhibited no effect on the rate of hydration step. In sharp contrast, both electronic and steric factors displayed detrimental effects on amidase in *Rhodococcus* sp. AJ270. Enantioselectivity of the strain, determined by the combination of selectivities of nitrile hydratase and of amidase, with the latter being a major contributor, seemed to be extremely sensitive to electronic factor too [114].

Rhodococcus sp. AJ270 also showed highly preference for aromatic and heteroaromatic nitriles, in which case the first step, hydration of the linear nitrile substituent, is not significantly hindered by steric or electronic factors, while the second amidase-mediated step showed high sensitivity to steric factors. It was observed that *para*- and *meta*-substituted benzonitriles were converted to the corresponding acids at a comparatively rapid rate in high yield irrespective of the electronic nature of the substituent. Nevertheless, benzonitriles with substituent at the *ortho*-position were rapidly and efficiently converted to amides, while conversion of amides to acids proceeded slowly, suggesting that the first step, is not significantly hindered by steric or electronic factors, while the second amidase-mediated step showed high sensitivity to steric factors [32].

It was evident that electronic nature showed a great effect on initial rate of nitrilase mediated hydrolysis of *para*-substituted phenylacetonitriles. The reaction was accelerated by electron-withdrawing substituents while slowed down by electron-donating substituents [67]. Similar results were achieved by Geresh et al. in whose study a Hammett-type linear free energy correlation with a ρ value (reaction constant) of 0.96 well described the relationship between the initial rates of the nitrilase catalyzed hydrolysis and the *para*-substituents [115].

5 Applications of Nitrile-Amide Converting Enzymes

Recently, a considerable number of products serving as the intermediates of pharmaceutical, fine chemical and food additives have been derived from enzyme-dependent reactions [116]. Biotransformation of amides or acids provides a feasible and valuable route. In fact, some have been successfully applied to industrial production, meanwhile, some are under active development.

5.1 Bioconversion of Various High-Value Amides and Acids

5.1.1 Industrial Production of Acrylamide and Enzymatic Manufacture of Acrylic Acid

Acrylamide, an important fine chemical with broad uses, is mainly applied in the synthesis of polyacrylamide. Achievements in bioconversion of acrylonitrile made it possible that enzyme-based manufacture would substitute its traditional production mode. *R. rhodochrous* J-1 nitrile hydratase has been well investigated as a powerful biocatalyst for the production of acrylamide, and was currently utilized in commercial synthesis of acrylamide in Japan [97]. In China, *Nocardia* sp. 163, a soil derived organism from Taishan Mountain in the 1980s, harbors nitrile hydratase activity active on acrylonitrile. Later it became the isolate with the highest NHase activity after optimization of culture conditions by Shanghai Pesticide Research Insititute. After that, a tens of thousand tons per year scale acrylamide production facility was set up [117].

As mentioned earlier, nitrile hydratase and amidase always coexist in the organism which leads to the formation of amides and acids in a certain portion. Due to this, a comparatively high ratio of nitrile hydratase/amidase activity is of great necessity to avoid contamination of the acrylamide by acrylic acid. In other words, effective measures should be taken to inhibit the amidase activity to a significant extent. Addition of urea, chloroacetone and phosphoramidate could sometimes efficiently inhibit the amidase activity, which in turn resulted in the accumulation of acrylamide [15, 74, 106].

As far as acrylic acid is concerned, it is a commodity chemical with an estimated annual production capacity of 4.2 million metric tons. Acrylic acid and its esters can be used in paints, coatings, polymeric flocculants, paper and so on. It is conventionally produced from petrochemicals. Currently, most commercial acrylic acid is produced by partial oxidation of propene which produces unwanted by-products and large amount of inorganic wasters [118]. Nowadays, there has occurred an innovative manufacture method using nitrile-amide converting enzymes. Being different from manufacture of acrylamide, its manufacture requires microorganisms with an excess of amidase over nitrile hydratase, namely the rate of amidase-mediated step markedly exceeds the first step. Alternatively, nitrilase possessing microorganisms affords acids as well. In our research, a nitrilase-producing strain *Arthrobacter nitroguajacolicu* ZJUTB06-99 was newly isolated from soil sample in order to develop a production process of acrylic acid by biotranformation of acrylonitrile. According to previous reports, ε-caprolactam induced *R. rhodochrous* J-1 cells containing abundant nitrilase were used in the manufacture of acrylic acid, 390 g L^{-1} of which were obtained under a periodic substrate feeding system [119].

5.1.2 Biotransformation of Aliphatic and Arylaliphatic Amides and Acids

Aliphatic and arylaliphatic amides and acids are readily available from the corresponding nitriles by the nitrile hydratase/amidase systems or nitrilase. A series of aliphatic nitriles, whether saturated or unsaturated, could be efficiently hydrolysed to the corresponding acids by the use of whole cells of *Rhodococcus* sp. AJ270. However, it was unsuitable for the preparation of amides, since the rate of the amide hydrolysis by amidase was greater than that of nitrile hydration. With one exception, unlike acrylonitrile and cinnamonitrile, methacrylamide was isolated after short activity due to the effect of adjacent subsistutent on the amidase-catalysed step [32].

In our study, a nitrilase from *B. subtilis* ZJB-063 was predominantly active on arylacetonitriles. It was observed that electronic nature had a great effect on the initial rate of nitrilase mediated hydrolysis of *para*-substituted phenylacetonitriles. The reaction was accelerated by electron-withdrawing substituents (Cl, NO_2) while slowed down by electron-donating substituents (OH, CH_3, OCH_3) [67]. Moreover, aliphatic and arylaliphatic amides could be provided by purified nitrile hydratase to avoid contaminated acids [19, 25].

5.1.3 Biosynthesis of Aromatic Acids and Amides

In previous studies there existed a hypothesis that aromatic nitriles were acceptable substrates for nitrilases while aliphatic nitriles were the same for nitrile hydratases. This hypothesis, however, was proved to be contradictory to many later observations which provided the evidence for the existence of nitrilases effective for aliphatic nitriles [120] and nitrile hydratases active on benzonitrile, its mono- or disubstituted derivatives [30, 32, 121, 122].

R. rhodocrous AJ270, a nitrile hydratase and amidase-containing microorganism, efficiently hydrolyzed all the seleceted aromatic nitriles including *para*-, *meta*- and *ortho*-substituted ones. Among these compounds, those with *para*-, *meta*-substituents gave acids at a fast rate, whereas conversion of aromatic nitriles bearing adjacent substituents almost ceased at the step of amides, and subsequent conversion of amides to acids proceeded rather slowly. The above finding clearly indicated that amidase was more sensitive to the electronic nature of the substituents. Disubstitution of benzonitrile with methoxy at positions 2 and 6 displayed a significantly adverse effect on nitrile hydratase. Displacement of methoxy by fluorin decreased the effect on nitrile hydratase but the step of hydrolysis of amides to acids still proceeded extremely slowly. This adjacent disubstitution phenomenon further confirmed that amide hydrolysis is stringently dependent on adjacent steric factors while nitrile hydration has a slight steric limitation [32].

5.1.4 Bioconversion of Heterocyclic Nitriles

In the past decades, commercial productions of acrylamide, nicotinamide and nicotinic acid have witnessed the significant use of nitrile-converting enzymes. Nicotinamide and nicotinic acid belong to heteroaromatic compounds. Nicotinamide and its acid are water-soluble B-complex vitamins (Vitamin B_3 or PP) used in pharmaceutical formulations, and as additives in food and animal feed; furthermore, their deficiency leads to pellagra. Moreover, nicotinic acid and its ester and amide derivatives have medical applications as antihyperlipidemic agents and peripheral vasodilators [123]. Currently, they are produced by chemical synthesis at high temperatures and pressures. Alternatively, they can be prepared under mild conditions by the bioconversion of 3-cyanopyridine with nitrile-converting containing microorganisms or enzymes.

R. rhodochrous J-1, a versatile nitrile-converting enzymes containing microorganisms, had high activity toward 3-cyanopyridine in the presence of crotonamide as an inducer (Fig. 4). This process straightforwardly transferred to industrial level and was soon developed into an industrial production scale of nicotinamide for the following reasons. First, 3-cyanopyridine could be converted to nicotinamide without formation of nicotinic acid. Namely, the strain displayed considerably high nitrile hydratase activity and it is noteworthy that nicotinamide is not contaminated by nicotinic acid, since nicotinamide is almost inert to the low amidase activity in this strain. Second, there seemed no occurrence of substrate inhibition as compared to conversion of acrylonitrile. Finally, the fermentation mode turned out to be pseudocrystal fermentation (namely, crystalline substrate, solution of substrate, solution of product, crystalline product). From the synthetic point of view, various useful amides other than nicotinamide and acrylamide can be obtained by using the

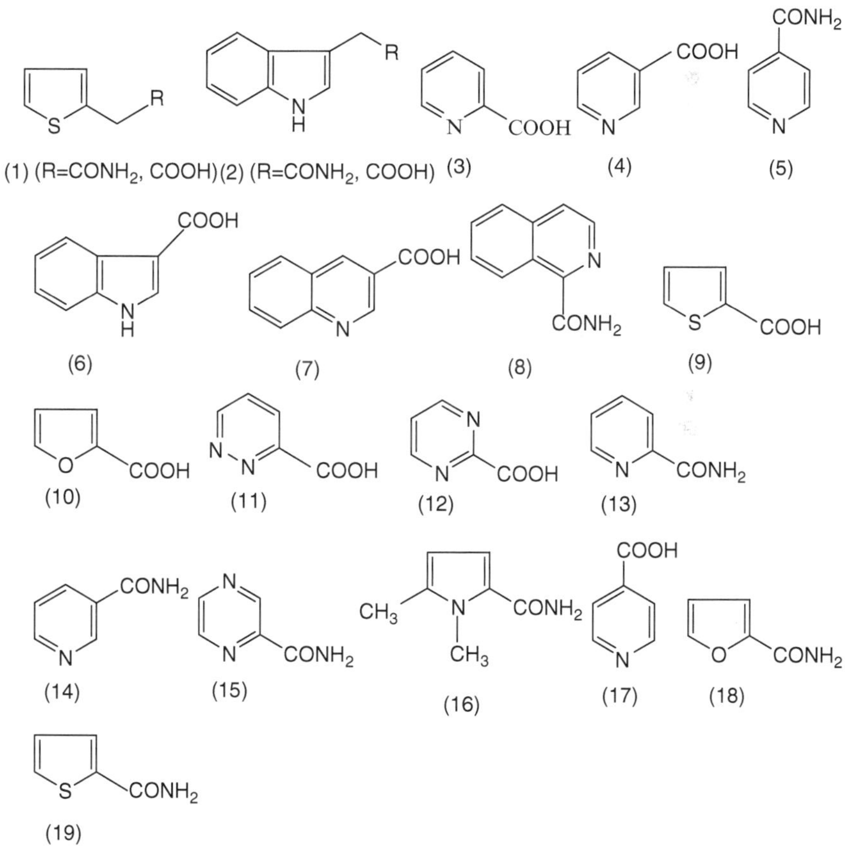

Fig. 4 Biotransformation of heterocyclic nitriles by the whole cells of *A. niger* K10 (**1**, **2**, **6–8**, **16**, **17**, **20**), *Rhodococcus* sp. AJ270 (**3–12**, **16–19**), *R. rhodochrous* NCIMB 11216 (**6**, **11**, **12**), *R. rhodochrous* J-1 (**5**, **13–15**), *Rhodococcus* sp. strain YH3-3 (**18**, **19**) and *R. equi* A4 (**5**, **13–15**)

R. rhodochrous J-1 cells cultured in the presence of crotonamide. Isonicotinamide and pyrazinamide, useful tuberculostatics, can be produced in this way as well [124]. As for nicotinic acid, a thermostable nitrilase produced by the thermophilic bacterium *B. pallidus* DAC521 catalyzed the direct hydrolysis of 3-cyanopyridine to nicotinic acid without detectable formation of nicotinamide [125].

In recent years, investigation was carried out on the preparation of nicotinamide with *Corynebacterium glutamicum* in China [126, 127]. As a robust bacterium, *Rhodococcus* sp. AJ*270* showed high nitrile hydratase activity against heterocyclic nitriles. The adjacent substituent impact existing in the hydrolysis of aromatic nitriles was encountered in the cases of heterocyclic nitriles. Those bearing an adjacent C=O/C=N remained intact after a long conversion time (Fig. 4) [32].

Our lab also provided an enzymatic route for the manufacture of nicotinamide. In addition, biosynthesis of 2-choloronicotinic acid is currently in active progress, which is a useful agricultural and pharmaceutical intermediate. Various other heteroaromatic amides and carboxylic acids procured currently via chemical synthesis could be produced biocatalytically from their nitriles. A cobalt-containing nitrile hydratase purified from *Rhodococcus* sp. strain YH3–3 was able to convert 2-cyanothiophene and 2-cyanofuran along with cyanopyrazine [128]. The nitrile hydratase from *R. equi* A4 also showed capacity for some heterocyclic nitriles, and similar results were observed that 2-cyanopyridine was transformed with a lower rate than 3- and 4-cyanopyridine (Fig. 4) [30]. In addition, bioconversion of some heterocyclic amides and acids could also be accomplished by *R. rhodochrous* J-1, *R. rhodochrous* NCIMB 11216 and *Aspergillus niger* K10 (Fig. 4) [25, 129].

5.1.5 Bioconversion of Alicyclic Nitriles

In the past few decades, a considerable amount of research concerning biotransformation of amides and acids has been carried out, the majority of which focused on the bioconversion of aromatic nitriles to its equivalent amides and acids and a paucity of which was concerned with the hydrolysis of alicyclic nitriles. Matoishi et al. conducted the hydrolysis of alicyclic mono- and dinitriles and amides mediated by *R. rhodochrous* IFO 15564, from which a variety of six-membered alicyclic cyano carboxylates, amido carboxylates, dicarboxylates from the corresponding nitriles were obtained (Fig. 5). The formation of these compounds was presumably ascribed to the stereochemistry of the substrate, the nature of substituents and presence of double bonds in alicyclic rings. These factors resulted in the rate difference of nitrile hydratase and amidase between enantiomers or enantiotopic groups, which in turn enabled kinetic resolution or asymmetrization [130].

Much attention has been paid in recent years to the preparation of enantiopure cyclopropyl compounds owing to the fact that enantiomers of these compounds often exhibit different biological activities. Hence, Wang and his coworker undertook the study of *Rhodococcus* sp. AJ270 mediated biotransformation of *trans*-2-arylcyclopropanecarbonitriles, by which not only optically active acids but also the amides were obtained (Scheme 5). In addition, bioconversions of 2-arylcyclopropanecarbonitriles

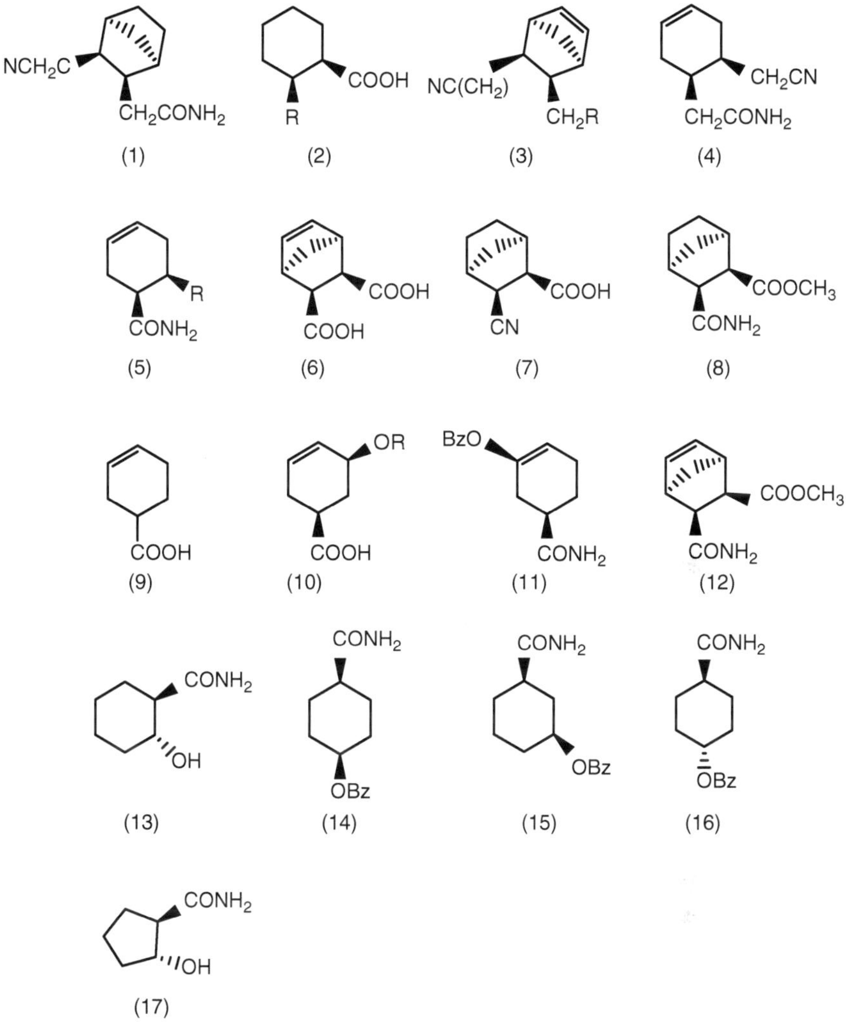

Fig. 5 Alicyclic mono-, di- and cyanocarboxylic acids and their amides prepared by the use of *Rhodococcus rhodochrous* IFO 15564 (R=$CONH_2$, COOH) (**1 – 12**) and R. equi **A4** (**13 – 17**)

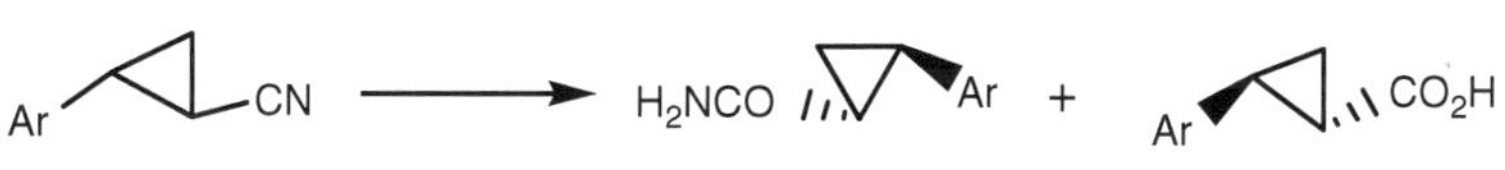

Racemic nitrile (–) (1R, 2R)-amide (+) (1S, 2S)-acid

(R = C_6H_5, 4-FC_6H_5, 4-ClC_6H_4 4-MeC_6H_4, 4-$MeOC_6H_4$)

Scheme 5 Enantioselective biotransformations of racemic *trans*-2-arylcyclopropanecarbonitriles by *Rhodococcus* sp. AJ270

Table 6 Bioconversions of 2-arylcyclopropanecarbonitriles and its analogs by *Rhodococcus* sp. AJ270

Ar	Time (h)	Amide (%)	Amide (*e.e.*)	Acid (%)	Acid (*e.e.* %)
C_6H_5	0.5	37	78	37	73(81)
C_6H_5	1	50	>99	39	55
C_6H_5	3	25	>99	61	19
4-FC_6H_4	2	44	89	48	75(91)
4-FC_6H_4	5	33	>99	66	30
4-ClC_6H_4	4	25	>99	43	52(78)
4-ClC_6H_4	5.5	29	>99	62	41
4-MeC_6H_4	2	46	70	32	75
4-MeC_6H_4	5	32	>99	53	71(82)
4-$MeOC_6H_4$	7	31	44	36	17
4-$MeOC_6H_4$	10	39	21	41	12

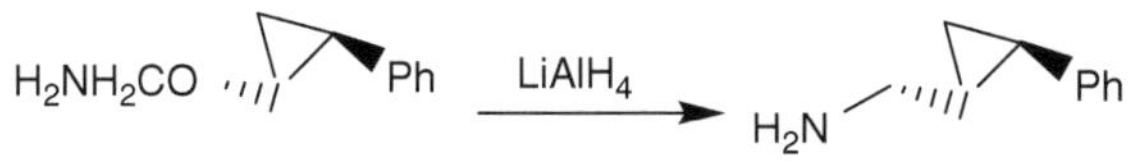

Scheme 6 Reduction of (–)-(1*R*,2*R*)-2-phenylcyclopropanecarboxamide using $LiAlH_4$

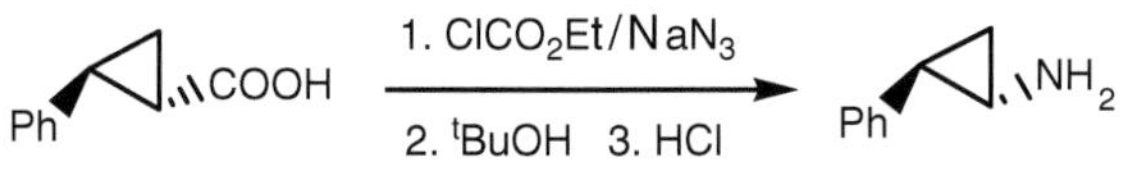

Scheme 7 Preparation of (+)-(1*S*,2*R*)-2-phenylcyclopropylmethylamine by a modified Curtius rearrangement

and its analogs with *para* substituents, a methyl, chlorine and fluorine, led to good to excellent enantioselectivity of the corresponding amides and acids, especially the amides. Exceptionally, comparatively low enantioselectivity of both amides and acids was detected in the reaction of 2-(4-methoxyphenyl)cyclopropanecarbonitrile, indicating that methoxy group caused a more dramatic effect than the above sterically smaller ones (Table 6). Moreover, (–)-(1*R*, 2*R*)-2-phenylcyclopropylmethylamine, a potential candidate for antihypertensive agent, could be provided by the reduction of (–)-(1*R*, 2*R*)-arylcyclopropanecarboxamide (Scheme 6). (+)-(1*S*, 2*R*)-2-Phenylcyclopropylmethylamine, antidepressant tranylcypromine, was obtained from (+)-2-arylcyclopropanecarboxylic acid via a modified Curtius rearrangement (Scheme 7) [131].

In addition, a recombinant nitrilase AtNITI from *A. thaliana* bore the capability of regio- and stereoselective hydrolysis of 3-(2-cyanocyclohex-3-enyl)propenenitriles, which consisted of four isomers numbered A–D prepared by Diels-Alder reaction of 1-cyano-1, 3-butadiene. Consequently, isomer D was exclusively hydrolyzed among the four dinitriles and thus (*E*)-*cis*-3-(2-cyanocyclohex-3-enyl)-propenoic acid was dominantly accumulated. Besides, isomer C was hydrolyzed to a small extent as well, which only occurred after complete conversion of D and when the enzyme concentration was high enough, while isomers A and B remained intact, in spite of higher enzyme concentration and prolonged conversion time. This outcome was considered to be explained by three factors: on one hand, the high regioselectivity, one the other hand, the high selectivity, including both (*E*)- and *cis* selectivity (Fig. 5) [132]. Similar discrimination between geometric isomers of substituted alicyclic nitriles was also accessed with the nitrile hydratase from *R. equi* A4. It was noted that *trans*-4-benzoyloxycyclohexanecarbonitrile, *cis*-3-benzoyloxy cyclohexanecarbonitrile, *trans*-2-hydroxycyclohexanecarbonitrile and *trans*-2-hydroxycyclopentanecarbonitrile were converted to the equivalent amides. In contrast, *cis*-2-hydroxycyclohexanecarbonitrile and *cis*-2-hydroxycyclo-pentanecarbonitrile gave a majority recovery of them. Although *cis*-4-benzoyl-oxycyclohexanecarbonitrile was an acceptable substrate for the enzyme, the reaction proceeded at a rather low rate (Fig. 6) [133].

Moreover, an impressive example of the application of nitrile-amide converting enzymes was enzymatic synthesis of optically active 2-methyl- and 2,2-dimethylcyclopropanecarboxylic acid and their derivatives. 2-Methyl- and 2,2-dimethylcyclopropanecarboxylic acid derivatives are key intermediates of curacin A, a potent antimitotic agent, and of cilastatin, a renal dehydropeptidase inhibitor commonly administered with penem and carbapenem antibiotics to prevent their degradation in the kidney, respectively. In the past, no direct asymmetric synthesis of these compounds has been reported. Optically active 2-methylcyclopropanecarboxylic acid and its amide derivatives have been prepared from the optical resolution or through the multistep transformations. Recently, nitrile-hydratase-associated amidases participated in the synthesis of optically active acids [110]. *S*-(+)-2,2-

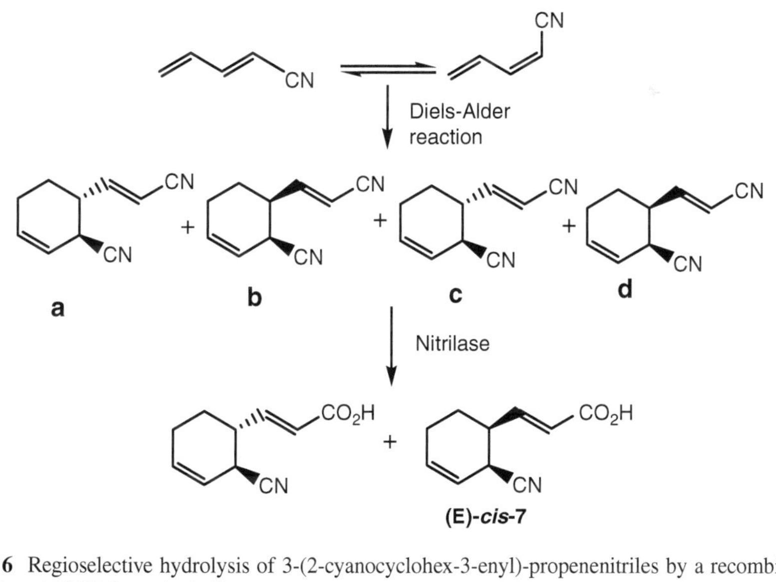

Fig. 6 Regioselective hydrolysis of 3-(2-cyanocyclohex-3-enyl)-propenenitriles by a recombinant nitrilase AtNITI from *A. thaliana*

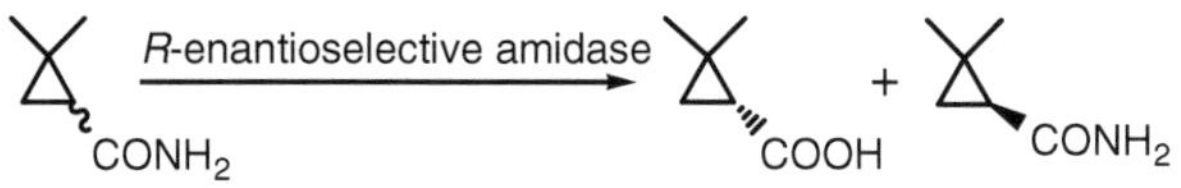

Scheme 8 Preparation of (*S*)-2,2-dimethylcyclopropanecarboxamide by (*R*)-enantioselective amidase in *D. tsuruhatensis* ZJB-05174

Dimethylcyclo-propanecarboxamide has been obtained from kinetic resolution of racemic 2,2-dimethylcyclopropanecarboxamide employing amidase-containing microbes or the amidases, which was a process adopted in Lonza for industrial operation. The above preparation involved in amidases exhibiting *R*-enantioselective. However, *Comamonas acidovorans* A18 was the only strain reported abroad, containing *R*-enantioselective amidase and has gained success in kinetic resolution of the racemic mixture [134]. Recently, *D. tsuruhatensis* ZJB-05174, capable of *R*-enantioselective degradation of 2,2-dimethylcyclopropanecarboxamide, was isolated employing a newly established colorimetric screening method. This isolate may be a suitable candidate for the production of (*S*)-2,2-dimethylcyclopropanecarboxamide from its racemic form after optimization of culture and biotransformation conditions (Scheme 8) [89].

5.1.6 Enantioselective Hydrolysis of Racemic Nitriles and Amides

Enantioselective Hydrolysis of Racemic Nitriles by Nitrile Hydratase

Although a significant number of nitrilases and amidases exhibited different degrees of enantioselectivity, nitrile hydratases were previously assumed to be relatively non-stereospecific. However, later on reports concerning enantioselective hydration of α-arylpropionitriles appeared. Unfortunately, an extreme paucity of nitrile hydratases exhibited enantioselectivity that gave rise to products with the ee value higher than 75%. *Pseudomonas putida* NRRL 18668, a soil derived Gram-negative bacterium, contained a nitrile hydratase capable of stereoselective hydrolysis of 2-(4-chlorophenyl)-3-methylbutyronitrile with more than 90% enantiomeric excess (ee) to the (*S*)-amide, which appeared to be the first description of a sterose-lective nitrile hydratase from a Gram-negative organism. This strain is also capable of a two-step hydrolysis of 2-(4-isobutylphenyl)-propionitrile and 2-(6-methoxy-2-napthyl)-propionitrile to the (*S*)-acids (ibuprofen and naproxen respectively) with stereoselectivity residing primarily in the aliphatic amidase [135]. Subsequently, another Gram-negative isolate *Agrobacterium tumefaciens* d3 appeared harboring enantioselective nitrile hydratase, which hydrated racemic 2-phenylpropionitrile, 2-phenylbutyronitrile, 2-(4-chlorophenyl)propionitrile, 2-(4-methoxy)propionitrile or ketoprofen nitrile, and the corresponding (*S*)-amides were formed with the highest enantiomeric excesses (ee >90% until about 30% of the respective substrates were converted) [136].

Former studies showed that in the nitrile hydratase-amidase system enantiomeric discrimination occured during amide hydrolysis. As an exception, *Rhodococcus* sp. 270, known as a nitrile hydratase and amidase containing organism, was effective for converting (*R*)-2-phenyl- butyronitrile to the corresponding amide with an *ee* value of 83%, suggesting that the amidase turned to be inactive for the amide and the formation of the chiral amide was due to the *R*-enantioselectivity of the nitrile hydratase [137].

In general, nitrile hydratase are coupled with amidase in organisms, and in most cases, formations of optically pure amides and acids are owing to the higher enantioselectivity of amidase compared with that of nitrile hydratase. A number of amides and acids were obtained via this method, for example, enzymatic synthesis of (*S*)-naproxen by *Rhodococcus* sp. C3II and *Rhodococcus erythropolis* MP50. From racemic naproxen nitrile, *Rhodococcus* sp. C3II formed *S*-naproxen amide and subsequently *S*-naproxen. Racemic naproxen amide was hydrolysed to *S*-naproxen. *Rhodococcus* sp. MP50 converted racemic naproxen nitrile predominantly to *R*-naproxen amide and racemic naproxen amide to *S*-naproxen. With both strains racemic naproxen amide was converted to *S*-naproxen with an enantiomeric excess >99% at a conversion rate up to 80% of the theoretical value (Scheme 9) [138]. By the hydrolysis of (*R*, *S*)-isopropyl-4 -chlorophenylacetonitrile using cells of *Pseudomonas* sp. B21C9, enantiopure (*S*)-2-isopropyl-4 -chlorophenylacetic acid, an intermediate of pyrethrins could be obtained. It appeared that the process proceeded via stepwise reactions by a nitrile hydratase exhibiting poor (*S*)-selectivity and an amidase exhibiting strict (*S*)-selectivity (Scheme 10). Heating resulted in the racemization of (*R*)-amide, which in turn caused enhanced yield of (*S*)-acid [139].

It was also described in China that *Rhodococcus* sp. AJ270 could be utilized for the preparation of enantiopure carboxylic acids and derivatives via stereoselective hydrolysis of a series of racemic α-substituted phenylacetonitriles and amides. The overall enantioselectivity of the process is mainly determined by the combination of selectivities of nitrile hydratase and amidase, with the latter being a major contributor, which was consistent with the above findings. Enantioselective synthesis

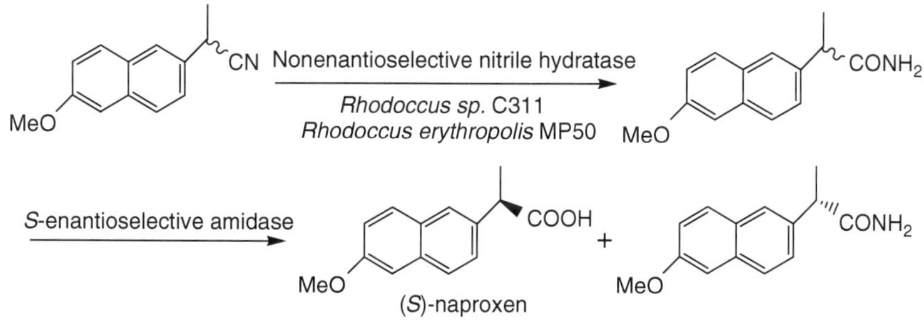

Scheme 9 Preparation of *S*-naproxen by *Rhodococcus* sp. C3II and *R. erythropolis* MP50

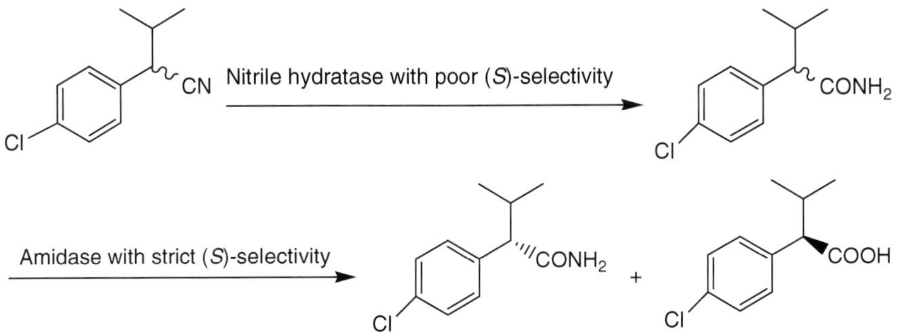

Scheme 10 Preparation of (*S*)-2-isopropyl-4′-chlorophenylacetic acid by *Pseudomonas* sp B21C9

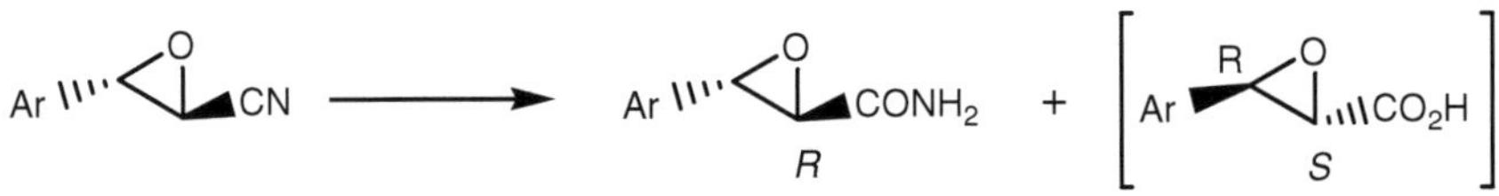

trans-racemic nitrile 2*R*, 3*S*-amide 2*S*, 3*R*-acid

(Ar=C_6H_4, 4F-C_6H_4, 4Cl-C_6H_4, 3-Cl-C_6H_4, 4-Br-C_6H_4, 2-Me-C_6H_4, 3, 4-$OCH_2OC_6H_3$)

Scheme 11 Biotransformations of racemic *trans*-2,3-epoxy-3-arylpropanenitriles by *R*hodococcus sp. AJ270

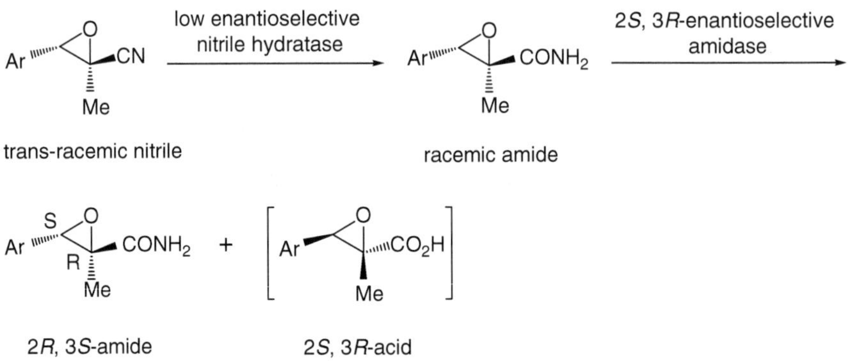

(Ar = C_6H_4, 4F-C_6H_4, 4Cl-C_6H_4, 3-Cl-C_6H_4, 4-Br-C_6H_4, 2-Me-C_6H_4, 3, 4-$OCH_2OC_6H_3$)

Scheme 12 Biotransformations of racemic *trans*-3-aryl-2-methyloxiranecarbonitrile by *Rhodococcus* sp. AJ270

of optically active 2-methyl-2,2-dimethylcyclopropanecarboxylic acids (see Sect. 5.1.5) and chiral cyclopropane compounds were successfully performed by the same strain (see Sect. 5.1.5). Very recently, it was observed that *Rhodococcus* sp. AJ270 are able to catalyze the hydrolysis of oxiranecarbonitrile to synthesize

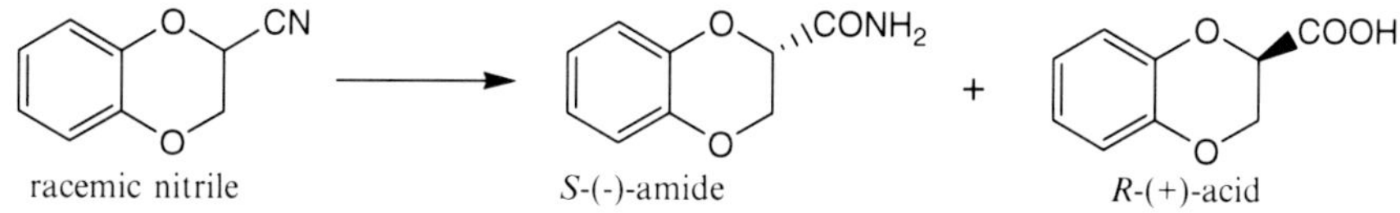

Scheme 13 Enantioselective biotransformations of racemic **1**, 4-benzodioxane-2-carbonitrile by *Rhodococcus* sp. AJ270

optically active 2*R*,3*S*-3-aryloxiranecarboxamides (Scheme 11) [140] and 2*R*,3*S*-3-aryl-2-methyloxiranecarboxamides (Scheme 12) [141]. As a useful microorganism, *Rhodococcus* sp. AJ270 is also able to rapidly catalyze enantioselective hydrolysis of racemic 1,4-benzodioxane-2-carbonitrile under very mild conditions, yielding 2*S*-1,4-benzodioxane-2-carboxamide and 2*R*-1,4-benzodioxane-2-carboxylic acid in high yields with excellent enantioselectivity, which are very important entities in medicinal chemistry for they are chiral building blocks in the design and synthesis of chiral therapeutic agents (Scheme 13) [142].

Enantioselective Hydrolysis of Racemic Nitriles by Nitrilase

The previously limited evidence for stereoselective nitrilases has been extended recently [143, 144] and their functions have been widely applied in the manufacture of various optically active carboxylic acids.

Amino acids are widely used as building blocks for the pharmaceutical industry, feed additives and human nutrition. Nitrilase participated manufacture of amino acids represents an attractive approach. In previous studies, formation of l-amino acids from α-aminonitriles was achieved using alginate immobilized *Acinetobacter* sp. APN containing nitrilase [145]. Later on, several optically active amino acids were produced from α-aminonitriles by *R. rhodochrous* PA-34 [146].

An intelligent screening method gave access to enantioselective nitrilases that are highly adapted to the production of high-value hydroxyl carboxylic acid derivatives, such as (*R*)-(−)-mandelic acid, which was described as a key intermediate of semi-synthetic cephalosporins and penillins, a chiral resolving agent, chiral synthon for the synthesis of anti-tumor agents and anti-obesity agents. Three isolates including *P. putida*, *Microbacterium paraoxydans* and *Microbacterium liquefaciens* gave rise to the desired product, (*R*)-(−)-mandelic acid. In terms of growth rate, enzyme activity, enantioselectivity and thermostability, *P. putida* was more suitable compared to the other two organisms. Thus it was selected for further use [95].

Noticeably, *Alcaligenes faecalis* ATCC 8750 was famous for the effective hydrolysis of mandelonitrile to (*R*)-(−)-mandelic acid. Interestingly, enantiomeric excess of (*R*)-(−)-mandelic acid formed from mandelonitrile was 100% and the yield attained was approximately 91%. Spontaneous racemization of *S*-mandelonitrile because of the chemical equilibrium accounted for the high yield. The observation was inconsistent

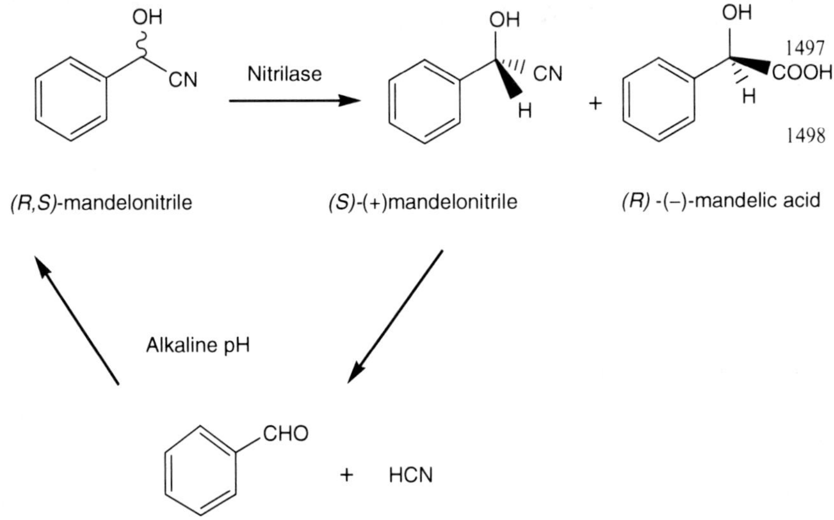

Scheme 14 Preparation of (*R*)-(−)-mandelic acid by *A. faecallis* ATCC 8750

with other enantioselective bioprocess where the yield attained was 50% at most. Later studies successfully completed the purification and characterization of the nitrilase, meanwhile, immobilization of the nitrilase (Scheme 14) [147, 148]. In particular, the immobilized nitrilase is useful for the production of hydroxy analogues of methionine derivatives that could have an interest in cattle feeding and in the transformation of compounds bearing other acid- or base-sensitive groups [149].

Researchers in China conducted the nitrilase-mediated manufacture of (*R*)-(−)-mandelic acid as well. He et al. screened a microbial strain identified as *Alcaligenes* sp. ECU0401 harboring a stereoselective nitrilase for the kinetic resolution of racemic mandelonitrile to (*R*)-(−)-mandelic acid with an enantiomeric excess of >99.9% [150]. Aiming at obtaining microorganisms with high enzyme activity and excellent selectivity, our lab has been undertaking bioconversion of (*R*)-(−)-mandelic acid and have made much progress. This convenient and practical approach to producing (*R*)-(−)-mandelic acid was developed into commercial application by Mitsubishi Rayon [151].

As mentioned earlier, optically active 2-arylpropionic acids like (*S*)-naproxen, (*S*)-ibuprofen and (*S*)-ketoprofen could be produced from the respective 2-arylpropionitrile by the aid of the sequential action of nitrile hydratase and amidase, besides, these acids could also be procured via (*S*)-enantioselective nitrilases. It was evident that interaction of racemic 2-(4′-isobutylphenyl) propionitrile with *Acinetobacter* sp. strain AK226 yielded *S*-(+)-2-(4′-Isobutylphenyl) propionic acid (*S*-(+)-ibuprofen) with a 95% enantiomeric excess (Scheme 15). The observation that *R*-enantiomer of the nitrile was inert to the organism and no detection of the

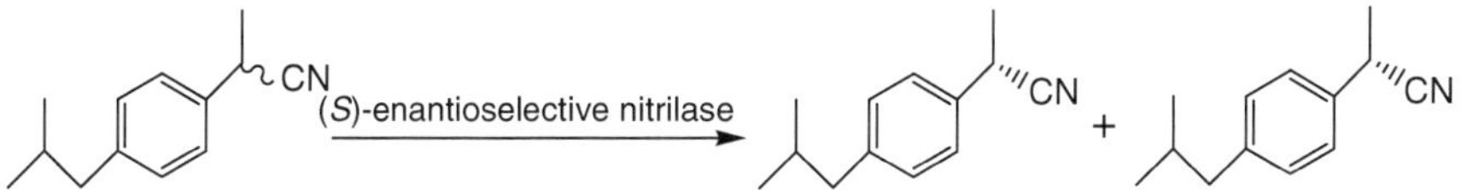

Scheme 15 Production of *S*-(+)-ibuprofen by (*S*)-enantioselective nitrilase in *Acinetobacter* sp. strain AK226

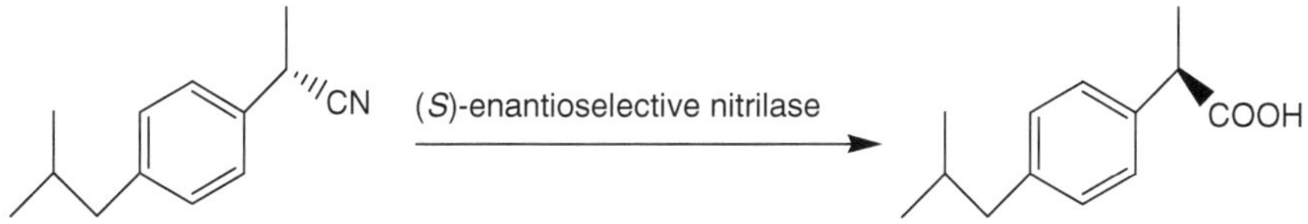

Scheme 16 Production of (*S*)-naproxen by (*S*)-enantioselective nitrilase in *R. rhodochrous* ATCC 21197

amide throughout the reaction suggested *Acinetobacter* sp. strain AK226 was a highly enantioselective nitrilase possessing bacterium [152]. Moreover, nitrilase in *R. rhodochrous* ATCC 21197 also appeared to be able to convert racemic 2-arylpropionitrile to optically active 2-arylpropionic acids. (*S*)-Naproxen could be obtained by this organism and excellent enantioselectivity could be achieved at the cost of conversion (Scheme 16) [153].

Enantioselective Hydrolysis of Racemic Amides by Amidase

Amidases with enantioselectivity are universal and these amidases catalyzed reactions offered optically pure compounds of pharmaceutical importance such as amino acids or 2-arylpropionic acids. Thus, they have attracted substantial interest and much progress has been made. A typical example of an enantioselective amidase catalyzed industrial biotransformation is the enzymatic chiral resolution of (*R*, *S*)-2,2-dimethylcyclopropane carboxamide providing the optically pure *S*-isomer. The process was developed by Lonza AG (see Sect. 5.1.5) (Lonza AG, CH). Similarly, using an *R*-enantioselective amidase producer, *D. tsuruhatensis* ZJB-05174, to produce (*S*)-2,2-dimethylcyclopropane carboxamide was investigated in our lab (See Sect. 5.1.5). The route, furthermore, is under progressive development and its potential of industrial application has been developed by Hisun Pharmaceutical Co., Ltd (Zhejiang).

As previously exemplified, efficient kinetic resolution of racemic piperazine-2-carboxamide and racemic piperidine-2-carboxamide to the corresponding enantiomerically pure carboxylic acids by the aid of whole cells of wild-type microorganisms harboring stereospecific amidases. The attained acids, (*S*)- and (*R*)-piperazine-2-carboxylic acid, and (*S*)-piperidine-2-carboxylic acid belong to non-proteinogenic amino acids and are used as precursors of numerous bio-active

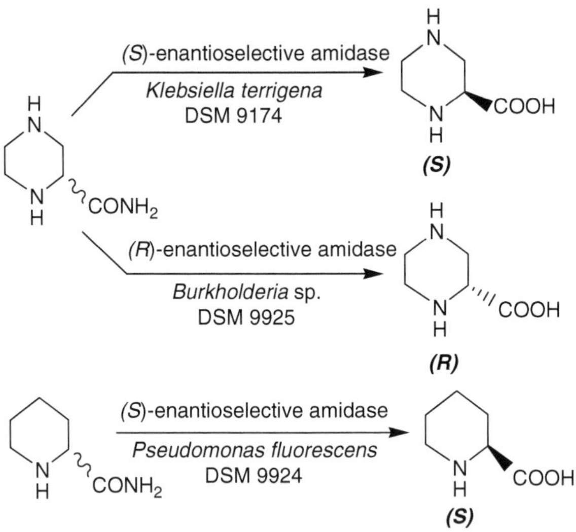

Fig. 7 Preparation of chiral non-proteinogenic amino acids by enantioselective hydrolysis of racemic amides

compounds. Due to the value of (*S*)-piperazine-2-carboxylic acid, which can be used for the synthesis of the HIV protease inhibitor Crixivan, its biosynthesis route employing *Klebsiella terrigena* DSM 9174 was developed into commercial application by Lonza AG [154] (Fig. 7).

Besides, amidase often coexists with the nitrile hydratase and formation of optically active acids is always ascribed to the combination of the poor selectivity of nitrile hydratase and excellent selectivity of amidase, which was discussed above (see "*Enantioselective Hydrolysis of Racemic Nitriles by Nitrile Hydratase*" above). As described above, in spite of the various applicability of the amidase, their application on a large or commercial scale has remained unexplored.

5.1.7 Regioselective Biotransformation of Di- and Trinitriles

It is rather difficult to conduct regioselective hydrolysis of dinitriles by chemical methods in a single step. An enzymatic approach, however, offers significant advantages since ammonium cyanocarboxylates from dinitriles could be obtained in nearly quantitative yields in one step. *Acidovorax facilis* 72W that possessed a regioselective nitrilase was successfully utilized for the conversion of an aliphatic α,ω-dinitrile to an ammonium salt of ω-cyanocarboxylic acid, which was then directly converted to the corresponding lactam by hydrogenation, without detection of the intermediate ω-cyanocarboxylic acid or ω-aminocarboxylic acid (Scheme 17) [155]. The process could also be achieved by the action of a combined nitrile hydratase and amidase present in *Comamonas testosteroni* 5-MGAM-4D, where an aliphatic α,ω-dinitrile is initially converted to an ω-cyanoalkylamide by the nitrile hydratase, and the ω-cyanoalkylamide is subsequently attacked by the amidase to the corresponding ω-cyanocarboxylicacid ammonium salt, implying that the regioselectivity of the system originated from nitrile hydratase (Scheme 17). Typically, a chemoenzymatic process for the preparation of 1,5-dimethyl-2-piperidone from 2-methylglutaronitrile with greater than 98% regioselectivity at 100% conversion has been demonstrated using immobilized *A. facilis* 72W cells [156] (Scheme 18). Besides, purification and characterization of the regioselective aliphatic nitrilase from this strain was carried out; furthermore, the nitrilase gene was cloned, sequenced and over-expressed in *Escherichia coli*, yielding a recombinant microorganism which more efficiently hydrolyzed aliphatic dinitriles to cyanocarboxylic acids in comparision with the wild type [157]. Thus, five- or six-membered ring lactams could be obtained from the sole product of regioselective conversion of dinitriles by means of nitrilase in *Acidovorax facilis*72W.

As exemplified, a biocatalytically commercial scale process for the highly regioselective hydration of adiponitrile to 5-cyanovaleramide, which is required in the first step of manufacture of a new herbicide, has been demonstrated using *Pseudomonas chlororaphis* B23 cells immobilized in calcium alginate beads[158]. In terms of enzyme stability and productivity of 5-cyanovaleramide, it was proved that it was superior to all other whole-cell microbial catalysts which were examined (Scheme 19)

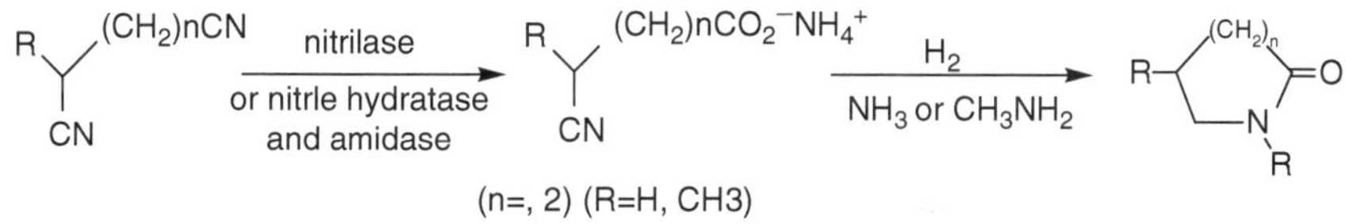

Scheme 17 Regioselective hydrolysis of aliphatic α,ω-dinitriles by nitrilase in *A. facilis* 72W or nitrile hydratase/amidase in *C. testosteroni* 5-MGAM-4D

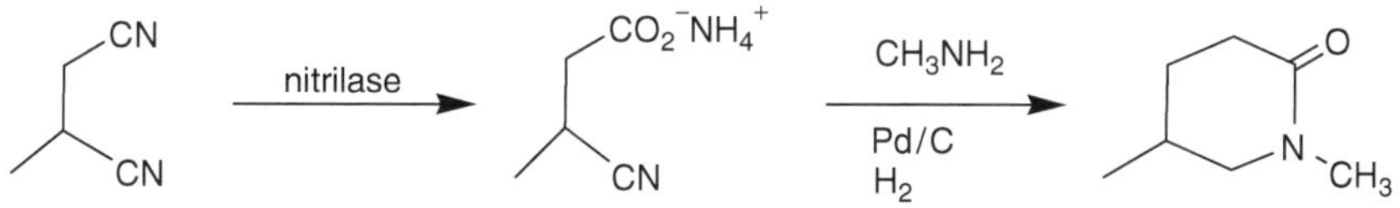

Scheme 18 Regioselective hydrolysis of 1,5-dimethyl-2-piperidone from 2-methylglutaronitrile by immobilized *A. facilis* 72W

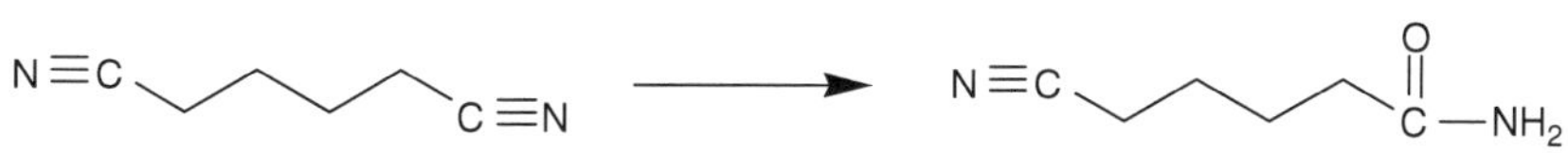

Scheme 19 Regioselective hydrolysis of adiponitrile by *P. chlororaphis* B23

[21]. Desymmetrization of prochiral dinitriles was also investigated by Crosby, Turner and their co-workers, in whose study, *O*-substituted-3-hydroxyglutaronitriles were converted using different *Rhodococcus* whole-cell catalysts and obtained the corresponding monocyanocarboxylic acids in good enantiomeric excess [159]. It is also interesting to note that biotransformation of a disubstituted malononitrile catalyzed by *R. rhodochrous* 21197 yielded an amido-acid with excellent enantioselectivity [160].

Regarding di- and tri-nitrile degradation, Asano et al. described it in the early 1980s [161–163]. Besides, *Rhodococcus* sp. AJ270 efficiently hydrolyzed a variety of dinitriles with excellent regioselectivity. Aliphatic dinitriles $NC[CH_2]_nCN$ underwent regieselective hydrolysis to give the mono acids with up to four methylenes between the nitrile groups, whereas those bearing $n > 4$ gave the diacids with good yield (Table 7). Significant regioselectivity was discovered using a range of α,ω-dinitriles $NC[CH_2]_nX[CH_2]_nCN$ with an ether or sulfide linkage as substrates. These compounds were efficiently transformed into the mono acids when an oxygen is placed β, γ or δ to the nitrile or β or γ-sulfur substituent is present (Table 8). As an efficient and robust

Table 7 Conversion of $NC[CH_2]nCN$ **1** into $NC[CH_2]_aCO_2H$ **2** and/or $HO_2C[CH_2]nCO_2H$ **3** by *Rhodococcus* sp. AJ270

Substrate (n)	Reaction condition		Product yield (%)	
1	Amount (mmol)	Time (h)	2	3
2	5	3	30	-
2	5	24	17	-
3	5	3	41	-
3	5	16	35	-
4	5	3	41	4
4	5	24	28	23
4	5	48	-	26
5	5	3	-	46
5	5	48	-	74
6	5	48	-	78
7	5	48	-	88
8	5	48	-	89

Table 8 Conversion of $NC[CH_2]nX[CH_2]nCN$ **4** into $NC[CH_2]nX[CH_2]_aCO_2H$ **5** and/or $HO_2C[CH_2]nX[CH_2]nCO_2H$ **6** by *Rhodococcus* sp. AJ270

Substract		Reaction conditions		Product yield (%)		
n	X	Amount (mmol)	Time (h)	5	6	4
2	O	3	48	61	-	-
3	O	5	1	35	Trace	35
3	O	3	72	-	70	-
4	O	3	2	73	-	-

(continued)

Table 8 (continued)

Substract		Reaction conditions		Product yield (%)		
n	X	Amount (mmol)	Time (h)	5	6	4
4	O	3	120	75	5	-
5	O	3	2	-	25	60
5	O	3	48	-	97.5	-
2	S	3	1	45	-	10
2	S	3	48	-	60	-
3	S	3	2	52	-	29
3	S	3	96	83	-	-
4	S	3	2	-	83	Trace
4	S	3	96	-	86	-
2	NPh	2	64	93	-	-
2	NC_6H_4Cl-*p*	2	168	71	-	26
2	NC_6H_4OMe-*p*	2	72	91	-	-
3	NPh	1.5	168	-	-	95
3	NC_6H_4OMe-*p*	1.5	168	-	-	92
4	NPh	1.5	168	-	-	94
5	NPh	1.5	168	-	-	99

nitrile-amide converting organism, the strain also effectively catalyzed the hydrolysis of a variety of miscellaneous aliphatic dinitirles into mono acids with the exception of *o*-phenylenediacetonitrile where *o*-phenylenediacetamide was derived as the major product (Table 9). Finally, efficient regiocontrol was also accessible when *m*- and *p*-dicyanobenzenes were used as substrates [164, 165] (Table 10).

Other nitrile-converting enzyme from *Rhodococcus* genus also exhibited regioselectivity. Nitrilase of *R. rhodochrous* NCIMB 11216 showed regioselectivity to some extent as well. The extent of conversion with saturated aliphatic dinitriles was very low, whereas hydrolysis of unsaturated aliphatic dinitriles occurred at similar rates to that of aromictic mononitriles. Greater structural rigidity of these compounds caused by the presence of a double or the aromatic ring could possibly accounted for the above observed phenomenon. In case of the hydrolysis of dicyanobenzenes, it proceeded also at a relatively low rate, among which 1, 4-dicyanobenzene bearing most spatially separated nitrile groups were converted to mononitrile at the highest rate. However, biotransformation of 1,3-dicyanobenzene turned out to be somewhat different with the addition of different inducers. The propionitriles-induced cells hydrolyzed not only 3-cyanobenzoate but also both the nitrile groups in 1,3-dicyanobenzene, yielding diacids. In sharp contrast, benzonitrile-induced cells were incapable of hydrolyzing 3-cyanobenzoate, so the process halted at the formation of mononitrile. Besides, 2-cyanophenylacetic acid could be prepared from α-cyano-*o*-tolunitrile by both of the nitrilases [99].

Dadd et al. assessed the biotransformation of 2-, 3- and 4-(cyanomethyl) benzonitriles employing *R. rhodochrous* LL100-21. As a result, 2-(cyanophenyl) acetic

Table 9 Conversions of miscellaneous dinitriles into acids by *Rhodococcus* sp. AJ270

Substrate	Condition: Concentration (mM)	Condition: T(h)	Product and yield
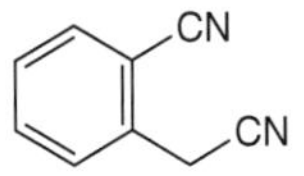	2	24	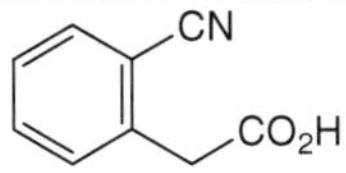 80%
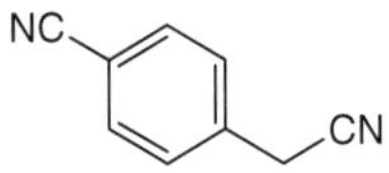	2	24	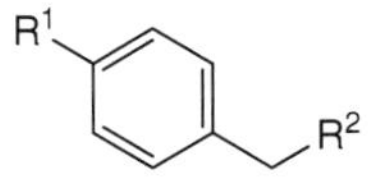 R^1=CN, R^2=COOH 65% R^1=COOH, R^2=CN 16%
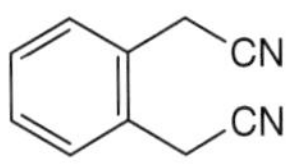	3	24	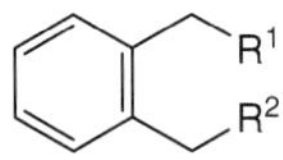 R^1=R^2=CN 13% R^1=CN, R^2=COOH 11% R^1= R^2=$CONH_2$ 65%
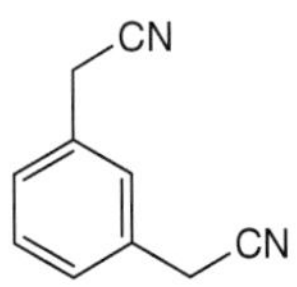	3	30	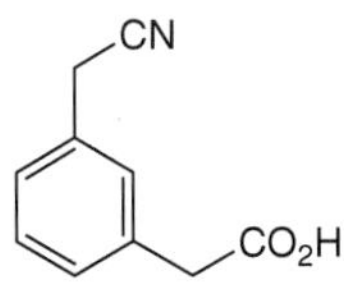 67%
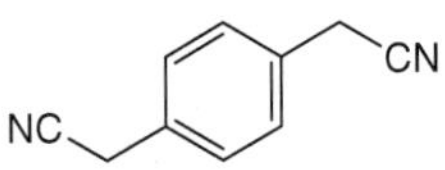	3	30	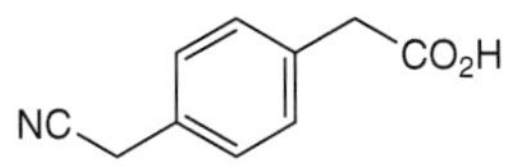 69%
	1.5	39	 99%
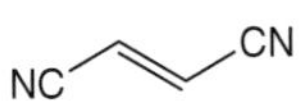	5	14	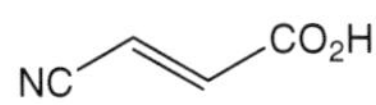 67%
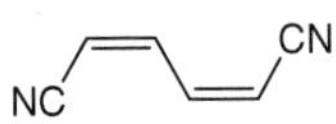	5	14	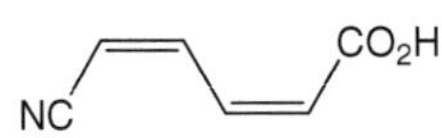 86%

acid was formed as the sole product by propionitrile or benzonitrile induced cells which may be explained by the hypothesis that aliphatic side chain of the compound prevented the hydrolysis of the aromatic cyano group owing to steric hindrance. Conversely, these cells led 3- and 4-(cyanomethyl) benzonitriles to 3- and 4-(cyanomethyl) benzoic acid as the major products with other products of relatively low concentrations. More interestingly, acetonitrile induced cells resulted in a mixture of

Table 10 Conversions of aromatic dinitriles into acids *Rhodococcus* sp. AJ270

Substrate	Conversion time (h)	Product and yield (%)
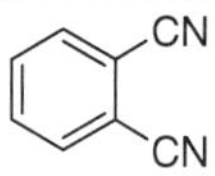	70	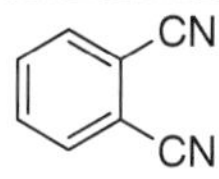 52%
		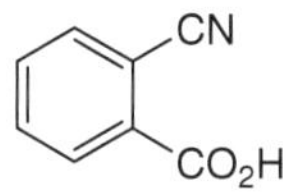 12%
		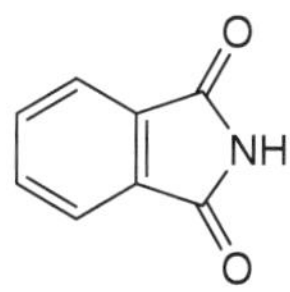 12%
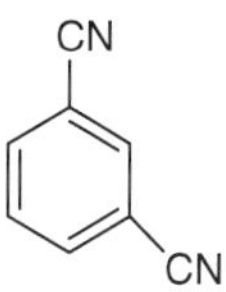	24	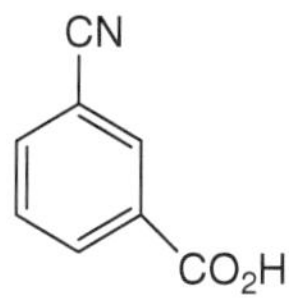 71%
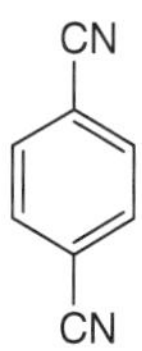	20	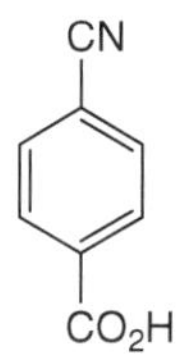 81%

different products from 2-, 3- and 4-(cyanomethyl) benzonitriles indicating less regioselectivity in such case [28].

5.1.8 Conversions of Nitriles Bearing Labile Functional Groups

Biocatalysts have been applied in the hydrolysis of nitriles containing other labile groups which are unable to survive under harsh chemical conditions. Attempts were made by Klempier et al. to selectively hydrolyze nitriles with acetal and ester groups in the molecule using *R. rhodochrous* NCIMB 11216 (Fig. 8). While the former functional moieties were inert to the catalyst during the hydrolysis, cleavage of the latter ones (leading to cyano acids and diacids) reduced the product yields, which was caused by the ubiquitous esterase activity [166]. Hence, suppressing the esterase activity of this strain was of significant importance, such as by modification of the growth media or employing RNAi technology. Alternatively, purified enzymes turned out to be more preferable.

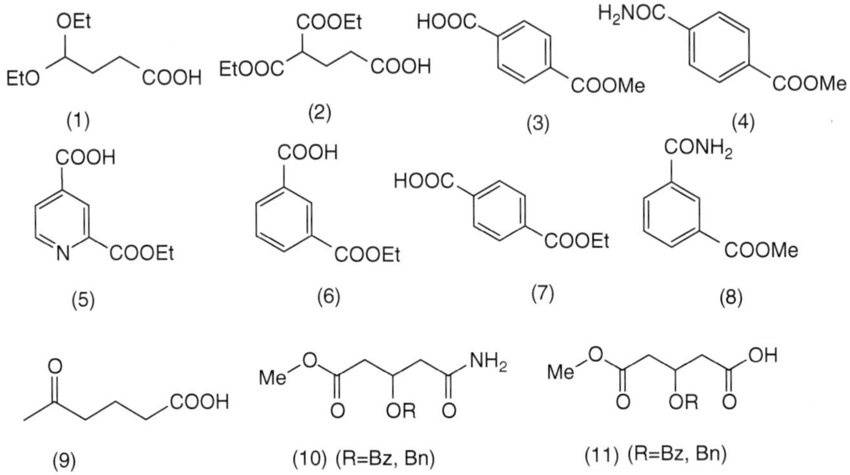

Fig. 8 Production of amides and acids from nitriles bearing labile functional groups by whole cells of *R. rhodochrous* NCIMB 11216 (**1–3**, **5**, **6**, **9**), *R. equi* A4 (**3**, **4**, **7**, **8**, **10**, **11**) and *Rhodococcus* sp. AJ270 (**3**)

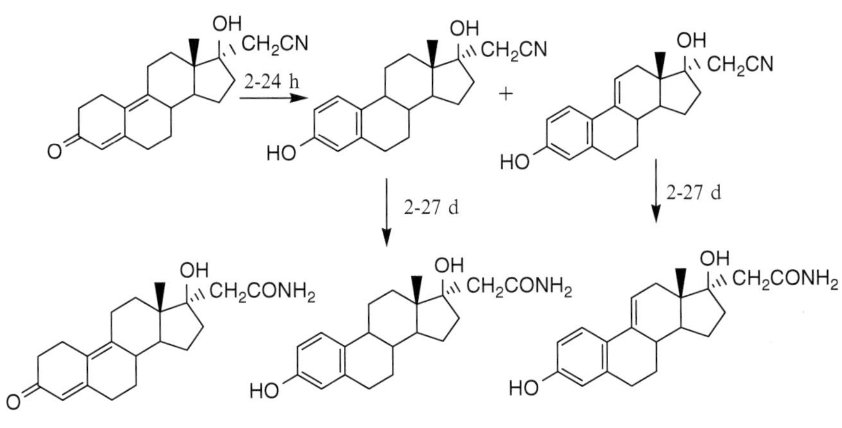

Fig. 9 Bioconversion of 17α-cyanomethyl-17β-hydroxy-estra-4,9-dien-3-one by whole cells of *R. erythropolis*

Excellent chemoselectivity was also observed for the hydrolysis of cyanobenzoates catalyzed by whole cells of *R. equi* A4. Monomethyl isophthalate and monomethyl terephthalate could be obtained from methyl-3-cyanobenzoate and methyl-4-cyanobenzoate, respectively, via resting cells of the strain [167]. Moreover, this strain are also capable of chemoselectively hydrolysing methyl (*R*,*S*)-3-benzoyloxy-4-cyanobutanoate and methyl (*R*,*S*)-3-benzyloxy-4-cyanobutanoate into monomethyl (*R*,*S*)-3-benzoyloxyglutarate and monomethyl (*R*,*S*)-3-benzyloxyglutarate, respectively [168] (Fig. 8). Hydrolysis of the progestin dienogest (17α-cyanomethyl-17β-hydroxy-estra-4, 9-dien-3-one) was performed by the nitrile hydratase-containing microorganism *R. erythropolis*. Along with the slow hydrolysis of cyano group, aromatization of ring A and hydroxylation occurred as well. After prolonged fermentation, the 17α-acetamido derivatives of estradiol and of 9(11)-dehydroestradiol were formed. Three of the metabolites were also prepared synthetically [169]. *Rhodococcus* sp. AJ270 also made it possible to synthesize some carboxylic acids and amides from the nitrile bearing sensitive groups (Fig. 9) [164].

5.2 *Biodegradation and Bioremediation*

Due to the intrinsic nature of nitriles as highly toxic, carcinogenic and mutagenic, it is necessary to control and monitor the discharge of these organonitriles into the environment. Typical examples of these compounds include acetonitrile, acrylonitrile and benzonitrile that are widely used in laboratories and industries as solvents and extractants, or used as an ingredient in pharmaceuticals, plastics, synthetic rubbers, drug intermediates (chiral synthons), herbicides and pesticides (e.g., dichlobenil, bromoxynil, ioxynil, buctril), etc. Recent awareness of environmental pollution caused by chemical-based industries has necessitated the development of enzyme-based process as alternatives to currently employed chemical processes. Moreover, bioconversion and biotransformation becomes partial or total replacement of currently employed toxic chemical process due to the distinct advantages of biotransformation [9]. Hence, their decomposition and detoxification by convenient and efficient methods is fairly urgent and challenging. More importantly, biodegradation and bioremediation, a convenient and cost-effective method, has the capability of eliminating these compounds by degrading them into harmless intermediates or, to a more desired form, carbon dioxide and water [170].

To the best of our knowledge, the huge potential of nitrile-converting enzymes has been explored in biodegradation of nitriles by many researchers. A variety of microorganisms, were demonstrated to be effective on metabolism of some organonitriles. As previously reported, a considerable number of nitrile-converting whole-cell biocatalysts have been applied to the removal of acrylonitrile waste effluents from the manufacture of acrylamide. A mixed culture of bacteria including different nitrile hydrolyzing enzymes that degrade effluent from the manufacture of acrylonitrile containing acrylonitrile, fumaronitrile, succinonitrile, etc. are grown in batch and continuous culture on these components of waste completely degraded all of the components [171].

Very recently, researchers from Singapore successfully enriched a whole consortium from the activated sludge of a pharmaceutical wastewater treatment plant and investigated its capability of biodegradation of saturated (acetonitrile), unsaturated (acrylonitrile), and aromatic (benzonitrile) organonitrile compounds [172].

Due to the potency and efficiency of biodegradation, similar studies were also conducted in China, especially in Taiwan, and moreover, some achievements were obtained [173–175]. Nitrile-converting enzymes have also participated in the cyano group-containing herbicide decomposition. A bromoxynil-degrading soil microorganism *Agrobacterium radiobacter* was used for degradation of the herbicide under nonsterile batch and continuous conditions. The efficacy of degradation was enhanced by addition of ferrous, cobaltous or cupric ions [176].

6 Cloning and Expression of Nitrile-Amide Converting Enzymes

Cloning and overexpression stands out as a promising method for the production of the desirable enzymes in sufficient quantities. So are the nitrile-amide converting enzymes, and there appeared numerous reports with respect to cloning and expression of these enzymes which are now accessible in sufficient quantities. As for nitrile hydratase, both the H- and L-nitrile hydratases genes, composed of two subunits of different sizes, have been cloned from *R. rhodochrous* J-1. The amino acid sequences of each subunit of the H- and L-NHases from *R. rhodochrous* J-1 showed generally significant similarities to those from *Rhodococcus* sp. N-774, but the arrangement of the coding sequences for two subunits is reverse. Each of the NHase genes was expressed in *E. coli* cells under the control of lac promoter only when they were cultured in the medium supplemented with $CoCl_2$ [177]. The stereoselective nitrile hydratase from *P. putida* 5B has been over-produced in *E. coli*. A clearly enhanced enzyme activity six times higher than the native strain and same stereoselectivity was observed [178].

In China, researchers have carried out similar work and some progress has been made. To obtain a recombinant *Rhodococcus* or *Nocardia* with not only higher enzymatic activity but also better operational stability and product tolerance for bioconversion of acrylamide from acrylonitrile, an active and stable expression system of nitrile hydratase (NHase) was tried to construct as the technical platform of genetic manipulations. Two NHase genes, NHBA and NHBAX, from *Nocardia* YS-2002 were successfully cloned into *E. coli* and *Pichia pastoris* system, however, expression level remained extremely low and the protein was unstable. To solve this problem, a possible genetic strategy, site-directed mutagenesis of the α-subunit of the NHase was carried out. After the successful mutagenesis, *E. coli* XL1-Blue (pUC18-NHBAM) was screened and the NHase activity was much higher than that of the prototype [179].

Along with the nitrile hydratase, some amidases have undergone the cloning and expression. Recently, Cheong et al. have undertaken the research concerning cloning of a wide-spectrum amidase from *Bacillus stearothermophilus* BR388 in *E. coli* and improving amidase expression using directed evolution. As a desired result, this mutant, prepared by PCR-based random mutagenesis which resulted in the substitution of arginine for histidine at position 26, demonstrated a 23-fold increase in amidase activity compared to the wild-type [88]. The amidase gene from the hyperthermophilic archaeon *Sulfolobus solfataricus* has been cloned, sequenced, and overexpressed in *E. coli* and the recombinant thermophilic protein was expressed as a fusion protein with an N-terminus six-histidine-residue affinity tag [86].

In the case of nitrilase, the cloning of this enzyme first occurred in *E. coli* and encoded a bromoxynil-degrading activity from *Pneumoniae* subsp. *ozaenae* [180]. In the previous studies, four nitrilases have been cloned from *A.thaliana* [181]. The corresponding gene of a regioselective aliphatic nitrilase from *A. facilis* 72W was cloned and over-expressed in *E. coli*, yielding a microorganism that efficiently and regioselectively catalyzes the conversion of aliphatic dinitriles to cyanocarboxylic acids. However, it was observed that, though a markedly increased quantity of nitrilase protein was obtained, the majority is present as inclusion bodies and is inactive. The phenomenon was consistent with the outcome obtained when nitrilase from *C. testosterone* was expressed in *E. coli*, although this was significantly improved with the co-expression of groESL chaperones [60]. *R. rhodochrous* J-1, appeared promising as a versatile nitrile-amide producing bacterium. Hence, it was investigated extensively, recent studies focusing on the molecular level. Komeda et al. demonstrated that the 1.4-kb downstream region from a nitrilase gene (nitA) was found to be required for the isovaleronitrile-dependent induction of nitrilase synthesis. Sequence analysis of the 1.4-kb region revealed the existence of an open reading frame (nitR) of 957 codes for a transcriptional regulator in nitA expression [182]. In China, researchers from Tongji University introduced a series of work concerning cloning and sequencing of nitrilase from an efficient degrader *Nocardia* sp. C-14-1. Southern blotting showed that there was a single nitrilase gene in the genome of C-14-1. Meanwhile, DNA sequencing and analysis suggested that there was a fragment of 1,143 bp DNA sequence encoded the nitrilase [183]. Besides, it was found that the expression of the *Vitreoscilla hemoglobin* (vgb) gene in vivo could improve the fermentation density and then contribute the extracellular secretion of the product of bromoxynil-specific nitrilase (bxn) gene. The recombination plasmid pPIC9K-vgb-bxn was constructed and transformed into *P. pastoris* GS115. The results of PCR and SDS-PAGE indicated that the vgb gene and bxn gene had integrated into the genome of *P. pastoris* GS115 and expressed in efficient level [184]. Subsequently, the bxn gene encoding bromoxynil-specific nitrilase was cloned from genomic DNA of *Klebsiella ozaenae* by PCR and over-expressed in *E. coli* DE3. The recombinant accessibility made it a promising candidate for eliminating bromoxynil to non noxious substances [185]. It could be concluded that the cloning and over-expression of the encoding genes resulted in a better understanding of enzyme function and the reaction mechanism, which in turn would lead to improvements in biotechnological applications.

Protein engineering of nitrilase have also been practiced to improve the substrate and product tolerance and specific activity. A high-activity biocatalyst based on an *A. facilis* 72W nitrilase was developed, where protein engineering and optimized protein expression in an *E. coli* transformant host were used to improve microbial nitrilase specific acticity for glycolonitrile by 33-fold compared to the wild-type strain [186]. Gene site saturation mutagenesis (GSSM) evolution technology was employed to improve enantioselectivity of nitrilase-catalyzed desymmetrization of

3-hydroxyglutaryl nitrile to afford (*R*)-4-cyano-3-hydroxybutyric acid, an intermediate to the cholesterol-lowering drug Lipitor [187]. Changing Ala to His in position 190 provided a 10% increase in the enantiomeric excess at the commercially relevant 3 M substrate reaction concentration.

7 Conclusions and Future Prospects

The past few decades have witnessed the fast development of nitrile-amide converting enzymes, both their reaction mechanisms and applications in manufacture of a series of pharmaceuticals and chemicals. Formerly, great contribution was made by hydrolases such as esterases and lipases in the production of enantiopure synthons. Nowadays, with the discovery of numerous nitrile-amide converting microorganisms and their extremely fast development, these enzymes are becoming more and more appreciated by organic chemists and are showing competency to compete with esterases and lipases. Besides their synthetic value, these enzymes also play an important role in protecting the environment due to the capability of removal highly toxic nitrile compounds which are rather detrimental to human beings, animals and plants. In order to exploit fully their biotechnological potential, researches concerning the following aspects should be carried out in the following ways: (1) Overcoming some disadvantages of the nitrile-converting biocatalysts, such as narrow substrate specificity, low thermostability and pH stability, low tolerance of substrate and product, undesired and unsatisfied enantioselectivity. (2) Screening and discovery of new nitrile-amide converting enzymes with promising and attractive properties. As previously demonstrated, microorganisms producing nitrile-amide converting enzymes turned out to be dominantly from prokaryotic ones and eukaryotic organisms constitute only a small part. The latter ones, however, were always neglected as an excellent source of nitrilase, nitrile hydratase and amidase. Moreover, these organisms may provide some different properties like excellent thermostability, regio-, enantio- selectivity, and improved stability in some acidic and alkaline media, which are favored by some process. (3) Employing genetic engineering to alter some undesired properties of wild type strain. Bearing these in mind, researchers would make great progress in techniques related to screening, cultivation, protein and genetic engineering and hence it is possible to isolate novel enzymes with extremophilic characteristics.

Despite the advantages of nitrile-amide converting enzymes catalyzed synthesis of a range of industrially useful acids and amides, a paucity of them have achieved success in industrial application, with the commercial production of acrylamide and nicotinamide being the most successful. Manufacture of many other substances has been proved to be accomplished by these enzymes, especially the processes involve the regioselective and enantioselective hydrolysis of some prochiral compounds and racemic nitriles. These compounds, nevertheless, are difficult if not impossible to convert by means of traditional chemical methods. Chemical hydrolysis of many nitriles with labile substituents catalyzed by acid or base is also virtually impossible because of the drastic reaction conditions required. Thus, their powerful potential in commercial application is being developed. In China, bioconversion of the high-value acids and amides has also been a hot issue. It is believed that an increasing number of novel nitrile-amide converting organisms will be screened and their potential in the synthesis of useful acids and amides will be further exploited in the near future. Furthermore, though cloning of the genes and expression of these versatile biocatalysts are presented in detail in many literatures and some of the recombinants exhibited rather high activity, it need to be largely exploited for the nitrile-converting enzymes. In a word, there is enough space for the development of these biocatalysts and there will turn on a bright and promising future of these enzymes.

Acknowledgements This work was supported by the Major Basic Research Development Program of China (973 Program) (No.2007CB714306), the Fund of the National High Technology Research and Development Program of China (863 Program) (No.2006AA02Z241).

References

1. Patel RN (2001) Curr Opin Biotechnol 12:587
2. Schmid A, Dordick JS, Hauer B, Kiener A, Wubbolts M, Withholt B (2001) Nature 409:258
3. Conn EE (1981) In: Vennesland B, Conn EE, Knowles CJ, Westly J, Wissing F (eds.) Cyanide in biology. Academic Press, London, p 183
4. Ahmed AE, Farooqui MYH (1982) Toxicol Lett 12:157
5. Pollak P, Romender G, Hagedorn F, Gelbke HP (1991) Ullman's encyclopedia of industrial chemistry. Wiley-VCH, Weinheim
6. Martínková L, Křen V (2002) Biocatal Biotransformation 20:73
7. Harper DB (1976) Biochem Soc Trans 4:502
8. Kobayashi M, Shimizu S (1994) FEMS Microbiol Lett 120:217
9. Banerjee A, Sharma R, Banerjee UC (2003) Appl Microbiol Biotechnol 60:33
10. Martínková L, Mylerová V (2003) Curr Org Chem 7:1279
11. Layh N, Hirrlinger B, Stolz A, Knackmuss HJ (1997) Appl Microbiol Biotechnol 47:668
12. Kobayashi M, Nagasawa T, Yamada H (1992) Trends Biotechnol 10:402
13. Mathew CD, Nagasawa T, Kobayashi M, Yamada H (1988) Appl Environ Microbiol 54:1030
14. Asano Y, Tani Y, Yamada H (1980) Agric Biol Chem 44:2251
15. Bauer R, Knackmuss HJ, Stolz A (1998) Appl Microbiol Biotechnol 49:89
16. Asano Y, Fujishiro K, Tani Y, Yamada H (1982) Agric Biol Chem 46:1165
17. Saroja N, Shamala TR, Tharanathan RN (2000) Process Biochem 36:119
18. Padmakumar R, Oriel P (1999) Appl Biochem Biotechnol 77:671
19. Takashima Y, Yamaga Y, Mitsuda S (1998) J Ind Microbiol Biotechnol 20:220
20. Alfani F, Cantarella M, Spera A, Viparelli P (2001) J Mol Catal, B Enzyme 11:687
21. Hann EC, Eisenberg A, Fager SK, Perkins NE, Gallagher FG, Cooper SM, Gavagan JE, Stieglitz B, Hennessey SM, DiCosimo R (1999) Bioorg Med Chem 7:2239
22. Chapatwala KD, Babu GRV, Armstead ER, White EM, Wolfram JH (1995) Appl Biochem Biotechnol 51–52:717
23. Yamaki T, Oikawa T, Ito K, Nakamura T (1997) J Ferment Bioeng 83:474
24. Yamada H, Kobayashi M (1996) Biosci Biotechnol Biochem 60:1391
25. Wieser M, Takeuchi K, Wada Y, Yamada H, Nagasawa T (1998) FEMS Microbiol Lett 169:17
26. Cull SG, Holbrey JD, Vargas-Mora V, Seddon KR, Lye GJ (2000) Biotechnol Bioeng 69:227

27. Dadd MR, Sharp DCA, Pettman AJ, Knowles CJ (2000) J Microbiol Methods 41:69
28. Dadd MR, Claridge TDW, Walton R, Pettman AJ, Knowles CJ (2001) Enzyme Microb Tech 29:20
29. Langdahl BR, Bisp P, Ingvorsen K (1996) Microbiology 142:145
30. Prepechalová I, Martínková L, Stolz A, Ovesná M, Bezouška K, Kopecký J, Křen V (2001) Appl Microbiol Biotechnol 55:150
31. Martínková L, Klempier N, Přepechalová I, Přikrylová V, Ovesná M, Griengl H, Křen V (1998) Biotechnol Lett 20:909
32. Meth-Cohn O, Wang MX (1997) J Chem Soc 8:1099
33. Wang C, Zhang G, Xu X, Li C (2007) Chin J Chem Eng 15:573
34. Wang YJ, Zheng YG, Xue JP, Shen YC (2007) World J Microbiol Biotechnol 23:355
35. Hu JG, Wang YJ, Zheng YG, Shen YC (2007) Enzyme Microb Technol 41:244
36. Endo T, Watanabe I (1989) FEBS Lett 243:61
37. Dias JCT, Rezende RP, Linardi VR (2001) Appl Microbiol Biotechnol 56:757
38. Linardi VR, Dias JCT, Rosa CA (1996) FEMS Microbiol Lett 144:67
39. Rezende RP, Dias JCT, Rosa CA, Carazza F, Linardi VR (1999) J Gen Appl Microbiol 45:185
40. Gerasimova T, Novikov A, Osswald S, Yanenko A (2004) Eng Life Sci 4:543
41. Kobayashi M, Shimizu S (1999) Eur J Biochem 261:1
42. Thimann KV, Mahadevan S (1964) Arch Biochem Biophys 107:62
43. Gavagan JE, DiCosimo R, Eisenberg A, Fager SK, Folsom PW, Hann EC, Schneider KJ, Fallon RD (1999) Appl Microbiol Biotechnol 52:654
44. Harper DB (1977) Biochem J 165:309
45. Piotrowski M, Schonfelder S, Weiler EW (2001) J Biol Chem 276:2616
46. Almatawah QA, Cramp R, Cowan DA (1999) Extremophiles 3:283
47. Goldlust A, Bohak Z (1989) Biotechnol Appl Biochem 11:581
48. Nagasawa T, Mauger J, Yamada H (1990) Eur J Biochem 194:765
49. Rezende RP, Dias JCT, Ferraz V, Linardi VR (2000) J Basic Microbiol 40:389
50. Rausch T, Hilgenberg W (1980) Phytochemistry 19:747
51. Yamamoto K, Fujimatsu I, Komatsu K (1992) J Ferment Bioeng 73:425
52. Vejvoda V, Kaplan O, Bezouška K, Martínková L (2006) J Mol Catal, B Enzyme 39:55
53. Ishikawa T, Okazaki K, Kuroda H, Itoh K, Mitsui T, Hori H (2007) Mol Plant Pathol 8:623
54. Kobayashi M, Nagasawa T, Yamada H (1989) Eur J Biochem 182:349
55. Kaplan O, Nikolaou K, Pišvejcová A, Martínková L (2006) Enzyme Microb Technol 38: 260
56. Harper DB (1985) Int J Biochem 17: 677
57. Rustler S, Stolz A (2007) Appl Microbiol Biotechnol 75: 899
58. Bhalla TC, Miura A, Wakamoto A, Ohba Y, Furuhashi K (1992) Appl Microbiol Biotechnol 37: 184
59. Kobayashi M, Yanaka N, Nagasawa T, Yamada H (1990) J Bacteriol 172: 4807
60. Lévy-Schil S, Soubrier F, Crutzlecoq AM, Faucher D, Crouzet J, Petre D (1995) Gene 161: 15
61. Layh N, Parratt J, Willetts A (1998) J Mol Catal B-Enzyme 5: 467
62. Hughes J, Armitage YC, Symes KC (1998) Antonie van Leeuwenhoek 74: 107
63. Yamamoto K, Komatsu K (1991) Agric Biol Chem 55: 1459
64. Stalker DM, Malyj LD, McBride KE (1988) J Biol Chem 263: 6310
65. Bandyopadhyay AK, Nagasawa T, Asano Y, Fujishiro K, Tani Y, Yamada H (1986) Appl Environ Microbiol 51: 302
66. Khandelwal AK, Nigam VK, Choudhury B, Mohan MK, Ghosh P (2007) J Chem Technol Biotechnol 82: 646
67. Zheng YG, Chen J, Liu ZQ, Wu MH, Xing LY, Shen YC (2008) Appl Microbiol Biotechnol 77:985
68. Ryuno K, Nakamura T (2003) Yuki Gosei Kagaku Kyokaishi/J Synth Org Chem 61: 517
69. Fournand D, Arnaud A (2001) J Appl Microbiol 91: 381
70. Hirrlinger B, Stolz A, Knackmuss HJ (1996) J Bacteriol 178: 3501
71. Kimani SW, Agarkar VB, Cowan DA, Sayed MFR, Sewell BT (2007) Acta Crystallogr D Biol Crystallogr 63: 1048

72. Suzuki Y, Ohta H (2006) Protein Expr Purif 45: 368
73. Hongpattarakere T, Komeda H, Asano Y (2005) J Ind Microbiol Biotechnol 32: 567
74. Asano Y, Tachibana M, Tani Y, Yamada H (1982) Agric Biol Chem 46: 1175
75. Kotlova EK, Chestukhina GG, Astaurova OB, Leonova TE, Yanenko AS, Debabov VG (1999) Biochemistry (Mosc) 64: 384
76. Inoue A, Komeda H, Asano Y (2005) Adv Synth Catal 347: 1132
77. Yamaoka JS, Pridgeon MG, Miner KD, Taylor SK, Liao MK, Goyne TE (2005) FASEB J 19: A303
78. Krieg L, Ansorge-Schumacher MB, Kula MR (2002) Adv Synth Catal 344: 965
79. Egorova K, Trauthwein H, Verseck S, Antranikian G (2004) Appl Microbiol Biotechnol 65: 38
80. Komeda H, Harada H, Washika S, Sakamoto T, Ueda M, Asano Y (2004) Eur J Biochem 271: 1580
81. Shaw NM, Naughton A, Robins K, Tinschert A, Schmid E, Hischier ML, Venetz V, Werlen J, Zimmermann T, Brieden W, de Riedmatten P, Roduit JP, Zimmermann B, Neumuller R (2002) Org Process Res Dev 6: 497
82. Baek DH, Kwon SJ, Hong SP, Kwak MS, Lee MH, Song JJ, Lee SG, Yoon KH, Sung MH (2003) Appl Environ Microb 69: 980
83. Komeda H, Ishikawa N, Asano Y (2003) J Mol Catal, B Enzym 21: 283
84. Neumann S, Kula MR (2002) Appl Microbiol Biotechnol 58: 772
85. Trott S, Bauer R, Knackmuss HJ, Stolz A (2001) Microbiol-Sgm 147: 1815
86. D'Abusco AS, Ammendola S, Scandurra R, Politi L (2001) Extremophiles 5: 183
87. Nawaz MS, Khan AA, Bhattacharayya D, Siitonen PH, Cerniglia CE (1996) J Bacteriol 178: 2397
88. Cheong TK, Oriel PJ (2000) Enzyme Microb Tech 26: 152
89. Zheng RC, Wang YS, Liu ZQ, Xing LY, Zheng YG, Shen YC (2007) Res Microbiol 158: 258
90. Fournand D, Arnaud A, Galzy P (1998) J Mol Catal, B Enzyme 4: 77
91. Buckles RE, Thelen CJ (1950) Anal Chem 22: 676
92. Zheng RC, Zheng YG, Shen YC (2007) Appl Microbiol Biotechnol 74: 256
93. Zhu Q, Fan A, Wang Y, Zhu X, Wang Z, Wu MH, Zheng YG (2007) Appl Environ Microb 73: 6053
94. Martínková L, Vejvoda V, Křen V (2008) J Biotechnol 133: 318
95. Kaul P, Banerjee A, Mayilraj S, Banerjee UC (2004) Tetrahedron Asymmetry 15: 207
96. Yazbeck DR, Durao PJ, Xie Z, Tao J (2006) J Mol Catal, B-Enzyme 39: 156
97. Nagasawa T, Takeuchi K, Yamada H (1988) Biochem Biophys Res Commun 155: 1008
98. Cheetham PJ, Knowles CJ (1988) J Gen Microbiol 134: 1099
99. Hoyle AJ, Bunch AW, Knowles CJ (1998) Enzyme Microb Technol 23: 475
100. Tauber MM, Cavaco-Paulo A, Robra KH, Gubitz GM (2000) Appl Environ Microb 66: 1634
101. Cramp RA, Cowan DA (1999) Biochim Biophys Acta 1431: 249
102. Bhalla TC, Kumar H (2005) Can J Microbiol 51: 705
103. Maier-Greiner UH, Obermaier-Skrobranek BMM, Estermaier LM, Kammerloher W, Freund C, Wulfing C, Burkert UI, Matern DH, Breuer M, Eulitz M, Kufrevioglu OI, Hartmann GR (1991) Proc Natl Acad Sci USA88: 4260
104. Endo I, Odaka M, Yohda M (1999) Trends Biotechnol 17: 244
105. Popescu VC, Munck E, Fox BG, Sanakis Y, Cummings JG, Turner IM, Nelson MJ (2001) Biochemistry 40: 7984
106. Graham D, Pereira R, Barfield D, Cowan D (2000) Enzyme Microb Technol 26: 368
107. Rustler S, Stolz A (2007) Appl Microbiol Biotechnol 75: 899
108. Wang MX, Lin SJ (2002) J Org Chem 67: 6542
109. Chen J, Zheng YG, Shen YC (2008) Biotechnol Appl Biochem 50:147
110. Wang MX, Feng GQ (2002) J Mol Catal, B-Enzyme 18: 267
111. Phillips RS (1996) Trend Biotechnol 14: 13
112. Prepachalova I, Martinkova L, Stolz A, Ovesna M, Bezouska K, Kopecky J, Kren V (2001) Appl Microbiol Biotechnol 55: 150
113. Wu ZL, Li ZY (2003) J Org Chem 68: 2479

114. Wang MX, Lu G, Ji GJ, Huang ZT, Meth-Cohn O, Colby J (1999) Tetrahedron Asymmetry 11: 1123
115. Geresh S, Giron Y, Gilboa Y, Glaser R (1993) Tetrahedron 49: 10099
116. Martinkova L, Mylerova V (2003) Curr Org Chem 7: 1279
117. Liu M, Jiao p, Cao ZA (2001) J Chem Ind Eng 52: 847
118. Straathof AJ, Sie S, Franco TT, van der Wielen LA (2005) Appl Microbiol Biotechnol 67: 727
119. Nagasawa T, Nakamura T, Yamada H (1990) Appl Microbiol Biotechnol 34: 322
120. Kobayashi M, Yanaka N, Nagasawa T, Yamada H (1990) J Bacteriol 172:4807
121. Kato Y, Tsuda T, Asano Y (1999) Eur J Biochem 263: 662
122. Bauer R (1997) Dissertation thesis. Stuttgart
123. Offermanns H, Kleeman A, Tanner H, Beschke H, Friedrich H (1984) In: Mark HF, Othmer DF, Overberger CG, Seaborg GT (eds.) Kirk–Othmer encyclopaedia of chemical technology. Wiley, New York p. 1
124. Nagasawa T, Mathew CD, Mauger J, Yamada H (1988) Appl Environ Microbiol 54: 1766
125. Almatawah QA, Cowan DA (1999) Enzyme Microb Technol 25: 718
126. Zou JH, Li MF, Liu XL (2007) Fangming Zhuanli Shenqing Gongkaishu CN 1952114 (in Chinese)
127. Shen YC, Xue JP, Li HB, Wang SL, Li Y, Zhu J (2006) Fangming Zhuanli Shenqing Gongkaishu CN 1730660 (in Chinese)
128. Kato Y, Tsuda T, Asano Y (1999) Eur J Biochem 263: 662
129. Kiempier N, Harter G, De Raadt A, Griengl H, Braunegg G (1996) Food Technol Biotechnol 34: 67
130. Matoishi K, Sano A, Imai N, Yamazaki T, Yokoyama M, Sugai T, Ohta H (1998) Tetrahedron Asymmetry 9: 1097
131. Wang MX, Feng GQ (2000) Tetrahedron Lett 41: 6501
132. Effenberger F, Osswald S (2001) Tetrahedron Asymmetry 12: 2581
133. Martínková L, Klempier N, Preiml M, Ovesná M, Kuzma M, Mylerová V, Křen V (2002) Can J Chem 80: 724
134. Shaw NM, Robins KT, Kiener A (2003) Adv Synth Catal 345: 425
135. Fallon RD, Stieglitz B, Turner I (1997) Appl Microbiol Biotechnol 47: 156
136. Bauer R, Hirrlinger B, Layh N, Stolz A, Knackmuss HJ (1994) Appl Microbiol Biotechnol 42: 1
137. Blakey AJ, Colby J, Williams E, O'Reilly C (1995) FEMS Microbiol Lett 129: 57
138. Effenberger F, Graef BW (1998) J Biotechnol 60: 165
139. Masutomo S, Inoue A, Kumagai K, Murai R, Mitsuda S (1995) Biosci Biotechnol Biochem 59: 720
140. Wang MX, Lin SJ, Liu CS, Zheng QY, Li JS (2003) J Org Chem 68: 4570
141. Wang MX, Deng G, Wang DX, Zheng QY (2005) J Org Chem 70: 2439
142. Liu J, Wang DX, Zheng QY, Wang MX (2006) Chin J Chem 24: 1665
143. Martinkova L, Kren V (2002) Biocatal Biotransformation 20: 73
144. Sugai T, Yamazaki T, Yokoyama M, Ohta H (1997) Biosci Biotechnol Biochem 61: 1419
145. Macadam AM, Knowles CJ (1985) Biotechnol Lett 7: 865
146. Bhalla TC, Miura A, Wakamoto A, Ohba Y, Furuhashi K (1992) Appl Microbiol Biotechnol 37: 184
147. Yamamoto K, Oishi K, Fujimatsu I, Komatsu K (1991) Appl Environ Microbiol 57: 3028
148. Yamamoto K, Fujimatsu I, Komatsu K (1992) J Ferment Bioeng 73: 425
149. Rey P, Rossi JC, Taillades J, Gros G, Nore O (2004) J Agric Food Chem 52: 8155
150. He YC, Xu JH, Xu Y, Ouyang LM, Pan J (2007) Chin Chem Lett 18: 677
151. Endo T, Yamagami T, Tamura K (1994) US Patent US 5326702
152. Yamamoto K, Ueno Y, Otsubo K, Kawakami K, Komatsu KI (1990) Appl Environ Microb 56: 3125
153. Kakeya H, Sakai N, Sugai T, Ohta H (1991) Tetrahedron Lett 32: 1343
154. Eichhorn E, Roduit JP, Shaw N, Heinzmann K, Kiener A (1997) Tetrahedron Asymmetry 8: 2533

155. Gavagan JE, Fager SK, Fallon RD, Folsom PW, Herkes FE, Eisenberg A, Hann EC, DiCosimo R (1998) J Org Chem 63: 4792
156. Cooling FB, Fager SK, Fallon RD, Folsom PW, Gallagher FG, Gavagan JE, Hann EC, Herkes FE, Phillips RL, Sigmund A, Wagner LW, Wu W, DiCosimo R (2001) J Mol Catal, B Enzyme 11: 295
157. Chauhan S, Wu S, Blumerman S, Fallon RD, Gavagan JE, DiCosimo R, Payne MS (2003) Appl Microbiol Biotechnol 61: 118
158. Asano Y, Yasuda T, Tani Y, H. Y (1982) Agric Biol Chem 46: 1183
159. Crosby JA, Parratt JS, Turner NJ (1992) Tetrahedron Asymmetry 3: 1547
160. Yokoyama M, Sugai T, Ohta H (1993) Tetrahedron Asymmetry 4: 1081
161. Asano Y, Ando S, Tani Y, Yamada H (1981) Agric. Biol. Chem 45: 57
162. Yamada H, Asano Y, Tani Y (1980) J Ferment Technol 58: 495
163. Asano Y, Ando S, Tani Y, Yamada H (1980) Agric Biol Chem 44: 2497
164. MethCohn O, Wang MX (1997) J Chem Soc [Perkin 1]: 3197
165. MethCohn O, Wang MX (1997) Chem Commun 1041
166. Klempier N, Harter G, DeRaadt A, Griengl H, Braunegg G (1996) Food Technol Biotechnol 34: 67
167. Martinkova L, Klempier N, Prepechalova I, Prikrylova V, Ovesna M, Griengl H, Kren V (1998) Biotechnol Lett 20: 909
168. Martínková L, Klempier N, Bardakji J, Kandelbauer A, Ovesná M, Podařilová T, Kuzma M, Přepechalová I, Griengl H, Kren V (2001) J Mol Catal, B Enzyme 14: 95
169. Kaufmann G, Dautzenberg H, Henkel H, Müller G, Schäfer T, Undeutsch B, Oettel M (1999) Steroids 64: 535
170. Nawaz MS, Davis JW, Wolfram JH, Chapatwala KD (1991) Appl Biochem Biotechnol 28: 865
171. Wyatt JM, Knowles CJ (1995) Int Biodeterior Biodegradation 35: 227
172. Li T, Liu J, Bai R, Ohandja DG, Wong FS (2007) Water Res 41: 3465
173. Kao CM, Chen KF, Liu JK, Chou SM, Chen SC (2006) Appl Microbiol Biotechnol 71: 228
174. Lee CM, Wang CC (2004) Water Sci Technol 49: 341
175. Wang CC, Lee CM, Chen LJ (2004) J Environ Sci Health Part A Tox Hazard Subst Environ Eng 39: 1767
176. Muller D, Gabriel J (1999) Folia Microbiol 44: 377
177. Kobayashi M, Nishiyama M, Nagasawa T, Horinouchi S, Beppu T, Yamada H (1991) Biochimica et Biophysica Acta 1129: 23
178. Wu S, Fallon RD, Payne MS (1997) Appl Microbiol Biotechnol 48: 704
179. Shi Y, Yu HM, Sun XD, Tian ZL, Shen ZY (2004) Enzyme Microb Technol 35: 557
180. Stalker DM, McBride KE (1987) J Bacteriol 169: 955
181. Piotrowski M, Schonfelder S, Weiler EW (2001) J Biol Chem 276: 2616
182. Komeda H, Hori Y, Kobayashi M, Shimizu S (1996) Proc Natl Acad Sci U S A 93: 10572
183. Fang P, Xu DQ, Zhang YL, Cao WH, Zhao JF, Zhu YM, Qin ZJ (2005) Huanjing Kexue Xuebao 25: 1414
184. Wang QL, Zhang R, Ni WC, Chen YQ, Guo SD (2004) Shengwu Gongcheng Xuebao 20: 730
185. Zhang JW, Xiong CR, Li JW, Wang WT, Meng YX, Chen ZH (2006) Acta prataculturae sinica 15: 87
186. Wu SJ, Fogiel AJ, Petrillo KL, Jackson RE, Parker KN, DiCosimo R, O'Ben-Bassat A, Keefe D, Payne MS (2008) Biotechnol Bioeng 99: 717
187. DeSantis G, Wong K, Farwell B, Chatman K, Zhu Z, Tomlison G, Huang H, Tan X, Bibbs L, Chen P, Kretz K, Burk MJ (2003) J Am Chem Soc 125: 11476

Adv Biochem Engin/Biotechnol (2009) 113: 79-150
DOI: 10.1007/10_2008_26

Published online: 26 May 2009

Secondary Metabolites from Higher Fungi: Discovery, Bioactivity, and Bioproduction

Jian-Jiang Zhong and Jian-Hui Xiao

Abstract Medicinal higher fungi such as *Cordyceps sinensis* and *Ganoderma lucidum* have been used as an alternative medicine remedy to promote health and longevity for people in China and other regions of the world since ancient times. Nowadays there is an increasing public interest in the secondary metabolites of those higher fungi for discovering new drugs or lead compounds. Current research in drug discovery from medicinal higher fungi involves a multifaceted approach combining mycological, biochemical, pharmacological, metabolic, biosynthetic and molecular techniques. In recent years, many new secondary metabolites from higher fungi have been isolated and are more likely to provide lead compounds for new drug discovery, which may include chemopreventive agents possessing the bioactivity of immunomodulatory, anticancer, etc. However, numerous challenges of secondary metabolites from higher fungi are encountered including bioseparation, identification, biosynthetic metabolism, and screening model issues, etc. Commercial production of secondary metabolites from medicinal mushrooms is still limited mainly due to less information about secondary metabolism and its regulation. Strategies for enhancing secondary metabolite production by medicinal

J.-J. Zhong (✉)
School of Life Sciences and Biotechnology, Key Laboratory of Microbial Metabolism
Ministry of Education,, Shanghai Jiao Tong University, 800 Dong-Chuan Road,
Shanghai 200240, China
e-mail: jjzhong@sjtu.edu.cn

J.-H. Xiao
Key Laboratory of Cell Engineering of Guizhou Province, Affiliated Hospital of Zunyi Medical College, 149 Dalian Road, Zunyi, 563003, China and
State Key Laboratory of Bioreactor Engineering, East China University of Science and Technology, 130 Meilong Road, Shanghai, 200237, China

mushroom fermentation include two-stage cultivation combining liquid fermentation and static culture, two-stage dissolved oxygen control, etc. Purification of bioactive secondary metabolites, such as ganoderic acids from *G. lucidum*, is also very important to pharmacological study and future pharmaceutical application. This review outlines typical examples of the discovery, bioactivity, and bioproduction of secondary metabolites of higher fungi origin.

Keywords Bioactive compound, Fermentation production, Higher fungi, Medicinal mushroom, Physiological and pharmacological activity

Contents

Abbreviations

ABTS	2,2'-Azinobis(3-ethylbenzothiazoline-6-sulfonate)
Abu	Aminobutyric acid
Aib	α-Aminoisobutyric acid
Ala	Alanine
BHA	Butyl hydroxyanisole
COX	Cyclooxygenase
DHNM	Dihydroxynaphthalene melanin
DMBA	7,12-Dimethylbenz[α]anthracene
DPPH	2,2-Diphenyl-1-(2,4,6-trinitrophenyl)hydrazyl
EBV-EA	Epstein–Barr virus early antigen
EC_{50}	50% Effective concentration
ESI-MS	Electrospray ionization mass spectrometry
Glc	Glucose

GLP	Ganoderma lucidum peptide
Gly	Glycine
HPLC	High-performance liquid chromatography
HRMS	High-resolution mass spectrometry
HyLeu	Hydroxyleucine
IC_{50}	50% Inhibitory concentration value
kDa	Kilo Dalton
Leu	Leucine
Lxx	*N*-methylleucine/*N*-methylisoleucine/*N*-methylalloisoleucine
MePro	Methylproline
mg	Milligram
mL	Milliliter
mmol	Millimolar
NE	Norepinephrine
NmePh	*N*-Methylphenylalanine
NmeVal	*N*-Methylvaline
Phe	Phenylalanine
Pro	Proline
PTP1B	Protein tyrosine phosphatase 1B
PTP1B	Protein tyrosine phosphatase 1B
RP-HPLC	Reversed phase HPLC
TCM	Traditional Chinese medicine
TPA	12-*O*-Tetradecanoylphorbol-13-acetate
UV	Ultraviolet
Val	Valine

1 Biodiversity of Higher Fungi

Although some fungi have been successfully domesticated for thousands of years without the realization of their existence, they have been recognized for little more than two centuries as an important and abundant group of organisms significant to humanity. Currently, the kingdom Fungi mainly includes four broad phyla, i.e., Chytridiomycota, Zygomycota, Ascomycota, and Basidiomycota, whose members mainly have chitinous cell walls, which clearly differentiates the fungi from plants which have cell walls strengthened with cellulose or hemicellulose. While Myxomycota and most members of Mastigomycotina (Oomycota, Hyphochytriomycota, Labyrinthulomycota, and Plasmodiophoromycota) used to be considered as fungi, they are now regrouped into the Protoctista as two separate kingdoms. Namely, members of Myxomycota and Plasmodiophoromycota are placed at the kingdom Protozoa, other members of Mastigomycotina that have been shown to have close relationship with the algae, with cellulosic cell walls, are assigned to independent phylums of the kingdom Chromista [1]. According to the evidence of fungal origin, evolution and phylogeny, Ascomycotina (the cup fungi, flask fungi and their allies, including moulds and most yeasts), Basidiomycotina (the smuts, rusts, jelly fungi, fairy clubs, bracket fungi, stinkhorns, bird's-nest fungi, puffballs and earthstars, toadstools and mushrooms) and their anamorphs (asexual fungi) are considered as typical representatives in so-called higher fungi divisions.

Members of the kingdom Fungi have been recognized as one of the largest biodiversity resources from both terrestrial and aquatic sources on Earth, fulfilling crucial ecological roles especially in terrestrial ecosystems. At present, estimates of the total number of fungal species on Earth are about 1.5 million species, whereas the number of fungi described worldwide is just about 7% of this number [2, 3]. Recently, Schmit and Mueller have conservatively estimated that there is a minimum of 712,285 extant fungal species worldwide according to the ratio of fungal species to plant species in the same region, 600,000 of which are fungi associated with terrestrial plants [4]. Unfortunately, only about 5–10% of fungi can be cultured artificially using current cultivation techniques [5]. According to related literatures, for the limitation of our knowledge base such as isolation and cultivation methods, fungal physiology, and fungal diversity, worldwide there are probably more than 20,000 fungal species already collected still awaiting formal description [6]. Therefore, the vast majority of fungi still remain hidden and need to be explored, identified, conserved and utilized for the benefit of humankind in particular, and the mycobiota and environment. As already stated, higher fungi (including asexual fungi) possess over 90% of recorded species of fungi [7]. Statistical data show that over 5,000 species belonging to approximately 1,200 genera of higher fungi have so far been reported only from southern China [8]. New report also indicates that approximately half of the recorded fungal names are of lichenized fungi and macrofungi with the other half corresponding to microfungi [4], while microfungi including aquatic fungi, soil-inhabiting fungi, terrestrial plant-associated fungi, and arthropod-associated fungi actually consist of mostly higher fungi and a few lower fungi. Therefore, higher fungi can be considered as one of the megadiversity bioresources in the world.

Chinese medicinal materials, possessing diversified physiological active substances, have been widely used for treatment of various diseases in China for thousands of years, which are looked at as an attractive source of new drug discovery for disease treatment and have received increasing interest around the world in recent years, while medicinal higher fungi play an important role in natural resources of Chinese medicinal materials. Shen Nong's Herbal Classic (Shen Nong Ben Cao Jing) of the Chinese Western Han Dynasty, the oldest book on herbal remedies, and Li Shi-zhen's Compendium of Materia Medica (Ben Cao Gang Mu) of the Ming Dynasty, recorded the beneficial effects of more than 20 species of medicinal higher fungi including *Cordyceps sinensis* (Dong Chong Xia Cao), *Ganoderma lucidum* (Ling Zhi), *C. cicadae*(Chan Hua), *Poria cocos* (Fu Ling), *Omphalia lapidescens* (Lei Wan), *Polyporus umbellatus* (Zhuling), etc., which have been widely used in many recipes and folk prescriptions for curing various diseases. Particularly *C. sinensis* and *G. lucidum* were ranked the superior medicines in the above two ancient books. Of course, today they are still considered as the most exalted medicines in traditional Chinese medicine in China and other East Asia regions [2, 9]. According to the Traditional Chinese Medical Database System of the Institute of Information on Traditional Chinese Medicine, China Academy of Traditional

Chinese Medicine (http://www.cintcm.com/default.htm), the large-scale investigation of China's resources of Chinese medicinal materials has shown that China has as many as 12,807 species of Chinese medicinal materials, among which 298 species are medicinal higher fungi belonging to 110 genera and 41 families of kingdom Fungi, and more than 70% of which are distributed at the following divisions including six families of basidiomycete, i.e., Polyporaceae, Tricholomataceae, Russulaceae, Boletaceae, Lycoperdaceae, and Agaricaceae, and three families of Ascomycete, i.e., Clavicepitaceae, Hypocreaceae, and Ustilaginaceae. Unfortunately, no new statistical data on China's resources of Chinese medicinal materials to date are available since the last investigation. However, recent survey also shows that ascomycetes, asexual fungi and basidiomycetes are the most frequent producers of bioactive secondary metabolites among fungal species, most species of which belong to the genera *Aspergillus*, *Penicillium*, *Fusarium*, *Trichoderma*, *Phoma*, *Alternaria*, *Acremonium* and *Stachybotrys*, *Ganoderma*, *Lactarius* and *Aureobasidium* [10]. In addition, with the great scientific and technical development of fungal isolation and taxonomy, more and more species of higher fungi, especially endophytic and marine fungi which are considered as a new and tremendous source of potent medicinal merits, have been found around the world since the 1990s [11, 12]. Medicinal higher fungi possess an extremely wide population with biodiversity.

2 Distribution and Chemodiversity of Secondary Metabolites in Higher Fungi

As well known, higher fungi have formed a special mechanism of metabolism which could produce diversified functional secondary metabolites possessing various properties of chemical structure and bioactivity during the long-term evolution, so as to resist unfavorable survivable environments and finish cell proliferation, differentiation and the entire life cycle to reach the purpose of self-defense and survival, which simultaneously provides an abundant and fascinating resource for new drug discovery. Thus the diversification of medicinal higher fungi represents a great potential for new drugs. Alexander Fleming published his observation on the inhibition of growth of *Staphylococcus aureus* on an agar plate contaminated with *Penicillium notatum* in 1929 and penicillin was finally exploited and clinical application was achieved during World War II, which is perhaps the most important discovery in the history of therapeutic medicine and shows an important route for the discovery of modern antibiotics. However, the first fungal-derived secondary metabolite is mycophenolic acid from *P. glaucoma* discovered in 1896 by Gosio in the history of natural product research. On the basis of these antibiotics discovery, the secondary metabolites of higher fungi have been received an increasing attention worldwide since the 1940s. Then the wide search for metabolites of higher fungi as potential new drugs was begun at Sandoz Ltd (Basel) in 1957, which discovered a number of new products with remarkable activities by Stoll (cultivation of fungi), Tamm (chemistry) and Stähelin (biological testing), such as cytochalasin B (phomin) from *Phoma exigua*, brefeldin A from *P. brefeldianum*, verrucarin A from *Myrothecium verrucaria*, anguidine (diacetoxyscirpenol) from *Fusarium diversisporium*, and famous immunosuppressive drug cyclosporin A from *Tolypocladium inflatum*, etc. [13]. Since the 1990s, with the extensive application of new screening, separation and characterization techniques, an exponential increase of the number of new bioactive secondary metabolites possessing wide bioactivities from higher fungi was emerged worldwide. However, only a relatively small number of higher fungi species are chemically investigated. A huge, untapped, and chemodiversity resource of secondary metabolites from higher fungi will provide more and more opportunities for finding new lead structures for medicinal chemistry, and a new era of higher fungi secondary metabolite research has appeared.

An interesting fact indicates that almost all models of drug screening can have corresponding active substances from higher fungi or other microbial secondary metabolites [14], and higher fungi are recognized as cell factories producing diversified bioactive compounds. More importantly, most secondary metabolites from higher fungi possess the drug-like characteristics of chemical structure, which can act as a major natural compounds library for new drug discovery [14, 15]. Generally speaking, the drug-likeness of secondary metabolites from higher fungi mainly includes the following: their molecular weight is in the range of 150–1,000 Da; the metabolites usually contain C, H, O, and N, even S, P and chlorine group atoms such as Cl, Br, and F; their chemical structure commonly contains some important functional groups such as hydroxyl, carboxyl, carbonyl, amino, etc., which can provide multipharmacophore points; their molecular properties such as relative molecular mass, logP value, and number of the donor and receptor of hydrogen-bonding meet the rules of drug-likeness. According to statistical data, the total number of bioactive microbial metabolites recognized has doubled every 10 years since the 1980s. As shown in Fig. 1, more than 22,000 species of bioactive secondary microbial metabolites have now been recognized, over 38% of which (approximately 8,600) were of fungal origin [10]. In addition, around 25,000 microbial secondary metabolites, over half of which were derived from fungi, need further assay about whether there are any bioactivities. At present, microbial secondary metabolites can be divided into over 20 major groups according to their chemical types [16]. In this review, roughly six groups of secondary metabolites reported in the past decade, which derive from higher fungi of aquatic, terrestrial and endophytic habitat environments (Table 1), are overviewed as follows.

2.1 Heterocyclics

Heterocyclic compounds are assigned as containing cyclic structures with at least two different atoms in the ring (as ring atoms or members of the ring), in which the ring itself is called a heterocycle. In principle, all elements except the alkali metals can act as ring atoms. The heterocycles constitute the largest group of organic compounds, and currently, of more than 20 million chemical compounds registered, about one half are heterocyclic [17]. Heterocyclics are important, not only because

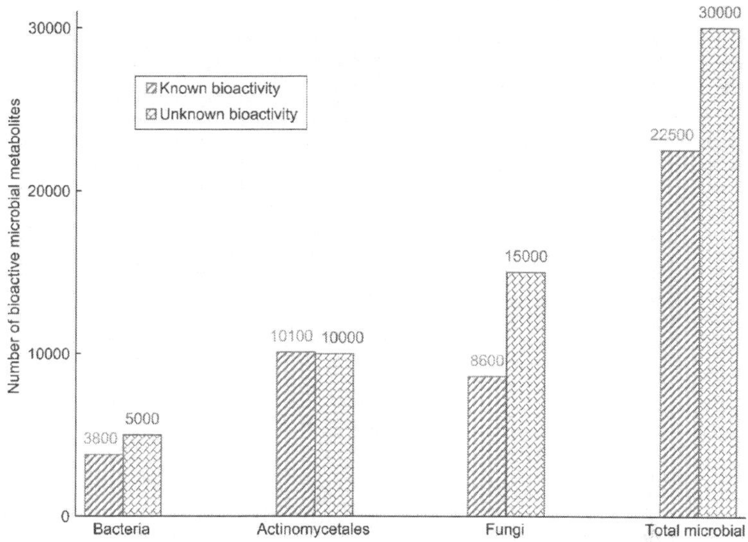

Fig. 1 Approximate number of known microbial secondary metabolites (according to [12])

of their abundance, but above all because of their chemical, biological and technical significance. Heterocyclic compounds ubiquitously occur and constitute almost one-third of the total number of known natural organic products, such as vitamins, hormones, antibiotics, alkaloids, as well as pharmaceuticals, herbicides, dyes, and other products of technical importance [17, 18].

Previous studies have shown that heterocyclic bioactive secondary metabolites produced by higher fungi are also major types of nitrogen heterocylics, oxygen heterocylics, and sulfur–nitrogen heterocyclics, etc. An overview on nitrogen-containing compounds of macromycetes was reported by Liu's group at the beginning of 2005 [18]. Subsequently, Liu's group isolated two new compounds, 21-(acetyloxy)-6,13, 14-trihydroxy-16,18-dimethyl-10-phenyl[11]cytochalasa-7,19-dien-1-one (**1**) [19] and a new derivative of benzofuran lactone, named concentricolide (**2**) [20], together with four known compounds including friedelin, cytochalasin L-696,474, armillaramide, russulamide, 2,3-dihydro-5-hydroxy-2-methyl-4*H*-1-benzopyran-4-one, 3,5-dihydroxy-2-(1-oxobutyl)-cyclohex-2-en-1-one, and pyroglutamic acid, from the fruiting bodies of the ascomycete *Daldinia concentrica*, of which the structures and relative configuration of the new compounds were elucidated by detailed spectroscopic analysis and confirmed by X-ray crystallography [20, 21]. Interestingly, compound **2** showed the potent blockage effect on syncytium formation between HIV-1-infected cells and normal cells; thus it was effective against HIV-1 [20]. More recently, they found a new nitrogen-containing heterocycle compound exhibiting a weak anti-HIV-1 activity, namely flazin (**3**), belonging to β-carboline derivatives, from the fruiting body of *Suillus granulatus* [22]. Based on the structure of **3**, flazinamide (**4**), 1-(5′-hydroxymethyl-2 -furyl)-*β*-carboline-3-carboxamide, was synthesized through converting the carboxyl

Table 1 Secondary metabolites from higher fungi and their bioactivities

Source (Family/species)	Denomination	Molecular formula	Bioactivity	Ref.
Aquatic fungi				
Massarina tunicata	massarilactones A(**58**)	$C_{11}H_{14}O_5$	antibacterial	[58]
	massarilactones B(**59**)	$C_{11}H_{14}O_5$	antibacterial	[58]
	massarinolin A (**200**)	$C_{15}H_{18}O_4$	antibacterial	[130]
	massarinolin B (**201**)	$C_{15}H_{22}O_4$	antibacterial	[130]
	massarinolin C (**202**)	$C_{15}H_{22}O_4$	NT	[130]
	massarigenin A (**60**)	$C_{11}H_{14}O_5$	antibacterial	[59]
	massarigenin B (**61**)	$C_{11}H_{14}O_5$	NT	[59]
	massarigenin C (**62**)	$C_{11}H_{12}O_5$	antibacterial	[59]
	massarigenin D (**63**)	$C_{11}H_{12}O_5$	antibacterial	[59]
	massarinin A (**64**)	$C_{21}H_{20}O_6$	antibacterial	[59]
	massarinin B (**65**)	$C_{20}H_{20}O_5$	antibacterial	[59]
Microsphaeropsis sp.	microsphaeropsisin(**56**)	$C_{16}H_{22}O_4$	antimicrobial	[57]
Coniothyrium sp.	(3S)-(3′,5′-dihydroxy phenyl)butan-2-one (**57**)	$C_{10}H_{12}O_3$	antimicrobial	[57]
Humicola fuscoatra	fuscoatrol(**198**)	$C_{16}H_{26}O_4$	antimicrobial	[129]
	11-epiterpestacin(**199**)	$C_{25}H_{35}O_4$	antimircrobial	[129]
Chaetomium sp.	chaetocyclinone A(**36**)	$C_{17}H_{16}O_8$	antifungal	[44]
	chaetocyclinone B(**37**)	$C_{16}H_{14}O_7$	NT	[44]
	chaetocyclinone C(**38**)	$C_{32}H_{24}O_{14}$	NT	[44]
Penicillium janthinellum	shearinine D (**8**)	$C_{37}H_{45}NO_6$	anticancer	[26]
	shearinine E (**9**)	$C_{37}H_{45}NO_6$	anticancer	[26]
	shearinine F(**10**)	$C_{37}H_{47}NO_4$	NT	[26]
P.aurantiogriseum	aurantiomide A(**11**)	$C_{19}H_{24}N_4O_4$	NT	[27]
	aurantiomide B(**12**)	$C_{18}H_{22}N_4O_4$	anticancer	[27]
	aurantiomide C(**13**)	$C_{18}H_{20}N_4O_3$	anticancer	[27]
Annulatascus triseptatus	annularin A (**66**)	$C_{10}H_{14}O_4$	antibacterial	[60]
	annularin B(**67**)	$C_{10}H_{14}O_4$	antibacterial	[60]
	annularin C(**68**)	$C_{10}H_{14}O_5$	antibacterial	[60]
	annularin D(**69**)	$C_{10}H_{14}O_3$	NT	[60]
	annularin E(**70**)	$C_9H_{12}O_3$	NT	[60]

	annularin F(**71**)	$C_{10}H_{10}O_5$	antibacterial	[60]
	annularin G(**72**)	$C_9H_{14}O_4$	NT	[60]
	annularin H(**73**)	$C_9H_{12}O_4$	NT	[60]
Phoma sp.	phomoxin (**79**)	$C_{15}H_{22}O_6$	NT	[62]
	phomoxide(**80**)	$C_{14}H_{20}O_4$	NT	[62]
	eupenoxide(**81**)	$C_{14}H_{22}O_4$	NT	[62]
Terrestrial fungi				
Chaunopycnis pustulata	compound A(**119**)	$C_{32}H_{43}NO_3$	antagonists of the calcium- gated potassium ion channel	[98]
	compound B(**120**)	$C_{37}H_{51}NO_5$	antagonists of the calcium- gated potassium ion channel	[98]
	compound C(**121**)	$C_{37}H_{47}NO_5$	antagonists of the calcium- gated potassium ion channel	[98]
	compound D(**122**)	$C_{28}H_{37}NO_5$	antagonists of the calcium- gated potassium ion channel	[98]
	compound G(**123**)	$C_{37}H_{49}NO_6$	antagonists of the calcium- gated potassium ion channel	[98]
Cha.alba	nalanthalide(**124**)	$C_{29}H_{44}O_5$	blocker of the voltage-gated potassium channel	[98]
Daldinia childiae	daldinal A (**28**)	$C_{17}H_{16}O_6$	anti-inflammtory	[42]
	daldinal B (**29**)	$C_{18}H_{18}O_6$	anti-inflammtory	[42]
	daldinal C (**30**)	$C_{20}H_{23}O_7$	anti-inflammtory	[42]
D.concentrica	concentricol(**165**)	$C_{30}H_{54}O_6$	NT	[110]
	concentriol B(**166**)	$C_{30}H_{52}O_7$	NT	[111]
	concentriol C(**167**)	$C_{32}H_{56}O_7$	NT	[111]
	concentriol D(**168**)	$C_{32}H_{54}O_6$	NT	[111]
	(17β,20*R*,22*E*,24*R*) -19-norergosta-1,3,5, 7,9,14,22-heptaene(**99**)	$C_{27}H_{34}$	NT	[73]
	(17β,20*R*,22*E*,24*R*) -1-methyl-19-nor-ergosta-1,3,5,7,9,14, 22-heptaene(**100**)	$C_{28}H_{37}$	NT	[73]
	1-isopropyl-2,7-di-methylnaphthalene(**236**)	$C_{15}H_{18}$	NT	[19]
	21-(acetyloxy)-6,13, 14-trihydroxy-16, 18-dimethyl-10	$C_{30}H_{41}NO_6$	NT	[19]

(continued)

Table 1 (continued)

Source (Family/species)	Denomination	Molecular formula	Bioactivity	Ref.
	- phenyl [11]cytochalasa-7, 19-dien-1-one (**1**)			
	concentricolide (**2**)	$C_{12}H_{10}O_3$	Anti-HIV	[20]
	3-alkyl-5-methoxy-2-methyl-1, 4-benzoquinones(**237**-**240**)	$C_{29\text{-}31}H_{50\text{-}54}O_3$	NT	[159]
	hept-6-ene-2,4,5-triols (**241**)	$C_7H_{15}O_3$	NT	[21]
Tyromyces fissilis	tyromycic acids B(**149**)	$C_{34}H_{50}O_7$	NT	[104]
	tyromycic acids C(**150**)	$C_{32}H_{48}O_5$	NT	[104]
	tyromycic acids D(**151**)	$C_{30}H_{44}O_4$	NT	[104]
	tyromycic acids E(**152**)	$C_{30}H_{44}O_3$	NT	[104]
	tyromycic acids F(**153**)	$C_{30}H_{42}O_3$	NT	[105]
	tyromycic acids G(**154**)	$C_{32}H_{46}O_5$	NT	[105]
	tyromycic acids(**148**)	$C_{30}H_{44}O_3$	NT	[105]
Creosphaeria sassafras	sassafrins A(**52**)	$C_{27}H_{32}O_7$	antifungal,antibacterial	[55]
	sassafrins B(**53**)	$C_{26}H_{30}O_7$	antifungal,antibacterial	[55]
	sassafrins C(**54**)	$C_{27}H_{30}O_7$	antibacterial	[55]
	sassafrins D(**55**)	$C_{27}H_{32}O_8$	antibacterial	[55]
Septocylindrium sp.	septocylindrins A(**221**)	$C_{94}H_{155}N_{23}O_{25}$	anticancer	[142]
	septocylindrins B(**222**)	$C_{94}H_{156}N_{24}O_{24}$	anticancer	[142]
Trichopezizella nidulus	trichoflectin (**49**)	$C_{17}H_{14}O_5$	inhibitors of the DHNM, antimicrobial	[54]
	6-deoxy-7-O-demethyl-3,4-anhydro-fusarubin (**50**)	$C_{14}H_{10}O_5$	inhibitors of the DHNM, antimicrobial	[54]
	6-deoxy-3,4-anhydro-fusarubin (**51**)	$C_{15}H_{12}O_5$	inhibitors of the DHNM, antimicrobial	[54]
Albatrellus dispansus	grifolin (**188**)	$C_{22}H_{32}O_2$	anti-fungal	[123]
A. confluens	(**188**)		antitumor / antioxidant /antiinflammatory	[124–128]
	albaconol (**190**)	$C_{22}H_{34}O_3$	antitumor	[127–128]
	emeheterone(**191**)	$C_{19}H_{16}N_2O_3$	NT	[127]
	5-Methoxy-3,6-bis(phenyl-methyl)pyrazin-2-ol(**192**)	$C_{19}H_{16}N_2O_2$	promote melanin synthesis	[127]
	aurovertin B(**235**)	$C_{25}H_{32}O_8$	bovine F1-ATPase inhibitor	[158]

	aurovertin E(**234**)	$C_{23}H_{30}O_7$	NT	[158]
A. ovinus	3-hydroxyneogrifolin(**193**)	$C_{22}H_{32}O_3$	antioxidant	[125]
	1-formylneogrifolin(**194**)	$C_{23}H_{32}O_3$	antioxidant	[125]
	1-formyl-3-hydroxy-neogrifolin (**195**)	$C_{23}H_{32}O_4$	antioxidant	[125]
A. caeruleoporus	neogrifolin (**189**)	$C_{22}H_{32}O_2$	antioxidant / antiinflammtory	[125-126]
	grifolinone A (**196**)	$C_{22}H_{30}O_3$	anti-inflammtory	[126]
	grifolinone B(**197**)	$C_{44}H_{54}O_7$	anti-inflammtory	[126]
Chaetomium subspirale	oxaspirodion (**39-42**)	$C_{13}H_{15}O_5$	anticancer,TNF-α inhibitor	[46]
Cha. brasiliense	chaetochalasin A(**35**)	$C_{27}H_{39}NO_2$	anticancer	[45]
Phycomyces blakesleeanus	phycomysterol A(**108**)	$C_{27}H_{40}O$	anticancer, anti-HIV	[79]
	phycomysterol B(**109**)	$C_{27}H_{42}O$	NT	[79]
	neoergosterol(**110**)	$C_{27}H_{40}O$	NT	[79]
Cordyceps sinensis	cordycedipeptide A(**204**)	$C_9H_{14}N_3O_3$	anticancer	[135]
C. sp. BCC1788	cordyheptapeptide A(**205**)	$C_{49}H_{65}N_7O_8$	anticancer	[136]
Paecilomyces tenuipes	beauvericin(**203**)	$C_{45}H_{57}N_3O_9$	insecticidal property	[132-134]
Beauveria bassiana	(**203**)		anticancer	[182]
C. cicade	(**203**)		antiangiogenic activity	[182]
C. heteropoda	cicadapeptin I(**206**)	$C_{50}H_{90}N_{10}O_{11}$	Antifungal antibacteral	[137]
	cicadapeptin II(**207**)	$C_{50}H_{90}N_{10}O_{11}$	Antifungal antibacteral	[137]
	myriocin(**208**)	$C_{21}H_{39}NO_6$	antifungal	[137]
Suillus granulatus	flazin (3)	$C_{17}H_{12}N_2O_4$	anti-HIV	[22]
Phellinus igniarius	phelligrin A (**15**)	$C_{22}H_{18}O_6$	NT	[32-33]
	phelligrin B (**16**)	$C_{22}H_{18}O_6$	NT	[32-33]
	phelligridin A(**17**)	$C_{13}H_8O_6$	NT	[34]
	phelligridin B(**18**)	$C_{15}H_{13}O_7$	NT	[34]
	phelligridin C(**19**)	$C_{20}H_{12}O_7$	anticancer	[35]
	phelligridin D(**20**)	$C_{20}H_{12}O_8$	anticancer	[35]
	phelligridin E(**21**)	$C_{25}H_{14}O_{10}$	NT	[35]
	phelligridin F(**22**)	$C_{26}H_{22}O_9$	NT	[35]
	phelligridin G(**23**)	$C_{32}H_{18}O_{12}$	antioxidant, anticancer	[36]
	phelligridin H(**24**)	$C_{33}H_{18}O_{13}$	antioxidant, inhibit PTP1B	[37]

(continued)

Table 1 (continued)

Source (Family/species)	Denomination	Molecular formula	Bioactivity	Ref.
	phelligridin I(**25**)	$C_{33}H_{21}O_{13}$	antioxidant, inhibit PTP1B	[37]
	phelligridin J(**26**)	$C_{13}H_{5}O_{8}$	anticancer	[37]
	phelligridimer(**27**)	$C_{52}H_{32}O_{20}$	antioxidant	[38]
Paxillus panuoides	paxillamide (**224**)	$C_{42}H_{85}NO_{6}$	NT	[153]
Russula cyanoxantha	(2S,3S,4R,2'R)-2-(2'-hydro-xytetracosanoylamino) oct-adecane-1,3,4-triol (**225**)	$C_{42}H_{85}NO_{5}$	NT	[118]
	5α,8α-epidioxy-(22E,24R) -ergosta-6,22-dien -3β-ol (**102**)	$C_{28}H_{44}O_{3}$	anticancer	[76,118]
	(22e,24r)-ergosta-4,6,8(**14**), 22-tetraen-3-one (**103**)	$C_{28}H_{40}O$	NT	[76,118]
Cyathus stercoreus	cyathusals A (**31**)	$C_{17}H_{14}O_{7}$	antioxidant	[43]
	cyathusals B (**32**)	$C_{17}H_{14}O_{8}$	antioxidant	[43]
	cyathusals C (**33**)	$C_{20}H_{20}O_{8}$	antioxidant	[43]
	pulvinatal (**34**)	$C_{18}H_{16}O_{8}$	anticancer/antioxidant	[43]
Poria cocos	15α-hydroxydehydro-tumulosic acid(**142**)	$C_{31}H_{48}O_{5}$	inhibiting EBV-EA	[103]
	16α,25-dihydroxydehydro-eburiconic acid(**143**)	$C_{31}H_{46}O_{5}$	inhibiting EBV-EA	[103]
	5α,8α-peroxydehydro-tumulosic acid(**144**)	$C_{31}H_{46}O_{6}$	inhibiting EBV-EA	[103]
	25-hydroxyporicoic acid H(**145**)	$C_{30}H_{48}O_{6}$	inhibiting EBV-EA	[103]
	16-deoxyporicoic acid B(**146**)	$C_{30}H_{44}O_{4}$	inhibiting EBV-EA, anticancer	[103]
	poricoic acid CM(**147**)	$C_{32}H_{48}O_{4}$	inhibiting EBV-EA	[103]
Sporormiella vexans	sporovexin A (**243**)	$C_{12}H_{14}O_{5}$	antimicrobial	[162]
	sporovexin B (**244**)	$C_{12}H_{14}O_{6}$	NT	[162]
	sporovexin C (**245**)	$C_{15}H_{19}NO_{6}$	NT	[162]

	3'-O-desmethyl-1-epipreussomerin C(**246**)	$C_{20}H_{14}O_8$	antimicrobial	[162]
R. lepida	rulepidol (**178**)	$C_{15}H_{22}O_2$	NT	[115]
	lepidamine (**179**)	$C_{15}H_{20}NO_3$	NT	[116]
	rulepidadiol (**177**)	$C_{15}H_{22}O_3$	NT	[114]
	rulepidatriol(**176**)	$C_{15}H_{22}O_4$	NT	[114]
	(24e)-3β-hydroxy-cucurbita-5,24-diene -26-oic acid (**173**)	$C_{30}H_{48}O_3$	NT	[114]
	(24e)-3,4-secocucur-bita-4, 24-diene-3,**26** -dioic acid (**174**)	$C_{30}H_{47}O_4$	NT	[114]
	(24e)-3,4-secocucurbita-4, 24-diene-3,26, 29-trioic acid (**175**)	$C_{30}H_{46}O_6$	NT	[114]
	lepidolide(**180**)	$C_{30}H_{40}O_6$	NT	[117]
	(2s,3s,4r,2'r)-2-(2'-hydro-xytetracosanoylamino) octadecane-1,3,4-triol (**225**)	$C_{42}H_{85}NO_5$	NT	[118]
Lactarius rufus	rufuslactone (**181**)	$C_{15}H_{21}O_3$	antifungal activity	[120]
L. hirtipes	2β,α-epoxy-6Z, 9Z-humuladien-8α-ol(**182**)	$C_{15}H_{24}O_2$	NT	[121]
L. mitissimus	mitissimol A (**183**)	$C_{15}H_{22}O_2$	NT	[122]
	mitissimol B (**184**)	$C_{15}H_{22}O_3$	NT	[122]
	mitissimol C (**185**)	$C_{15}H_{22}O_3$	NT	[122]
	mitissimol A oleate(**186**)	$C_{33}H_{54}O_3$	NT	[122]
	mitissimol A linoleate(**187**)	$C_{33}H_{52}O_3$	NT	[122]
Tylopilus	tylopiol A (**106**)	$C_{28}H_{44}O_2$	NT	[78]
plumbeoviolaceus	tylopiol B (**107**)	$C_{28}H_{44}O_3$	NT	[78]
Monascus.purpureus	monascodilone (**44**)	$C_{15}H_{12}O_4$	NT	[51]
	monascopyridine A (**45**)	$C_{21}H_{25}NO_4$	NT	[52]
	monascopyridine B (**46**)	$C_{23}H_{30}NO_4$	NT	[52]

(continued)

Table 1 (continued)

Source (Family/species)	Denomination	Molecular formula	Bioactivity	Ref.
	monascopyridine C (**47**)	$C_{20}H_{27}NO_3$	anticancer	[53]
Tuber indicum	monascopyridine D (**48**)	$C_{22}H_{31}NO_3$	anticancer	[53]
	tuberoside (**104**)	$C_{34}H_{56}O_8$	NT	[77]
	(22e, 24r)-ergosta- 7, 22-dien-3β, 5α, 6β-triol (**105**)	$C_{28}H_{46}O_3$	NT	[77]
	(2s, 3s, 4r, 2'r)-2-N-(2'-hydroxytricosanoyl) -octadecan-1, 3,4-triol (**226**)	$C_{41}H_{83}NO_5$	NT	[154]
	(2s,2'r,3s,4r)-2-(2'-D-hydroxyalkanoylamino) octadecane-1,3,4-triol (**227**)	$C_{40-42}H_{81-85}NO_5$	NT	[155]
	(2s,3s,4r)-2-(alkanoylamino) octadecane-1, 3,4-triol (**228**)	$C_{34-41}H_{69-83}NO_4$	NT	[155]
	(2s,3r,4e)-2-(alkanoy-lamino)-4-octadecene 1-,3-diol (**229**)	$C_{34-36}H_{67-71}NO_3$	NT	[155]
	9,10,11-trihydroxy-(12Z)-12-octadecenoic acid (**230**)	$C_{18}H_{34}O_5$	NT	[156]
Fusarium oxysporum	6-epi-oxysporidinone (**94**)	$C_{28}H_{42}NO_5$	antifungal	[68]
	the dimethyl ketalof oxysporidinone (**95**)	$C_{30}H_{49}NO_7$	antifungal	[68]
	N-demethylsambutoxin (**96**)	$C_{27}H_{37}NO_4$	antifungal	[68]
Leucopaxillus gentianeus	cucurbitacins B (**156**)	$C_{32}H_{46}O_8$	anticancer	[107]
	deoxycucurbitacin B (**160**)	$C_{32}H_{46}O_7$	anticancer	[107]
	cucurbitacins D (**161**)	$C_{30}H_{44}O_6$	anticancer	[107]
	leucopaxillone A (**162**)	$C_{34}H_{54}O_7$	anticancer	[108]
	deoxyleucopaxillone A (**164**)	$C_{34}H_{54}O_6$	anticancer	[108]

	leucopaxillone B (**163**)	$C_{34}H_{53}O_8$	anticancer	[108]
	15-deoxyoxalicine B (**5**)	$C_{30}H_{33}NO_6$	antiinsectan activity	[25]
P. decaturense	decaturin A (**6**)	$C_{30}H_{35}NO_6$	antiinsectan	[25]
	decaturin B (**7**)	$C_{30}H_{35}NO_6$	antiinsectan	[25]
Ganoderma lucidum	lucidenic acid N (**127**)	$C_{27}H_{40}O_6$	anticancer	[101]
	methyl lucidenate F (**128**)	$C_{28}H_{38}O_6$	NT	[101]
	lucialdehyde A (**129**)	$C_{30}H_{46}O_2$	NT	[102]
	lucialdehyde B (**130**)	$C_{30}H_{44}O_3$	anticancer	[102]
	lucialdehyde C (**131**)	$C_{30}H_{46}O_3$	anticancer	[102]
	3β,5α-dihydroxy-(22e,24r)-ergosta -7,22-dien-6β-yl oleate (**132**)	$C_{46}H_{78}O_4$	NT	[74]
Tricholomopsis rutilans	3β,5α-dihydroxy-(22e,24r)-ergosta -22-en-7-one-6β-yl oleate (**133**)	$C_{46}H_{78}O_5$	NT	[74]
Xylaria sp.	kolokoside A (**169**)	$C_{36}H_{59}O_{10}$	antibacterial	[113]
	kolokoside B (**170**)	$C_{36}H_{58}O_9$	NT	[113]
	kolokoside C (**171**)	$C_{36}H_{58}O_{10}$	NT	[113]
X. euglossa	kolokoside D (**172**)	$C_{36}H_{58}O_{10}$	NT	[113]
Cortinarius umidicola	xylactam (**134**)	$C_{23}H_{31}NO_6$	NT	[56]
Cor. vibratilis	3-aldehyde-2-amino-6-methoxypyridine (**137**)	$C_7H_8N_2O_2$	NT	[40]
Cor. sp.	vibratilicin (**138**)	$C_{43}H_{80}N_2O_7$	NT	[41]
	unsymmetrical disulfidecortamidine oxide (**231**)	$C_{13}H_{17}O_4N_3S_2$	antimicrobial, anticancer	[157]
	2,2'-dithiobis(pyridine N-oxide) (**232**)	$C_{10}H_{10}O_2N_2S_2$	antimicrobial, anticancer	[157]
	symmetrical disulfide cortamidine oxide (**233**)	$C_{16}H_{26}O_6N_4S_2$	NT	[157]
Sarcodon scabrosus	scabronine G (**139**)	$C_{27}H_{36}O_5$	NT	[96]
	scabronine H (**140**)	$C_{27}H_{36}O_5$	NT	[96]

(continued)

Table 1 (continued)

Source (Family/species)	Denomination	Molecular formula	Bioactivity	Ref.
	sarcodonin I (**141**)	$C_{20}H_{28}O_4$	NT	[97]
S. laevigatum	sarcodan (**223**)	$C_{24}H_{17}O_{10}$	NT	[152]
Endophytic fungi				
Coniothyrium sp	massarilactone C (**82**)	$C_{12}H_{18}O_6$	NT	[64]
	massarilactone D (**83**)	$C_{11}H_{14}O_6$	NT	[64]
	massarilactone E (**87**)	$C_{11}H_{12}O_5$	NT	[65]
	massarilactone F (**88**)	$C_{12}H_{16}O_6$	NT	[65]
	massarilactone G (**89**)	$C_{12}H_{16}O_6$	NT	[65]
	massarilactone acetonide (**90**)	$C_{14}H_{18}O_5$	NT	[65]
	massarigenin E (**84**)	$C_8H_{12}O_4$	NT	[64]
	coniothyrenol (**85**)	$C_{22}H_{36}O_6$	NT	[64]
	graphislactone A (**86**)	$C_{16}H_{14}O_6$	NT	[64]
Taxomyces andreanae	paclitaxel (**118**)	$C_{47}H_{51}NO_{14}$	anticancer	[84]
Pestalotiopsis	(**118**)		anticancer	[85]
microspora	(**118**)		anticancer	[86]
Sporormia minima	(**118**)		anticancer	[86]
Trichothecium sp	(**118**)		anticancer	[87]
Tubercularia sp	(**118**)		anticancer	[88–89]
Nodulisporium sylviforme	(**118**)		anticancer	[91]
Seimatoantlerium tepuiense	podophyllotoxin (**242**)	$C_{22}H_{20}O_8$	anticancer, antioxidant,	[160]
Phialocephala fortinii	(**242**)		anticancer, radioprotective	[161]
Trametes hirsute	camptothecin (**14**)	$C_{20}H_{16}N_2O_4$	immunomodulatory	[28]
Entrophospora	3β,5α-dihydroxy-6β-acetoxy-ergosta-7,22-diene (**111**)	$C_{30}H_{48}O_4$	anticancer,antibacterial, antifungal activity	[80]
Infrequens	3β,5α-dihydroxy-6β-phenylacetyloxy-ergosta-7,22-diene (**112**)	$C_{36}H_{52}O_4$	antibacterial and antifungal activity	[80]

Colletotrichum sp.	CR377 (**93**)	$C_{12}H_{16}O_4$	antifungal	[67]
F. sp.	Radicicol (**43**)	$C_{18}H_{17}O_6Cl$	anticancer, Hsp90 inhibitor	[47]
Cha. chiversii	Phomoxin B (**91**)	$C_{15}H_{22}O_6$	NT	[66]
Eupenicillium sp.	Phomoxin C (**92**)	$C_{15}H_{22}O_6$	NT	[66]
	Xyloketal G (**135**)	$C_{15}H_{18}O_4$	NT	[69]
X. sp.	Xyloketal H (**136**)	$C_{13}H_{15}O_4$	NT	[70]

group in 3-position of compound **3** to the formamido group, which showed more effective anti-HIV-1 activity and lower cytotoxicity than **3** [23].

Fungi of the genus *Penicillium* are promising for new biologically active compounds, which approximately 400 secondary metabolites produced by these fungi have been reported up to now [24], some of which belong to nitrogen-containing heterocyclics. In recent years, nine novel heterocyclics including antiinsectan oxalicine alkaloids, 15-deoxyoxalicine B (**5**) and decaturins A (**6**) and B (**7**) from *P. decaturense* and *P. thiersii*, respectively [25], three new indole alkaloids, shearinines D (**8**), E (**9**) and F (**10**), along with the previously reported shearinine A from the marine fungus *P. janthinellum* Biourge [26], three new quinazoline alkaloids, aurantiomides A (**11**), B (**12**) and C (**13**) from another marine *Penicillium* strain of *P. aurantiogriseum* SP0-19 [27], were isolated by bioassay-guided fractionation. Their structures were also elucidated by a combined spectroscopic and chemical method. Both compounds **6** and **7** are members of a rare structural class with a new polycyclic ring system [25]. Further bioassay results suggested that compounds **8**, **9**, **12** and **13** could inhibit the proliferation of various tumor cells [26, 27].

Previous investigations showed that endophytic higher fungi, as a new source of bioactive products, may yield many potentially useful medicinal compounds including heterocyclics. For example, camptothecin (**14**), a pentacyclic quinoline alkaloid, is an effective anticancer drug lead structure first isolated from *Camptotheca acuminata*, a native plant of China. Puri and coworkers obtained a camptothecin-producing endophytic fungus *Entrophospora infrequens* by isolating from an important Indian medicinal tree *Nothapodytes foetida*, which is commonly known as 'Kalgur' in India [28]. The fungus *E. infrequens* produced the compound **14** when grown in a synthetic liquid medium (Sabouraud broth) in shake flask fermentation, and the identity of **14** in the fungal culture broth was also confirmed by optical rotation, UV, IR, CD, LC/MS, LC-MS/MS, HRMS, and ^{1}H and ^{13}C NMR spectra [28]. A total synthesis of compound **14** has been reported, but the too complex synthetic route and relatively low yield is not able to meet the industrialization and commercial demands. Hence, the discovery of endophyte *E. infrequens* producing **14** may provide an alternative source for its production by fermentation.

2.2 Polyketides

Polyketides, originating from a polyketone and/or the polyketide chain and containing at minimum one acetate group, comprise a huge class of chemicals with a wealth of structural variety [29]. In general, the polyketide chain is formed by successive addition of simple carboxylic acids like acetate. During each extension step of one unit, the polyketide chain is elongated with two carbon atoms, where the β-carbon is a keto group. The vast diversity of polyketides is due to the stepwise reduction of a part or all of these keto groups to hydroxyls or enoyls, which are in some compounds finally completely reduced to an alkyl chain [30]. Polyketides are the most abundant medicinal sources that have been shown to display a wide range of potentially useful therapeutics values due to their antibiotic, anticancer, antifungal, hypolipidemic and immunosuppressive properties among natural products [31]. Higher fungi are recognized as prolific producers of new polyketide natural products over the decades and they continue to be the source of new structural and/or bioactive polyketide chemistry.

Phe. igniarius, a basidiomycete belonging to the family Polyporaceae and known as Sang Huang in China and considered as a famous traditional Chinese medicine, has been used to treat wounds, abdominalgia, and bloody gonorrhea since ancient times [32]. In recent years, Shi's group has obtained over 20 secondary metabolites with interesting chemical structures and significant bioactivities from the fungus *Phe. igniarius* [32–37]. Two new benzyl dihydroflavones, phelligrins A (**15**) and B (**16**), were isolated from the EtOAc soluble fraction of the ethanolic extract of *Phe. igniarius* fruiting body, where their structures were identified as 5,7,4′-trihydroxy-6-*O*-hydroxybenzyldihydroflavone and 5,7,4′-trihydroxy-8-*O*-hydroxybenzyldihydroflavone, respectively, by means of spectral methods [32, 33], and bioassay indicated that both of them had no significant inhibitory effects on some cancer cell lines at a concentration of 10 μmol L^{-1} [32]. Subsequently, the EtOAc soluble portion of the ethanolic extract of the same fungus was further subjected to chromatography on normal phase silica gel, Sephadex LH-20 and reverse phase HPLC to yield phelligridins A-G (**17–23**) [34–36]. Herein both compounds **17** and **18** were pyrone derivatives and were elucidated as 8,9-dihydroxy-3-methyl-1*H*,6*H*-pyrano[4,3-*c*][2]benzopyran-1,6-dione and 4-hydroxy-6-(3′,4′-dihydroxystyryl)-3-methoxycarbonyl-2-pyrone, respectively, by spectroscopic methods including IR, MS, and 1D and 2D NMR [34]. Interestingly, compounds **19–22**, three unique pyrano[4,-*c*][2]benzopyran-1,6-dione derivatives and a new furo[3,2-*c*]pyran-4-one, were characterized as 3-(4-hydroxystyryl)-8,9-dihydroxypyrano[4,3-*c*]isochromene-4-one, 3-(3,4-hydroxystyryl)-8,9-dihydroxypyrano[4,3-*c*]isochromene-4-one, 8,9-dihydroxy-3-[5′,6′-dihydroxy-5″- methyl-3″-oxo-spiro[fural-2″(3″*H*),1′-indene]-2′-yl]-1*H*, 6*H*-pyrano[4,3-*c*][2]benzopyran-1,6-dione, and (3*Z*)-3-(3,4-dihydroxybenzylidene)-6-(3,4-dihydroxystyryl)-2,3-dihydro-2-methoxy-2-(2-oxo-propyl)furo[3,2-*c*] pyran-4-one, respectively, of which all possessed potent cytotoxicity against several human cancer cell lines, especially compounds **18** and **19** had significant selective cytotoxicity against A549 human lung cancer cell and Bel7402 human liver cancer cell [35]. Compound **23**, containing an unprecedented carbon skeleton, was a unique pyrano[4,3-*c*][2]benzopyran-1,6-dione derivative, which showed not only moderate cytotoxic activities against human cancer cells but also antioxidant activity inhibiting rat liver microsomal lipid peroxidation [36]. Recently, further investigation of the EtOAc soluble portion of the ethanolic extract of this fungus has resulted in the isolation and structural elucidation of three novel members of pyrano-[4,3-*c*] isochromen-4-one derivatives, designated as phelligridins H-J (**24–26**), together with the known compounds davallialactone, scopolin, nebularine, uridine, glucitol, trehalose, and ethyl glucoside [37]. Both compounds **24** and **25** possessed unprecedented carbon skeletons, and **26** was a derivative of **17** in which the methyl group is oxidized as a carboxyl group [37]. With another *n*-BuOH soluble portion of the ethanolic extract of *Phe. igniarius* fruiting body, these investigators obtained a highly oxygenated and new unsaturated macrocyclic compound with an unprecedented 26-membered ring system via middle-pressure liquid chromatography over reversed-phase silica gel and chromatography over Sephadex LH-20. The macrocyclic compound possessing a symmetric structure was characterized as a dimmer of hypholomine B that was isolated from the fungus *Hypholoma*, and designated as phelligridimer A (**27**) [38]. Additionally, for the related cooccurring compounds **17–27**, a possible biosynthetic pathway involving the fungal metabolite precursor 4-hydroxy-6-methyl-2-pyrone that couples with activated 3,4-dihydroxybenzoyl-SCoA or the cooccurring 3,4- dihydroxybenzaldehyde and/or 4-hydroxybenzaldehyde was also proposed by the authors [36–38].

Cortinarius is one of the largest genera in the subdivision Basidiomycotina, comprising hundreds of species widely distributed in the world [39], and most investigations carried out on the chemical constituents of *Cortinarius* have been focused on toadstools in Europe and Australia [40]. Liu's group recently investigated the chemical constituents of the mushroom *Cortinarius* that grows in a mountainous region in Southwest China [40, 41], from which 3-aldehyde-2-amino-6-methoxypyridine (**137**) [41] and vibratilicin(**138**) [40] were obtained from the fruiting bodies of *Cor. umidicola* and *Cor. vibratilis*, respectively. Compound **138**, identified as 3-[3-(dimethylamino)-4-(hydroxyamino)-4-oxobutoxy]-2-(palmitoyloxy) propyl(9*E*,12*E*)-octadeca-9,12-dienoate based on spectroscopic data, is a representative of the rare natural products containing hydroxamic acid moieties, and can be viewed as a derivative of neoengleromycin [40].

Three benzophenone derivatives, daldinals A–C (**28–30**), from the inedible mushroom *D. childiae*, were obtained by Asakawa's group, which strongly suppressed the LPS-induced production of NO through inhibition of iNOS mRNA expression [42]. More recently, three new polyketide-type antioxidative compounds, cyathusals A (**31**), B (**32**) and C (**33**), and the known pulvinatal (**34**) were isolated from fermented products of the basidiomycete *Cyathus stercoreus* which belongs to the bird's nest fungi; their structures were identified, and they showed higher free radical scavenging activities than those of Trolox and BHA [43].

Members of the ascomycete genus *Chaetomium* occur widely in nature, of which a remarkable variety of chemical diverse metabolites are known, e.g., chaetomin, chaetoglobosins, chaetoquadrins, chaetospiron and orsellides [44]. Oh et al. obtained a new metabolite exhibiting antimicrobial and cytotoxicity bioactivity from an EtOAc extract of *Cha. brasiliense*, whose structure, a cytochalasin-type compound with a novel ring system was determined by NMR and single-crystal X-ray diffraction, and named chaetochalasin A (**35**) [45]. More recently, several polycyclic polyketide derivatives, chaetocyclinone A to C (**36–38**) from a marine fungus *Chaetomium* sp. [44], a mixture of four oxaspirodion isomers (**39–42**) from the *Cha. subspirale* [46], and radicicol (**43**) from *Cha. chiversii* [47], have been isolated by bioassay-guided fractionation, and their structures were also identified by combined spectroscopic techniques. Biosynthetically, chaetocyclinone A (**36**) and B (**38**) should be generated by a fungal polyketide synthase in a one-chain heptaketide folding process. For chaetocyclinone C (**37**), it was proposed as a dimerization step of two heptaketides [44]. The spectroscopic analysis showed that all four compounds **39–42** have the same planar structure, and they should therefore have different configurations [46]. Compound **43** displayed Hsp90 inhibitory and in vitro anticancer activities, but it was found to be devoid of

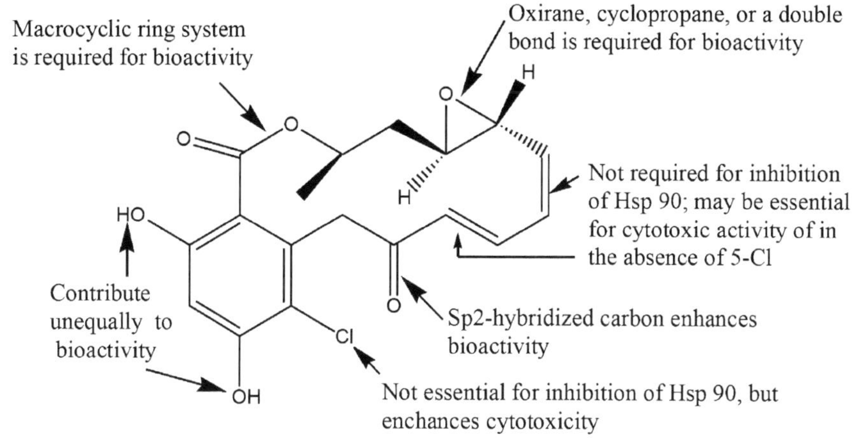

Fig. 2 Partial structure-bioactivity relationships for radicicol from *Chaetomium chiversii* (adapted from [47])

any in vivo activity in animal models [47]. As shown in Fig. 2, partial structure–activity relationships for **43** were summarized by Turbyville and coworkers [47].

Members of the ascomycete genus *Monascus*, mainly *M. purpureus* and *M. anka*, are used for the production of red fermented rice, which is used as a natural food colorant and medicine in China for over millennium [48]. Historic data showed that the use of red fermented rice in China was first documented in the Tang Dynasty in 800 AD, and a complete and detailed description of its manufacture was exhibited in the ancient Chinese pharmacopoeia, *Ben Cao Gang Mu*, published during the Ming Dynasty (1368–1644). Due to its cholesterol lowering properties, red fermented rice has recently gained an increasing interest in the Western countries [49]. About the *Monascus* derived secondary metabolites, Juzlová and coworkers well summarized the progress in the chemistry, bioactivity and biosynthesis [50]. In recent years, Wild and Humpf's group has obtained several new compounds, i.e., monascodilone (**44**) [51], monascopyridines A–D (**45–48**) [52, 53], respectively, from red fermented rice produced by *M. purpureus*. The monascopyridines have a unique aromatic pyridine ring which has previously not been found in metabolites of *Monascus*. Compound **45** contains a γ-lactone, propenyl group, hexanoyl side chain, and a pyridine ring, whereas the more lipophilic compound **46** was a higher homologue of **45** with the more lipophilic octanoyl instead of the hexanoyl side chain [52]. Herein, compound **48** with a C_7H_{15} octanoyl side chain was also a higher homologue of **47** with a C_5H_{11} side chain. Compared with compounds **45–46**, their structural difference was the missing lactone ring [53].

Some investigators also reported polyketides and their derivatives from other ascomycetes. From culture broth of the ascomycete *Trichopezizella nidulus*, Thines et al. obtained a novel azaphilone with the basic structure as deflectins, named as trichoflectin (**49**), and two known fusarubin metabolites including 6-deoxy-7-*O*-demethyl-3,4-anhydrofusarubin (**50**) and 6-deoxy-3,4-anhydrofusarubin (**51**) using bioactivity-guided fractionation, of which all showed antimicrobial activity and inhibited dihydroxynaphthalene melanin (DHNM) biosynthesis in fungi [54]. The structure of **49** was also elucidated by spectroscopic methods [54]. Asakawa's group, using silica gel column chromatography and reversed-phase HPLC, obtained four novel azaphilones named as sassafrins A–D (**52–55**), exhibiting broad-spectrum antimicrobial activity, from the methanol extract of the stromata of the ascomycete *Creosphaeria sassafras*, of which **54** indicated the largest inhibition zones of 22 mm against *S. aureus, Pseudomonas aeruginosa, Klebsiella pneumoniae,* and *Escherichia coli* 95, and a possible biosynthetic pathway for compound **55** was also proposed [55]. Additionally, Liu's group obtained a novel nitrogen-containing polyketide, named as xylactam **(134)**, along with two known alkaloids, penochalasin B2 and neoechinulin A from extracts of the fruiting bodies of *Xylaria euglossa*, which mainly exists on stumps and fallen branches of forested areas in Southwestern China [56].

Aquatic and endophytic fungi are recognized as novel sources of potentially useful medicinal compounds. From the marine-derived higher fungi, two new compounds, microsphaeropsisin (**56**) and (3*S*)-(3′, 5′-dihydroxyphenyl)butan-2-one **(57)** were isolated from *Microsphaeropsis* sp. and *Coniothyrium* sp, respectively, which exhibited potent antimicrobial properties in agar diffusion assays [57]. Gloers's group reported that two new polyketide-derived antibacterial lactones, massarilactones A (**58**) and B (**59**), were isolated from cultures of the higher freshwater aquatic fungus *Massarina tunicate* [58]. The compound **59** is an isomer of **58**, but both of which also had significant structural difference, i.e., **58** contains a methanofuro [3,4-*b*]oxepin ring system, but **59** possesses a furo [3,4-*b*] pyran ring system. Both of them are novel and unusual ring systems [58]. Later the group reported six additional new compounds, including four new rosigenin analogues, massarigenins A–D (**60–63**), and two new aromatic polyketide metabolites, massarinins A and B (**64, 65**), from the same fungal species [59], and eight new polyketide metabolites, annularins A–H (**66–73**) from another freshwater fungus *Annulatascus triseptatus* [60]. Compound **61** was determined to be an isomer of **60**, and **63** was an isomer of **62** on the basis of spectroscopic data [59]. Compounds **66–71** are 3,4,5-trisubstituted α-pyrones, and the fused bicyclic pyrone–furanone system in **71** has not been reported previously among natural products, while other two compounds, namely **72** and **73**, are 3,4-disubstituted α, β-unsaturated γ-lactones [60]. Among eight compounds, only compounds **66–68** and **71** exhibited antibacterial activity [60].

Irish research group obtained four novel polyketides including two hexaketide compounds *iso*-cladospolide B (**74**) and *seco*-patulolide C (**75**), two 12-membered macrolides, pandangolide 1 (**76**) and pandangolide 2 (**77**), along with the known terrestrial fungal metabolite, cladospolide B (**78**), from the fermentation of an unknown marine fungal species; all of these compounds had no significant activity against a panel of Gram-positive and Gram-negative bacteria and yeast at a concentration of 250 μg per well [61]. Recently, Liu and coworkers isolated two highly oxygenated polyketides, named phomoxin (**79**) and phomoxide (**80**), as well as a previously synthesized antibiotic eupenoxide (**81**), from the marine-derived *Phoma* sp., of which **79** contains an unusual cyclic carbonate moiety that is rare among natural products [62]. Compounds **79** and **80** represent new carbon skeletons that appear to be derived via polyketide pathway, which appear to be produced by the cyclization of a polyketide intermediate to form the cyclohexene ring [62]. Interestingly, an investigation suggested that marine and terrestrial *Phoma* species differed significantly with respect to their secondary metabolite content [63].

More recently, from the ethyl acetate extracts of the endophytic fungus *Coniothyrium* sp. associated with *Carpobrotus edulis*, four previously unknown polyketide-derived natural products including massarilactone C (**82**) and D (**83**), massarigenin E (**84**), and coniothyrenol (**85**), together with the known graphislactone A (**86**) and massarilactone A (**58**), were isolated and identified by Krohn's group [64]. Two of these compounds, massarilactone C and D, are related to the massarilactones, first isolated by Oh et al. from the aquatic fungus *Massarina tunicate* [58]. Compound **84**, a cyclohexene derivative, shows a substitution pattern similar to the rosigenin analogue massarigenin A (**60**), which was later isolated from the same fungus [59]. Therefore, there may be certain genetic relationship between the aquatic fungus *Mas. tunicate* and the endophytic fungus *Coniothyrium* sp according to the related chemotaxonomic theory. Compound **85** has a structural hexadecahydro-1*H*-benzo[α]xanthene skeleton, which to date is unknown as a natural product [64]. Later, four new massarilactones including massarilactones E to G (**87–89**) and massarilactone acetonide (**90**) were isolated from the same endophytic fungus *Coniothyrium* sp. from another plant species *Artimisia maritime* and characterized [65]. The relative structure of the parent compound **87** was determined by X-ray single crystal diffraction analysis, and its absolute configuration was analyzed by the solid-state CD-TDDFT approach [65]. Davis and colleagues isolated and identified two new natural polyketides, named phomoxins B (**91**) and C (**92**), along with a previously reported antibiotic eupenoxide, from the culture broth of an endophytic fungus *Eupenicillium* sp. [66], both of which are an isomer of the previously isolated fungal metabolite compound **79** from the marine fungus *Phoma sp*. Although compound **81** is reported as an antibiotic, the authors here did not observe significant antibacterial and antifungal activity or cancer cell cytotoxicity for any of these metabolites [62, 66]. Additionally, Brady and Clardy reported the isolation and characterization of CR377 (**93**), a potent antifungal activity possessing pentaketide derivative, from an endophytic fungus *Fusarium* sp. [67]. Three new *N*-methyl-4-hydroxy-2-pyridinone analogues, 6-*epi*-oxysporidinone (**94**), dimethyl ketal of oxysporidinone (**95**), and *N*-demethylsambutoxin (**96**), along with the known compounds oxysporidinone (**97**), sambutoxin (**98**), wortmannin, enniatin A, enniatin A1, and enniatin B1 were isolated from another *Fusarium* species *F. oxysporum* by a bioassay-guided fractionation [68], of which these compounds were all shown to have antifungal activities, and herein wortmannin was a powerful inhibitor of phosphatidylinositide 3-kinase [68].

Among marine microorganisms, the marine mangrove fungus has attracted much research due to its importance in ecology. Recently, Lin's group successively reported that two representatives of a new family of xyloketals, xyloketal G (**135**) [69] and xyloketal H (**136**) [70], together with a known xyloketal D were isolated from the marine mangrove endophytic ascomycetes *Xylaria* sp. Chemical investigation suggested that **135** is the stereoisomer of xyloketal D, while the synthetic route of the family xyloketals via *ortho*-quinone methide was also successfully performed by Wilson's group [71, 72].

However, in the preliminary bioassay, compound **136** was not bioactive against Hep-2 cell line and Gram-positive bacterium *S. aureus* in standard disk assay [70].

2.3 Sterols

Liu's group reported the characterization of two new aromatic steroids, (17β,20*R*,22*E*,24*R*)-19-norergosta-1,3,5,7,9,14,22-heptaene (**99**) and (17β,20*R*,22*E*,24*R*)-1-methyl-19-norergosta-1,3,5,7,9,14,22-heptaene (**100**) isolated from the ascomycete *D. concentrica*, and proposed that the origin of these compounds is derived from the transformation undergone by their precursor due to microbial action [73]. Commonly, steroid skeleton occurs in sediment and crude oil, and diaromatic and 1-methyl diaromatic steroid hydrocarbons have never been found from any living organisms, which could be the long-sought, biological precursor steroids from living organisms for organic matter in Earth's subsurface. Therefore, these compounds that could be potentially used as biological markers for the contribution of microorganisms to sediments give a link between biological marker compounds or fossil molecules and biological origin [73]. Recently, from the fruiting bodies of the basidiomycete *Tricholomopsis rutilans*, two steryl esters with a polyhydroxylated ergostane-type nucleus, 3*β*,5*α*-dihydroxy-(22*E*,24*R*)-ergosta-7,22-dien-6*β*-yl oleate (**132**) and 3β,5α-dihydroxy-(22*E*,24*R*)-ergosta-22-en-7-one-6β-yl oleate (**133**) were also obtained by that group[74].

Agrocybe aegerita, an edible mushroom, is an important valuable source possessing varieties of bioactive secondary metabolites such as indole derivatives with free radical scavenging activity, cylindan with anticancer activity, and agrocybenine with antifungal activity, etc. Recently, two known sterols with new cyclooxygenase (COX) inhibitory and antioxidant activities including ergosterol (**101**) and 5*α*,8*α*-epidioxy-(22*E*,24*R*)-ergosta-6,22-dien-3*β*-ol (**102**) were isolated from the fruiting body of *Agr. aegerita* [75]. In the course of chemical investigation on the basidiomycete *Paxillus panuoides*, Gao and colleagues isolated two ergosteroide compounds, i.e., compound **102** and (22*E*,24*R*)-ergosta-4,6,8(14),22-tetraen-3-one (**103**), in which **102** possessed potent anticancer activity [76]. They also obtained a new polyhydroxy sterol glycoside, named as tuberoside (**104**), together with additional four known ergostane-type compounds, brassicasterol, (22*E*,24*R*)-ergosta-7,22-dien-3*β*,5α, 6*β*-triol (**105**), (22*E*, 24*R*)-ergosta-4,6,8(14),22-tetraen-3-one, and **102** from the fruiting body of the edible truffle *Tuber indicum*, in which the structure of **104** was identified as 3-*O*-*β*-*D*-glucopyranosyl-(22*E*,24*R*)-ergosta-7,22-dien-5*α*,6*β*-diol on the basis of spectroscopic and chemical means [77]. Additionally, Wu et al. reported that two novel secoergosterols, tylopiols A (**106**) and B (**107**), were isolated from the fresh fruiting bodies of *Tylopilus plumbeoviolaceus*. Their structures were reported as 3β-hydroxy-8α,9α-oxido-8,9-secoergosta-7,9(11),22-triene and 3β-hydroxy-8α,9α-oxido-8,9-secoergosta-7,22-dien-12-one, respectively, where the bond between C-8 and C-9 is cleaved to form an enol ether oriented in the α-position [78].

In the course of searching for new bioactive products from filamentous fungi, three novel naturally occurring sterols with a 19-norergostane skeleton and an aromatic

B ring, from mycelium of *Phycomyces blakesleeanus*, were isolated and analyzed using semipreparative HPLC, GC-MS, and NMR techniques, and proposed as phycomysterol A (**108**), phycomysterol B (**109**),and neoergosterol (**110**), respectively [79]. Herein **109** was assigned as an isomer closely related to **108** that possesses anti-HIV and antitumor activities [79]. Later, Tan and coworkers reported that, by combining spectroscopic methods, two new sterols metabolites produced by an endophytic fungus *Colletotrichum* sp. in *Artemisia annua* were isolated and characterized as 3β,5α-dihydroxy-6β-acetoxy-ergosta-7,22-diene (**111**) and 3β,5α-dihydroxy-6β-phenylacetyloxy-ergosta-7,22-diene (**112**), respectively, together with several known ergosterol, 3β,5α,6β-trihydroxyergosta-7,22-diene, 3β-hydroxy-ergosta-5-ene, 3-oxo-ergosta-4,6,8(14),22-tetraene, 3β-hydroxy-5α,8α-epidioxy-ergosta-6,22-diene, 3β-hydroxy-5α,8α-epidioxy-ergosta-6,9(11),22-triene and 3-oxo-ergosta-4-ene [80]. Among these metabolites characterized, the new compounds and two known ergosterol derivatives (3β-hydroxy-ergosta-5-ene and 3β-hydroxy-5α,8α-epidioxy-ergosta-6,22-diene) possessed antibacterial and antifungal activities, while another sterol 3-oxo-ergosta-4,6,8(14),22-tetraene was only inhibitory against bacteria [80].

The ascomycete *Cordyceps* and related fungi also contain rich sterols with various bioactivities. From crude methanol extracts of *Cordyceps sinensis* mycelia, Bok et al. obtained two novel antitumor sterol derivatives, i.e., 5α,8α-epidioxy-24(*R*)-methylcholesta-6,22-dien-3β-d-glucopyranoside (**113**) and 5α,6α-epoxy-24(*R*)-methylcholesta -7,22-dien-3β-ol (**114**) by bioassay-guided fractionation, of which the glycosylated form of ergosterol peroxide was found possessing a greater inhibitory effect on the proliferation of K562, Jurkat, WM-1341, HL-60 and RPMI-8226 tumor cell lines [81]. Nam et al. [82] reported that both ergosterol peroxide 5α,8α-epidioxy-24(*R*)-methylcholesta-6,22- dien-3β-ol (**115**) and acetoxyscirpenediol 4β-acetoxyscirpene-3α,15-diol (**116**) from artificial culture of *Paecilomyces tenuipes* could markedly inhibit the proliferation of tumor cells in vitro. Additionally, Chung's group observed that a trichothecene derivative, 4-acetyl-12,13-epoxyl-9-trichothecene-3,15-diol isolated (**117**) from the methanolic extract of the fruiting body of *Isaria japonica*, was a potent inducer of apoptosis in various cancer cells [83].

2.4 Terpenes

Terpenes are known as an important variety of naturally occurring bioactive metabolites produced by many higher fungi species. Especially diterpenoid, triterpenoid, and sesquiterpenoid are the typical representatives of terpenes with interesting biological activities.

Paclitaxel (**118**), a famous diterpene anticancer drug, is also produced by some endophytic fungi. Stierle and coworkers found that *Taxomyces andreanae*, a higher endophytic fungus belonging to hyphomycetes from the phloem (inner bark) of the Pacific yew (*Taxus brevifolia*), could produce it and its analogue taxane [84]. At present, the compound **118** is mainly produced in extremely poor yield from the yew bark due to difficulties of total synthesis on a large scale, while **118** is not abundant and is only available in a very low content in the bark of *Taxus* species, which unfortunately faces the fact that yew belongs to a rare tree species with low distribution and poor populations worldwide and cannot satisfy the market demand, while over-exploitation and destruction of wild populations of the primary source yew has brought a series of ecological environment and biodiversity problems. Accordingly, endophytic fungi from *Taxus* species were regarded as a potential source producing **118** in an economic scale and received increasing attention worldwide in last decade. As a result, a global survey on endophytes producing **118** from various *Taxus* species indicated that *Pestalotiopsis microspora*, *Sporormia minima* and *Trichothecium* sp from *Taxwallichiana* [85, 86], *Tubercularia* sp from *Tax. mairei* [87], *Nodulisporium sylviforme* from *Tax. cuspidase* [88, 89], etc. could produce **118** to a certain extent. Furthermore, *Pestalotiopsis microspora* from bald cypress was also shown to produce **118** [90], which was the first observation that endophytic fungus residing in plants other than *Taxus* spp. could also produce **118**. More surprisingly, *Maguireothamnus speciosus*, a rubiaceous plant from southwestern Venezuela, yielded a novel fungus *Seimatoantlerium tepuiense* producing **118** [91]. Thus, the distribution of those endophytic fungi producing **118** is worldwide and not confined to endophytes of yews. Unfortunately, none of endophytic fungi producing **118** as yet has been used to produce in large-scale fermentation. Currently **118** production by all endophytic fungi in fermentation is in the range of submicrograms to micrograms per liter [92]. Although many investigators have tried to utilize some elicitors to enhance the **118** production by endophytic fungi in attenuated cultures, the yield of **118** is as yet unable to reach the commercial fermentation requirement.

Sarcodon scabrosus is a basidiomycete belonging to the family Thelephoraceae and has a strong bitter taste due to its abundant bitter diterpenoids such as sarcodonins A to H, scabronines A to F, and scabronines L and M, where all these compounds possess a cyathane skeleton consisting of angularly condensed five-, six- and seven-membered rings and show stimulating effects on nerve growth factor synthesis in vitro [93–95]. Recently, Liu and coworkers obtained three novel cyathane-type diterpenes, named as scabronine G (**139**), H (**140**) [96], and sarcodonin I (**141**) [97], from the fruiting bodies of the same higher fungus, but their bioactivity was unknown. According to a detailed comparative study on NMR and ROESY of compounds **139** and **140**, **140** was recognized as the 11-epimer of **139** [96].

In addition, Bills and coworkers isolated several indole diterpene alkaloids including compounds A–D (**119–122**), G (**123**), and anlanthalide (**124**) from a new clavicipitalean anamorph *Chaunocycnis pustulata*, which exhibited potassium channel antagonist activity and thus may be useful in treating Alzheimer's disease and other cognitive disorders [98]. While these compounds have not yet been reported from the insect or fungal parasitic lineages of the Clavicipitaceae, it suggested that delineation of a *Chaunopycnis* clade has revealed a previously unrecognized lineage of indole diterpene alkaloid-producing fungi among the subfamily Cordycipitoideae [98].

Triterpenoids and related compounds are fairly common among fungal metabolites. *G. lucidum*, known as Ling Zhi in China and Reishi in Japan as a well-known traditional Chinese medicine (TCM) in eastern Asia, is used as a folk remedy for the treatment of cancer, hepatitis, chronic bronchitis, asthma, hemorrhoids, and fatigue symptoms, and has provided over 130 highly oxygenated and pharmacological active lanostane type triterpenoids from its fruiting bodies, mycelia, and spores, many of which exhibited cytotoxic activity against various tumor cell lines [13]. Most recently, our research group has established a fast and efficient method for the separation and purification of triterpenes GA-T (**125**) and GA-Me (**126**) from extracts of *G. lucidum* mycelia, i.e., using RP-HPLC on a semi-preparative C18 column with an acidified methanol–water mobile phase in combination with UV detection and ESI-MS [99]. We have found that GA-T (**125**) could induce mitochondria mediated apoptosis in lung cancer cells [100]. Additionally, there are yet additional new triterpenoids isolated from *G. lucidum* and published in the recent literature. Two triterpenes lucidenic acid N (**127**) and methyl lucidenate F (**128**), and three new lanostane-type triterpene aldehydes lucialdehydes A–C (**129–131**), were successively isolated from *G. lucidum*, of which **127**, **130** and **131** showed significant cytotoxic effects against cancer cells [101, 102]. Therefore, the mushroom *G. lucidum* is considered as a rich mine of triterpenoids bioresource.

Like *G. lucidum*, *Poria cocos* (called Fu Ling in China) also belongs to the family Polyporaceae, and is a famous TCM and rich source of lanostane-type triterpene acids. Seventeen lanostane-type triterpene acids, including six novel compounds 15α-hydroxydehydrotumulosic acid (**142**), 16α,25-dihydroxydehydroeburiconic acid (**143**), 5α,8α-peroxydehydrotumulosic acid (**144**), 25-hydroxyporicoic acid H (**145**), 16-deoxyporicoic acid B (**146**), and poricoic acid CM (**147**), and 11 known compounds, have been recently isolated from an acidified $CHCl_3$-soluble fraction of a MeOH extract of the epidermis of *Poria cocos* sclerotia, of which all were identified based on spectroscopic methods [103]. In addition, all the compounds except eburicoic acid, 3-epidehydrotrametenolic acid, dehydroeburicoic acid and dehydroeburiconic acid exhibited inhibitory effects on the EBV–EA activation induced by TPA in Raji cells. Compounds **146** and **147** exhibited inhibitory effects on skin tumor promotion in an in vivo two-stage carcinogenesis test using DMBA as an initiator and TPA as a promoter [103].

Previous studies suggest that *Tyromyces* species belonging to the family Polyporaceae are rich sources of lanostane and rearranged lanostane carboxylic acids. From a methanol-soluble extract of the fruit bodies of the higher inedible mushroom *Tyr. fissilis*, Asakawa's group isolated six novel triterpenoids, tyromycic acids B to G (**149–154**), whose structures were similar to that of tyromycic acid previously described, together with two known triterpenoids, tyromycic acid (**148**) and trametenolic acid B. Compounds **149–151** and **154** possess a lanostane skeleton, while both **152** and **153** are based on a rare 14(13→12) *abeo*-lanostane skeleton [104, 105]. Many lanostane-type triterpene carboxylic acids show various bioactivities, such as cholesterol biosynthesis inhibitory activity and antinociceptive effects, but results of bioassay on compounds **149–152** showed neither antioxidant nor anti-HIV activities [104], although they may possess other unknown bioactivities.

The fungi of the genus *Hebeloma* belongs to another basidiomycete family Cortinariaceae comprising many inedible or toxic species. Liu and colleagues investigated chemical components of *Heb. versipelle*, where a cytotoxic lanostane triterpenoid, 24(*E*)-3β-hydroxylanosta-8,24-dien-26-al-21-oic acid (**155**), possessing a previously unknown α,β-unsaturated aldehyde group at the side chain of lanostanoids, was isolated from its fruiting bodies [106].

In searching for bitter constituents of the mushroom *Leucopaxillus gentianeus*, Clericuzio and coworkers obtained nine cucurbitane triterpenoids, namely, cucurbitacin B (**156**), the corresponding cucurbitacin B 16-oleyl, 16-linoleyl, and 16-palmityl esters (**157–159**), 16-deoxycucurbitacin B (**160**), cucurbitacin D (**161**), leucopaxillones A–B (**162–163**), and 18-deoxyleucopaxillone A (**164**) from the fruiting bodies or mycelial biomass [107, 108]. Compound **156** imparts a bitter taste to the flesh of the fungus; however, it occurs in the fruiting bodies mainly esterified as a mixture of tasteless fatty acid esters of **156** [107]. Compounds **156** and **161**, as well as **162** and **163**, were isolated from both sources of fruiting bodies and mycelia. However, compounds **157–160** were absent in the mycelia, while **164** was only detected in the mycelia [108]. The antiproliferative bioassay of these triterpenes isolated against cancer cell lines suggested that all compounds showed potent effects except cucurbitacin B esters **157–159** [107, 108]. Interestingly, compound **156** seems to represent a nice example of secondary metabolite convergence between distant taxa such as fungi and vascular plants, where they likely exert a similar role of protection [107].

Ascomycetes of the genus *Daldinia* are rich in secondary metabolites. Hashimoto and Asakawa have obtained more than 20 new metabolites from two Japanese *Daldinia* spp. [109]. Recently, four new squalene-type linear triterpenoids, named concentricol (**165**) and concentricols B–D (**166–168**), were isolated from the fungus *D. concentrica* by means of repeated silica gel and Sephadex LH-20 column chromatography, and preparative HPLC, whose structures were also elucidated by a combination of NMR, MS, IR, and UV spectra [110, 111]. The authors thought that the production of concentricol was possibly related to the morphogenesis of the sexual stage in *D. concentrica* [110]. In general, linear triterpenoids are quite rare in fungi, as compared to cyclic triterpenes and steroids. More intriguingly, the occurrence of secondary metabolites was correlated with both morphology and genetic fingerprints [112], while concentricol, as a major metabolite, has been recognized as a taxonomically significant marker in the genus *Daldinia* according to HPLC-based secondary metabolite fingerprinting methodology [110, 112].

Most recently, Gloer's group found that extracts from cultures of one isolate of ascomycete *Xylaria* sp. showed moderate antifungal and antiinsect activity. Subsequent chemical studies of this extract led to the isolation of four triterpenoid glycosides named as kolokosides A–D (**169–172**), respectively, where the kolokosides appear to be members of the fernane class of triterpenoids [113]. The compound **169** has the same carbon skeleton as **172**, differing only in the presence of oxygen at C-29, and the NMR and MS data for **172** were very similar to those of **171**, but the second oxygen atom is located on the opposite side of the carbonyl carbon in comparison to **171**. Among these compounds, only **169** exhibited a marked activity against Gram-positive bacteria [113].

Russula, one of the largest genera in the subdivision Basidiomycotina fungi, consist of hundreds of species. Liu's group has investigated secondary metabolites in the fruiting bodies of a few *Russula* species since 2000 [114–119]. Five new terpenes, named (24*E*)-3β-hydroxycucurbita-5,24-diene-26-oic acid (**173**), (24*E*)-3,4-secocucurbita-4,24-diene-3,26-dioic acid (**174**), (24*E*)-3,4-secocucurbita-4,24-diene-3,26,29-trioic acid (**175**), rulepidadiol (**176**), and rulepidatriol (**177**) were isolated from the EtOH and $CHCl_3$/MeOH 1:1 extract of the fruiting bodies of basidiomycete *R. lepida*, in which compounds **174** and **175** are the first report of naturally occurring seco-ring-A cucurbitane triterpenoids, while **176** and **177**, belonging to the aristolane-type sesquiterpenoids, are a rather rare type in nature, especially among fungal species [114]. Other three terpene compounds viz. two new aristolane-type sesquiterpenes rulepidol (**178**) [115] and lepidamine (**179**) [116], and lepidolide (**180**) [117], a novel *seco*-ring-A cucurbitane triterpenoid, were successively isolated from the fruiting body of the same fungus using repeated column chromatography and preparative TLC methods. Interestingly, compound **179** was the first naturally occurring aristolane-type sesquiterpene alkaloid containing nitrogen atom [116]. Compounds **174**, **175**, and **180** which were previously described [114], completely belong to the same structure cluster, in which the big difference was only that a heterocycle was formed between C-28 and C-6 in lepidolide [117]. Unfortunately, the bioactivities of these terpenes from *R. lepida* were not reported by the group, notwithstanding the fungus has been used as a food and medicinal agent in China, and extract of its fruiting bodies showed antitumor activity [114].

The genus *Lactarius* and *Russula* belong to the family Russulaceae, Basidiomycotina. However, most members of the genus *Lactarius* abundantly contain a milky juice, which can be observed when the fruiting bodies are injured. Sesquiterpenes play an important biological role in the great majority of *Lactarius* species, being responsible for the pungency and bitterness of the milky juice and the change in color of the latex on exposure to air and constituting a chemical defense system against various invaders such as bacteria, fungi, animals, and insects. In the course of bioactive metabolites investigation for *Lactarius* sp., Liu's group isolated a new lactarane sesquiterpene, named as rufuslactone (**181**), from the fungus *L. rufus* [120], and six novel humulane-type sesquiterpenes, namely 2β,α-epoxy-6Z,9Z-humuladien-8α-ol (**182**) from the fruiting body of *L. hirtipes* [121], mitissimols A–C (**183–185**), a mixture of mitissimol A oleate (**186**), and mitissimol A linoleate (**187**) from the fruiting body of *L. mitissimus* [122], in which their structures were elucidated by comprehensive spectroscopic techniques and necessary chemical methods. Compound **181** is an isomer of a previously described lactarane 3,8-oxa-13-hydroxylactar-6-en-5-oic acid γ-lactone from *L. necata*, and possessed potent antifungal activity [120]. Compound **182** was the first purified sesquiterpene with a humulene skeleton from higher fungi [121]. Generally, sesquiterpenes isolated from *Lactarius* species including lactaranes, secolactaranes, marasmanes, isolactaranes, norlactaranes, or caryophyllanes are believed to be biosynthesized from humulene [121, 122].

Grifolin (**188**), neogrifolin (**189**) and their derivatives, possessing many interesting biological activities such as antioxidant, antimicrobial, and anticancer etc., are naturally occurring substances isolated from the fruiting bodies of members of the genus *Albatrellus* of the basidiomycete Polyporaceae family [123–126]. Liu's group obtained a new compound albaconol (**190**), which possesses the skeleton of a drimane-type sesquiterpenoid and is directly connected to a resorcinol (benzene-1,3-diol) moiety, together with three known **188**, emeheterone (**191**), and 5-methoxy-3,6-

bis (phenylmethyl) pyrazin-2-ol (**192**), a pyrazinol derivative from the fresh fruiting bodies of the inedible basidiomycete *Albatrellus confluens* [127]. Herein, the prenylated resorcinol of **190** represents a new C skeleton, and the **188** is recognized as the precursor of **190** [127]. Interestingly, the **190** exhibits significant anticancer activity through inhibiting the DNA topoisomerase II activity [128], and has a high content in fruiting bodies of *A.confluens* [127]. Another Asakawa's research group successively obtained three novel neogrifolin derivatives including 3-hydroxyneogrifolin (**193**), 1-formylneogrifolin (**194**) and 1-formyl-3-hydroxyneogrifolin (**195**) from *A. ovinus* [125], and two new grifolin derivatives named grifolinones A (**196**) and B (**197**) from *A. caeruleoporus* [126], and their structures were established by a combination of NMR, MS, IR, and UV spectra, in which compound **196** was characterized as 2,6-dihydroxy-4-methylphenyl-1-(3,7,11-trimethyldodeca-2*E*,6*E*,10-trien-4-one-1-yl), and **197** was determined to be a dimer of grifolin derivatives. Interestingly, small structural differences in these compounds caused significant changes in activity. The presence of the conjugated ketone group at C-16 in **196** and **197** increased their activities, as compared to **188**. In addition, the location of phenolic hydroxyls also clearly affected the activity, of which **189** was stronger than **188**. However, the presence of the furane ring and para-quinone in **197** did not support the inhibitory properties [126]. Additionally, **188** and related derivatives were isolated not only from the above two members of *Albatrellus* but also from other *Albatrellus* species such as *A. dispansus*, *A. confluens*, etc. as the major products, for which the distribution of **188** and related compounds were considered as the most important chemical markers for the *Albatrellus* species [125].

A new caryophyllenic sesquiterpene, fuscoatrol A (**198**) along with two known compounds, bicyclic sesterterpene 11-epiterpestacin (**199**) and β-nitropropionic acid, were isolated from the ethyl acetate extract of marine fungus *Humicola fuscoatra* using column chromatography on silica gel and Sephadex LH-20 chromatography in succession [129]. And all three compounds showed markedly antimicrobial activities [129]. In addition, from the higher aquatic fungus *Massarina tunicate*, Oh et al. isolated and identified three new bioactive sesquiterpenoids with unusual tetracyclic and tricyclic ring systems including Massarinolins A to C (**200–202**), which were the first secondary metabolites reported from any members of the genus *Massarina* [130]. Compounds **200–202** seem to be sesquiterpenoids biosynthesized from a farnesyl-type precursor, and both **200** and **201** were presumably cyclized products of **202**. In standard disk assays at 200 μg per disk, compounds **200** and **201** were active against *Bacillus subtilis* (ATCC 6051), and the former was also active against *S. aureus* (ATCC 29213) [130].

2.5 Miscellaneous

Naturally occurring small molecular peptides is always recognized as an important source for new drug discovery due to their higher plasma clearance rate, shorter half-lifetime, and lower side effect [131]. A few literature surveys indicated that entomogenous fungi, a special higher fungal family, produce diversified cyclic peptides and depsipeptides with various bioactivities [131]. Beauvericin (**203**), a cyclohexadepsipeptide antibiotic representative consisting of three l-*N*-methylphenylalanine units connected alternatively with three d-2-hydroxyisovaleric acid residues, was successively isolated from entomopathgenic fungi *Beauveria bassiana*, *Paecilomyces tenuipes*, and *C. cicadae*, etc. [132, 133], which could regulate intracellular ion such as Ca^{2+}, K^+ and Na^+ and cyto-homeostasis [134] and significantly induce cell death in tumor cells due to triggering Ca^{2+} release from endoplasmic reticulum [132]. From culture broth of *C. sinensis*, Jia and colleagues isolated a cyclodipeptide named as cordycedipeptide A (**204**) showing a significant cytotoxic effect on cancer cells [135]. Cordyheptapeptide A (**205**), a new cycloheptapeptide, was isolated from fermentation mycelia of the entomopathogenic fungus *C.* sp. BCC 1788, and possessed cytotoxicity against Vero cell line [136]. Recently, Krasnoff and coworkers found that *C. heteropoda* yielded two nonribosomal linear peptides of complex microheterogeneous family, i.e., cicadapeptins I (**206**) and II (**207**), together with a known antifungal compound, myriocin (**208**), where both novel compounds are acylated at the N-terminus by *n*-decanoic acid and amidated at the C-terminus by 1,2-diamino-4-methylpentane, and the amino acid sequence of cicadapeptin I was N-terminus-Hyp-Hyp-Val-Aib-Gln-Aib-Leu-C-terminus, Ile substitutes for Leu in cicadapeptin II, namely each cicadapeptin contained two residues of *R*-aminoisobutyric acid (Aib) [137].

More recently, from organic-solvent extraction of hyphal stands of the fungus *Isaria* species grown in solid media, Balaram's group isolated 12 new cyclohexadepsipeptides belonging to members of both known isariin and unknown isaridin family using a reverse-phase HPLC method and characterized by LC–ESI–MS/MS and NMR [138, 139]. Generally, the isariins possess a *β*-hydroxy fatty acid and five α-amino acid residues, while the isaridins contain an α-hydroxy acid and a *β*-amino acid, with a preponderance of N-methylated residues. The sequences of most new cyclic depsipeptides have been obtained. Isariin C2 (**209**), E (**210**), F2 (**212**), and G1 (**213**) possess similar sequence cyclo($^{D}\beta$-HA-Gly-Val-^{D}Leu-Ala-Val), of which only the number of methylene units (*n*) of side chain $-(CH_2)_n-CH_3$ of $^{D}\beta$-HA residue is different, and the n value of side chain is 3, 2, 4, and 5, respectively. The sequence of isariin F1 (**211**) is cyclo($^{D}\beta$-HA-Gly-Val-^{D}Leu-Ala-Abu/Aib), where the side chain of $^{D}\beta$-HA residue is $-(CH_2)_5$-CH3, but a residue, the presence of Abu or Aib, still need to be further analyzed. The sequence of isariin G2 (**214**) contains two Val residues viz. cyclo($^{D}\beta$-HA-Gly-Val-^{D}Leu-Ala-Ala), of which the $^{D}\beta$-HA residue possesses $-(CH_2)_8-CH_3$ side chain [139]. The sequences of new isaridins including isaridin A, B, C1, C2, D and E (**215–220**) were also identified to be cyclo(HyLeu-Pro-Phe-NMeVal-NMePhe-*β*Gly), cyclo(HyLeu-*β*MePro-Phe-NMeVal-NMePhe-*β*Gly), cyclo-(HyLeu-Pro-Phe-NMeVal-Lxx-*β*Gly), cyclo-(HyLeu-*β*MePro-Phe-NMeVal-NMeVal-*β*Gly), cyclo-(HyLeu-*β*MePro-Phe-NMeVal-Lxx-*β*Gly), and cyclo-(HyLeu-Pro-Phe-NMeVal-NMeVal -*β*Gly), respectively. Thereby *β*Gly, Phe, NMeVal, and HyLeu are the conserved amino acid residues in all of these isaridins, the microheterogeneity arises only due to the permutations of certain residues including Pro and *β*-MePro, Lxx and NMeVal, and N-MePhe and Lxx [138, 139].

In addition, single crystals of both **215** and **216** have been obtained, and their 3D structures were also elucidated by X-ray diffraction [138].

Peptaibols, nonribosomally synthesized peptides with typically 15–20 residues, comprise a large class of peptides characterized by a high number of unusual aminoisobutyric acid (Aib) residues, an acetylated N-terminus, and a hydroxylated C-terminal amino acid [140]. Aminoisobutyric acid has a high tendency to form helices and this is caused by the helical structures of the peptaibols. Other nonnatural amino acids occurring in peptaibols are iso-valine (IVA) and hydroxy-proline (HYP). According to the related database on naturally occurring peptaibols (http://www.cryst.bbk.ac.uk/peptaibol/home.shtml), peptaibols known to date have 317 species and are divided into 9 subfamilies, which are mainly derived from fungi of the genera *Trichoderma* and *Emericellopsis* and generally exhibit antimicrobial properties, which is thought to be due to their ability to form ion-channels in lipid membranes [141]. Recently, from a terrestrial fungus *Septocylindrium* sp., Summers and colleagues obtained two new peptaibols, septocylindrin A (**221**) and septocylindrin B (**222**), related to the well-studied membrane-channel-forming peptaibol alamethicin [142]. According to a combined data analysis of NMR and HRMS, the linear sequences of both **221** and **222** were found to be Ac-Aib-Pro-Aib-Ala-Aib-Ala-Gln-Aib-Val-Aib-Gly-Leu-Aib-Pro-Val-Aib-*Glu*-Gln-Phaol$^+$, and Ac-Aib-Pro-Aib-Ala-Aib-Ala-Gln-Aib-Val-Aib-Gly-Leu-Aib-Pro-Val-Aib-*Gln*-Gln-Phaol$^+$, respectively [142]. The results indicated that both **221** and **222** are linear 19-amino acid peptides with a modified phenylalanine C-terminus, but the HRMS data indicated that there is a slight difference in the 18th residue, where **221** contains Glu and **222** contains Gln. Zhang's group recently isolated three peptaibols antimicrobial peptides with full length of 20 amino acids viz. trichokonin VI, VII and VIII, produced by *Trichoderma koningii* [143]. Their linear sequences were confirmed as follows: trichokonin VI: Ac-Aib-Ala-Aib-Ala-Aib-*Ala*-Gln-Aib-Val-Aib-Gly-Leu -Aib-Pro-Val-Aib-*Aib*-Gln-Gln-Phaol$^+$; trichokonin VII: Ac-Aib-Ala-Aib-Ala-Aib-*Ala*-Gln- Aib-Val-Aib-Gly-Leu-Aib-Pro-Val-Aib-*Iva*-Gln-Gln-Phaol$^+$; trichokonin VIII Ac-Aib-Ala- Aib-Ala-Aib-*Aib*-Gln-Aib-Val-Aib-Gly-Leu-Aib-Pro-Val-Aib-*Aib*-Gln-Gln-Phaol$^+$ [143]. Among three trichokonins, they differ only in the 6th and 17th residue, namely their 6th residue was Ala, Ala, and Aib, respectively, and their 17th residue was Aib, Iva, and Aib, respectively.

Terphenyls are aromatic hydrocarbons consisting of a chain of three benzene rings. There are three isomers in which the terminal rings are ortho-, meta-, or para-substituents of the central ring. Natural terphenyls consisted of major *p*-terphenyl derivatives and very few *m*-terphenyl derivatives, while *o*-terphenyls have not been found in nature until now and no *m*-terphenyls have ever been reported from the kingdom of fungi. The chemical investigation of *p*-terphenyls as one class of the pigments of mushrooms began in 1877. Two recent reviews from Liu's and Asakawa's groups gave detailed information on natural terphenyls including the isolation, structure elucidation, biological activities, transformation, and total synthesis of terphenyl derivatives from nature [144, 145], but it is still necessary for us to pay close attention to two basidiomycete families Thelephoraceae and Paxillaceae due to their abundant antioxidative *p*-terphenyls metabolites. *Thelephora ganbajun*,

The. aurantiotincta, *The. terrestris* and *Paxillus curtissii* are typical representatives possessing rich *p*-terphenyls secondary metabolites. For example, *The. ganbajun*, locally called as Gan-Ba-Jun, is one of the most favorite edible mushrooms in southwestern China yielded rare natural poly (phenylacetyloxy)-substituted *p*-terphenyl analogues ganbajunins A–G [146, 147]. Approximately 20 *p*-terphenyls, named as curtisians A–Q, were successively isolated from *Pax. curtissii* growing widely in East Asia and north America on decayed pine trees [148–151]. Furthermore, most of the above-mentioned *p*-terphenyl compounds exhibited attractive antioxidant activities against lipid peroxidation or radical-scavenging activity against DPPH. More recently, Liu and coworkers isolated a new *p*-terphenyl derivative, sarcodan (**223**) from the fruiting bodies of *Sarcodon leavigatum* belonging to the same family as members of the genus *Thelephora*, but have not yet detected its bioactivity [152].

Gao and colleagues have obtained several new ceramides from various higher fungi since 2000. Namely, three phytosphingosine-type ceramides including paxillamide (**224**) with a 2,3-dihydroxytetracosanoic acid moiety from the basidiomycete *Pax. panuoides* [153], (2*S*,3*S*,4*R*,2′*R*)-2-(2′-hydroxytetracosanoylamino)octadecane-1,3,4-triol (**225**) containing a 2-hydroxy fatty acid from the basidiomycete *R. cyanoxantha* [118], and (2*S*,3*S*,4*R*,2′*R*)-2-*N*-(2′-hydroxytricosanoyl)-octadecan-1, 3, 4-triol (**226**) from the ascomycete *Tuber indicum* [154], and three other ceramides (2*S*,2′*R*,3*S*,4*R*)-2-(2′-d-hydroxyalkanoylamino) octadecane-1,3,4-triol (**227**), (2*S*,3*S*,4*R*)-2-(alkanoylamino)octadecane-1,3,4-triol (**228**), and (2*S*,3*R*,4*E*)-2-(alkanoylamino)-4-octadecene-1,3-diol (**229**) from the same ascomycete *T. indicum* [155], were successively isolated and determined unequivocally by means of spectroscopic and chemical methods. Interestingly, Gao and coworkers isolated the same phytosphingosine-type ceramide **225** from the fruiting bodies of another basidiomycete *Pax. panuoides* [76]. From the fruiting bodies of *T. indicum*, a new trihydroxylated monounsaturated fatty acid, named 9,10,11-trihydroxy-(12*Z*)-12-octadecenoic acid (**230**) was additionally obtained by the research group [156].

Many species in the basidiomycete *Cortinarius* genus are known to produce biologically active natural products, including pigments and toxins. Munro and coworkers found that crude extracts (both organic and aqueous) of a *Cortinarius* sp. showed significant antimicrobial activity and cytotoxicity against the P388 murine leukaemia cell line [157]. Subsequent fractionation of further extracts yielded three disulfide metabolites, i.e., the unsymmetrical disulfide cortamidine oxide (**231**), 2,2′-dithiobis(pyridine *N*-oxide) (**232**), and the symmetrical disulfide (**233**), utilizing repeated reverse-phase column chromatography, where both **231** and **232** exhibited potent antimicrobial activity and cytotoxicity [157]. According to spectral data, it is suggested that the two symmetrical dimers **232** and **233** arise via disulfide exchange from the compound **231**. Both compounds **231** and **232** contain a 2-thiopyridine *N*-oxide functionality. This functionality is consistent with the biological activity (cytotoxicity and antimicrobial activity) observed for compounds **231** and **232**. A secondary metabolite containing a pyridine *N*-oxide functionality has been reported from *Cor. orellanus* and *Cor. speciosissimus* before, namely, the toxin orellanine [157].

Ellagic acid and its derivatives are widely distributed in plants, but are rare in fungi. Liu's group isolated nigricanin from the fruiting bodies of the basidiomycete *R. nigricans* using repeated extraction and column chromatography, whose structure was identified as a phenolic compound based on the ellagic-acid skeleton by spectroscopic and chemical methods, and considered as the first ellagic acid like compound found in higher fungi [119]. Later, they found interesting metabolites that were a new polyene pyrone named as aurovertin E (**234**), and a known aurovertin B (**235**) from the mycelia biomass of another basidiomycete *A. confluens*, which was the first example of the occurrence of aurovertins in basidiomycetes [158]. In addition, the structures of both aurovertins **234** and **235** were elucidated on the basis of spectroscopic studies, and the **234** consists of a substituted pyrone ring linked by a rigid spacer containing conjugated double bonds to a substituted dioxabicyclo[3.2.1]octane and as a product of the reaction from **235** by alkaline hydrolysis [158]. From the fruiting bodies of the ascomycete *D. concentrica*, this research group also successively isolated and identified five new compounds, including 1-isopropyl-2,7-dimethylnaphthalene (**236**) [19], three homologs of 3-alkyl-5-methoxy-2-methyl-1,4-benzoquinoneswith chain lengths of C21–C23 (**237–239**) [159] and a pair of heptentriol stereoisomers hept-6-ene-2,4,5-triols (**240**) [21].

Podophyllotoxin (**242**), an aryl tetralin lignan, is biosynthesized by the plant *Podophyllum* species and is in great demand worldwide due to their use as a precursor of synthesis of some topoisomerase inhibitors including three anticancer drugs, etoposide, teniposide, and etoposide phosphate. However, the sustained production of **242** requires large-scale harvesting from the natural environments, which has resulted in the plant-endangered status, while their total chemical synthesis also has too many difficulties. From rhizomes of the plant *Pod. peltatum*, Eyberger and coworkers recently obtained two isolates of endophytic Ascomycete *Phialocephala fortinii* producing compound **242** [160]. Another laboratory also found an alternative source of **242** production, in which the methodology for isolation, identification, and characterization of a novel endophytic fungus *Trametes hirsute* that produces **242** and its glycoside from the dried rhizomes of *Pod. hexandrum* was established [161]. This strategy promises to improve the production of these therapeutically important compounds at lower cost.

Members of the genus *Sporormiella* have afforded a number of interesting bioactive metabolites. Here silica gel chromatography of the EtOAc extract from the coprophilous fungus *S. vexans* liquid cultures, followed by semipreparative reversed-phase HPLC, afforded three new *p*-hydroxybenzoic acid derivatives sporovexins A–C (**243–245**) and a new preussomerin analogue, 3′-*O*-desmethyl-1-epipreussomerin C (**246**) [162]. During bioassay, compounds **243** and **246** showed antifungal and antibacterial activities [162].

3 Bioactivity of Secondary Metabolites from Higher Fungi

Actually, it is difficult to estimate either the number of compounds or the chemical space represented by microbial secondary metabolites, but for evolutionary reasons, the range of bioactivities will be correspondingly large. Chadwick and Whelan proposed that all small molecules or secondary metabolites produced by microbes have biological functions, although these compounds may be as suggested simply waste products of cellular metabolism [163]. Indeed, secondary metabolites synthesized by microbes have been evolved to be specific functional regulators within complex microbial communities in the environment; roles of which are of ecological and evolutionary significance. That is to say, microorganisms have evolved the ability to biosynthesize secondary metabolites due to the selectional advantages they obtain as a result of the functions of the compounds.

Interestingly, based on the ecological points of view, Bérdy thought that the microbial secondary metabolites represent a kind of chemical communication between microbes and other microbes or nonmicrobial systems including higher plants, lower animals or mammalian (humans) systems, which reflects antagonistic, synergistic, regulatory or modulatory and any other biochemical or either biophysical interactions [12]. Bérdy summarized these interactions as illustrated in Table 2 covering the whole area of known biological activities of microbial metabolites, which also represented their possible practical applications and helped us to understand the reason why the microbes exhibit so wide range of bioactivities via their chemical products [12]. It is without doubt that secondary metabolites from higher fungi possess a vast array of biologically activities (Table 1). Of course physiological functions and acting mechanism of many known and unknown higher fungal metabolites still need to be revealed. In other words, all secondary metabolites have certain kinds of inherent activities but in many cases those activities have not yet been discovered. Here, major biological activities of higher fungi-derived secondary metabolites and their acting mechanisms are summarized as follows.

3.1 Antimicrobial Activity

Following Fleming's discovery, the tremendous success attained in the battle against disease with penicillin not only led to the development of a new field of antibiotic research but also created an entirely new industry [164]. In the course of chemical investigation on higher fungi, antimicrobial activities of secondary metabolites

Table 2 Microbial interactions (adapted from [12])

Microbe–microbe	Antimicrobial antibiotics, microbial regulators, growth factors, signaling compounds, mating hormones, etc.
Microbe–lower animals (invertebrates)	Insecticides, miticides, antiparasitic compounds, algicides, antifeedants, (invertebrates) repellents, molluscicides, anti-worm agents, etc.
Microbe–higher plants	Herbicides, phytotoxins, plant growth regulators, chlorosis inducers, phytoalexins, etc.
Microbe–mammalians (humans)	Anticancer antibiotics, pharmacologically active agents, enzyme inhibitors, (humans) immunoactive, CNS-active etc., agent etc., feed additives, etc.

were widely analyzed due to a relative simple screening system, for which it is not surprising that antimicrobial metabolites become one of major population among presently known higher fungal secondary metabolites. For example, an estimated 75% of polypore fungi that have been tested show strong antimicrobial activity, and these may constitute a good source for developing new antibiotics [165]. Actually, over 22,000 bioactive compounds had been discovered from microbes by 2002, which included 20,000 antibiotics [12].

The antimicrobial activities of secondary metabolites produced by higher fungi involve mainly antibacterial, antifungal, and/or antiviral activities. Many chemotherapeutic agents isolated from higher fungi and used to treat bacterial infections have contributed greatly to the improvement of human health during the past century. The most classical example of antibacterial fungal metabolite is the penicillin that kills susceptible bacteria by specifically inhibiting the transpeptidase that catalyzes the final step in cell wall biosynthesis, the crosslinking of peptidoglycan. In addition, the mechanisms of action for higher fungal-derived antibacterial metabolites include: to inhibit the synthesis of protein synthesis, nucleic acid and folic acid, interfere the metabolic processes of biochemistry, and enhance the permeability of cytoplasma membrane, etc. Gloers and coworkers reported that the polyketide metabolites **35**, **66–68**, and **71** exhibited antibacterial activity in standard disk assays against *B. subtilis*, and only **35** and **68** also displayed activity against *S. aureus* [45, 60]. Regretfully, most similar studies that the vast number of polyketides [55, 57–59, 70], sterols [80], terpenes [113, 129, 130], peptides [142, 143, 166], and other types of compounds [157, 162] produced by higher fungi possessing broad-spectrum antibacterial activities, have not provided further reports on their mechanisms of action against various pathogenic bacteria. More recently, peptaibols **221** and **222** from *Septocylindrium* sp. exhibited significant antibacterial and antifungal activities, where the antimicrobial activity was thought to arise from the peptaibols' membrane activity and their ability to form pores in lipid membranes [142]. The pores so formed were able to conduct ionic species, and this conductance led to the loss of osmotic balance and cell death [142].

Fungal infections range from superficial conditions of the skin and nails to disseminated life-threatening diseases. Systemic fungal infections caused by *Candida* spp., *Cryptococcus neoformans*, *Aspergillus* spp., and *Pneumocystis carinii*, show a significant increasing threat to human health during the past decade, while only a limited number of antifungal agents are currently available for the treatment of life-threatening fungal infections. Therefore, on the basis of the struggle for existence among fungi, the antifungal activities of fungal natural products especially higher fungal metabolites received much attention for the discovery of new antifungal agents. Compounds **49–51** from the ascomycete *Trichopezizella nidulus* were inhibitors of DHN-melanin biosynthesis. Using *Lachnellula* sp. A32–89 as a test organism, 50 μg per paper disk (6 mm) of compound **49** resulted in a pigment inhibition zone of 19 mm in diameter, while the same amount of compounds **50** and **51** gave inhibition zones of 14 mm. Compound **49** still exhibited moderate activity against *Mucor miehei* with an inhibition zones of 23 mm at 50 μg per paper disk in the standard plate diffusion assay [54]. Asakawa's group found that four new azaphilones **52–55** yielded by *Creosphaeria sassafras* showed relatively strong antifungal activities with inhibition zone ranging from 17 to 20 mm against *Asp. niger* and *Can. albicans* [55]. Compound **93**, from an endophytic fungus *Fusarium* sp., also indicated potent antifungal activity against *Can. albicans* [67]. Unfortunately, these investigations did not show further antifungal acting mechanism. Generally, the antifungal agents are classified according to their mechanisms of action, covering inhibitors of synthesis of cell wall components (glucan, chitin and mannoproteins), of sphingolipid synthesis (serine palmitoyltransferase, ceramide synthase, inositol phosphoceramide synthase and fatty acid elongation) and of protein synthesis (sordarins) [167].

The search for antiviral compounds from marine fungi has yielded some promising secondary metabolites. Compound **2** yielded from the ascomycete *D. concentric*$_a$ showed the potent blockage effect on syncytium formation between HIV-1-infected cells and normal cells; thus it was suggested it might be effective against HIV-1 [20]. More recently, **3** from the fruiting body of *Sui. granulatus* exhibited a weak anti-HIV-1 activity [22]. The compound **4** derived from **3** showed more effective anti-HIV-1 activity and lower cytotoxicity than **3**, the TI value increased from 12.1 to 312.2 [23]. Moreover, Liu's group found that **4** displayed anti-HIV activity via interfering in the early stage of HIV life cycle [23]. HIV-1 integrase is one of the three critical enzymes for viral replication; its inhibition is therefore one of the most promising new drug strategies for antiretroviral therapy, with potentially significant advantages over existing therapies [168]. Based on the viral targets, Singh and coworkers screened a series of HIV-1 inhibitors that inhibited the coupled and strand-transfer reaction of HIV-1 integrase with an IC_{50} value of 0.5–120 μM from the organic extract of fermentations of some terrestrial fungi through bioassay-directed isolation [169].

3.2 Antiinflammatory Activity

Inflammation is the human body's first response to irritation, and is characterized by redness, heat, swelling, pain, and organ dysfunction. Inflammation is good phenomenon to a certain extent due to the heat produced during an inflammatory response killing off bacteria that may have invaded. Antiinflammatory is a function of a substance to reduce inflammation. Unfortunately, uncontrolled inflammation can cause more damage to the tissue and organ of organisms. Therefore, it is valuable to discover antiinflammatory agents from natural sources. Previous studies demonstrated that higher fungi showed significant antiinflammatory actions by inhibiting inflammation mediators. For example, methanolic extracts of *C. pruinosa* (CPME) caused a significant downregulation of inflammation mediators' gene expression including IL-1β, TNF-α, inducible nitric oxide synthase (iNOS), and COX-II via the inhibition of NF-κB activation in RAW264.7 cells and mice stimulated with LPS, which resulted in a corresponding suppression of inflammatory mediators IL-1β, TNF-α, NO and PGE2 production in vitro and in vivo [170]. CPME therefore possessed a potential antiinflammatory activity by inhibiting NF-κB-dependent inflammatory gene expression, suggesting that the CPME may be beneficial to the treatment of endotoxin shock or sepsis; unfortunately the authors did not conduct further chemical investigations on CPME. More recently, however, Asakawa and coworkers observed that compounds **28–30** derived from *D. childiae* strongly suppressed the LPS-induced production of NO with IC_{50} values of 15.2, 4.6, and 6.4 μM, respectively [42]. Further experimental results indicated that the inhibition of the LPS-induced NO production of **29** was due to the inhibition of iNOS mRNA synthesis [42]. Compounds **188**, **189**, **196** and **197** produced by inedible mushroom *A. caeruleoporus* also exhibited inhibitory activity against nitric oxide (NO) production stimulated by lipopolysaccharide (LPS) in RAW 264.7 cells with IC_{50} values of 23.4, 22.9, 29.0, and 23.3 μM, respectively [126], in which preliminary investigation on acting mechanism of compound **196**, similar to **29**, showed that the NO production stimulated by LPS in RAW cells was suppressed due to the downregulation of iNOS gene expression [126]. TNF-α is the main proinflammatory cytokine in inflammatory diseases like septic shock, rheumatoid arthritis and Crohn's disease. Sterner's group identified compounds **39–42** as four isomers of oxaspirodion from *Cha. subspirale*, while oxaspirodion could inhibit the TNF-α driven luciferase reporter gene expression with an IC_{50}-value of 2.5 μg mL^{-1} (10 mM) in TPA/ionomycin stimulated Jurkat T-cells by interfering with signal transduction pathways involved in the inducible expression of many proinflammatory genes [46]. Additionally, in the inflammatory process, both COX-I and COX-II involve in the conversion of arachidonic acid to prostaglandins [171], and inducible COX-II is associated with inflammatory conditions, whereas extensively expressed COX-I is responsible for the cytoprotective effects of prostaglandins [172]. Compounds **101** and **102** from the fruiting body of *Agr. aegerita* showed COX inhibitory activities. The inhibition value of COX-I enzyme by **101** and **102** at a dose of 100 μg mL^{-1} was 19 and 57%, respectively. Similarly, COX-II enzyme activity was reduced by **101** and **102** (both at 100 μg mL^{-1}) with 28 and 22%, respectively [75].

3.3 Antioxidant Activity

Active oxygen, and in particular free radicals, are considered to induce oxidative damage in biomolecules and to play an important role in aging, cardiovascular diseases, cancer, atherosclerosis, and inflammatory diseases, while antioxidants help protect healthy cells from damage caused by free radicals. Synthetic antioxidants have begun to be restricted because of their health risks and toxicity [173]. Therefore, the discovery of natural antioxidants from various sources such as plant and higher fungi to replace synthetic antioxidants has attracted more and more attention in the last decade [145].

G. lucidum has been shown to possess potent antioxidant activity in multiple research studies with little or no side effects [174]. Sun et al. [175] evaluated the antioxidant activity of *G. lucidum* peptide (GLP) using different oxidation systems. Compared to butylated hydroxytoluene, GLP showed a higher antioxidant activity in the soybean oil system. Soybean lipoxygenase activity was blocked by GLP in a dose-dependent manner with an IC_{50} value of 27.1 μg mL^{-1}. GLP showed scavenging activity toward hydroxyl radicals produced in a deoxyribose system with an IC_{50} value of 25 μg mL^{-1}, and GLP effectively quenched superoxide radical anion produced by pyrogallol autoxidation in a dose-dependent manner. GLP also showed substantial antioxidant activity in the rat liver tissue homogenates and mitochondrial membrane peroxidation systems, and the auto-hemolysis of rat red blood cells was also blocked by GLP in a dose-dependent manner. It is therefore suggested that GLP is the major constituent responsible for the antioxidant activity of *G. lucidum* and GLP could play an important role in the inhibition of lipid peroxidation in biological systems through its antioxidant, metal chelating, and free radical scavenging activities.

Phe. igniarius, belonging to the family Polyporaceae, has long been used for the treatment of fester, abdominalgia, and bloody gonorrhea in China. Shi and coworkers successively obtained more than 20 metabolites in the course of chemical and biological investigation on fruiting bodies of *Phe. igniarius*, of which the compounds **23–27** exhibited potent antioxidant activity inhibiting rat liver microsomal lipid peroxidation with IC_{50} values of 3.86, 4.8, 3.7, 6.5, 8.2, and 10.2 μM, respectively [36–38]. More recently, Kim and colleagues isolated and identified the four antioxidants **31–34** from the fermented mushroom *Cya.stercoreus*, where the antioxidant activity of **31–34** was evaluated by DPPH and ABTS radical scavenging activity assays and compared with that of reference antioxidants, BHA and Trolox [43]. The compounds **31–34** showed free radical scavenging activities on the DPPH radical with EC_{50} values of 41.6, 46.0, 26.6, and 28.6 μM, respectively, and on the ABTS cation radical with EC_{50} values of 7.9, 11.1, 9.1, and 8.4 μM, respectively, where EC_{50} (μM) was defined as an amount of antioxidant necessary to decrease the initial DPPH and ABTS radical concentration by 50% [43]. Moreover, DPPH radical scavenging activities of **33** and **34** were higher than those of Trolox and BHA, while compounds **31** and **32** showed almost the same activity as those of the reference antioxidants. In the ABTS radical scavenging assay, compounds **31–34** showed higher activity than those of Trolox and BHA [43]. Liu's and Asakawa's groups observed that many *p*-terphenyl compounds from the basidiomycete families Paxillaceae and Thelephoraceae exhibited attractive antioxidant activities. For example, Yun et al. [148] isolated curtisians A–D from *Pax. curtissii* and reported strong radical-scavenging activity against DPPH, and Asakawa and coworkers obtained curtisians E–Q from the same fungus, of which curtisians I–Q were shown to be of moderate to strong free-radical-scavenging activities as compared with the standards [149–151].

3.4 Anticancer Activity

As to anticancer mechanism of higher fungal secondary metabolites, Xiao and Zhong have reviewed higher fungi *Cordyceps*-derived anticancer metabolites [13]. Cancer chemotherapy has relied mostly on cytotoxic drugs, which inhibit tumor cell proliferation and cause cell death. As mentioned above, cytotoxic activities against various tumor cells lines of higher fungi secondary metabolites were widely investigated in the past decades. *Shiraia bambusicola*, an ascomycete parasitic on bamboo twigs and named as 'Zhu Huang' in China, is recorded only in China and Japan and used to treat rheumatism and pneusomia in TCM. Fang et al. reported that hypocrellin D from *Shi. bambusicola*, a cytotoxic fungal pigment, significantly inhibited the growth of tumor cell lines Bel-7721, A-549 and Anip-973 with IC_{50} values of 1.8, 8.8, and 38.4 μg mL^{-1}, respectively [176]. The cytotoxicities of compounds **11–13** isolated from marine fungus *P. aurantiogriseum* were evaluated against the P388, BEL-7402, A-549, and HL-60 cell lines by the MTT assay [27]. Compound **12** exhibited moderate cytotoxic activities against HL-60 and P388 cell lines with IC_{50} values of 52 and 54 μg mL^{-1}, respectively, while compound **13** inhibited BEL-7402 and P388 cell lines with IC_{50} values of 62 and 48 μg mL^{-1}, respectively [27]. Compounds **8** and **9** from another marine *Penicillium* species *P. janthinellum* induced apoptosis in human leukemia HL-60 cells at a dose of 100 μM by 10, 39, and 34% of the apoptotic cells compared to control cells, respectively, and compound **9** also inhibited EGF-induced malignant transformation of JB6 P^+ Cl 41 cells in a soft agar [26]. Shi and coworkers isolated over decade compounds from *Phe. igniarius* by following a specific bioassay-guided separation protocol, whose most compounds indicated in vitro cytotoxicity against several human cancer cell lines [35–37]. Of those compounds, **18** and **19** showed potent selectivity against A549 and Bel7402 with IC_{50} values of 0.012, 0.016, 0.010, and 0.008 μM, respectively [35], while **23** showed moderate cytotoxic activities against human ovary cancer A 2,780 cell line and human colon cancer HCT-8 cell line with IC_{50} values of 20.4 and 30.2 μM, respectively [36]. Compounds **24** and **25** from *Phe. igniarius* were inactive (IC_{50} > 10 μM) against the cell lines tested, but they could inhibit protein tyrosine phosphatase 1B with IC_{50} values of 3.1 and 3.0 μM, respectively, while **26** possessing moderate cytotoxic activities against these human cancer cell lines was inactive against PTP1B (IC_{50} > 10 μM) [37].

Hsp90 is abundant and important in maintaining the conformation, stability, and function of many proteins involved in cell survival. Furthermore, Hsp90 is recognized to play a critical role in the cancer phenotype and provide a particularly effective target for cancer chemotherapy because of its importance in maintaining the function of key oncogenic client proteins [177, 178]. Actually, Hsp90 inhibitors have shown promising antitumor activity in vitro and in vivo preclinical model systems, of which a Hsp90 inhibitor 17-allylamio-17-desmethoxygeldanamycin is currently in clinical trial [178]. Recently, Gunatilaka and coworkers have developed an effective strategy for discovering natural product-based Hsp90 inhibitors with potential anticancer activity, and they obtained two Hsp90 inhibitors compound **43** and monocillin I by bioassay-guided fractionation from *Cha. chiversii* and *Paraphaeosphaeria quadriseptata* [47].

Liu et al. observed that compound **190** from *A. confluens* significantly inhibited tumor growth in S180 and H22 tumor-bearing mice with the inhibitory rates of 47.4 and 37.8% at a dose of 3.46 mg kg^{-1}, respectively, and compound **190** obviously influenced DNA topoisomerase II activity, stimulated DNA cleavage and inhibited DNA reunion mediated by topoisomerase II [128]. Its acting mechanism, therefore, is similar to a famous anticancer agent camptothecin (**14**) produced by plant *Cam. acuminate* and/or some endophytic fungus like *Ent. infrequens* [28]. Another most successful anticancer agent paclitaxel (**118**) was originally discovered in plants, but has also been found to be a metabolite of endophytic fungi including *Taxomyces andreanae*, *Pestalotiopsis microspora*, *Sporormia minima*, *Trichothecium* sp, *Tubercularia* sp, and *Nodulisporium sylviforme* [84–89]. Compound **118** is approved for breast and ovarian cancer and is the only antitumor drug known to act by blocking depolymerization of microtubules [179]. Compound **188**, another metabolite of *A. confluens*, significantly inhibited the proliferation of cancer cell lines, but its antitumor mechanism differed from compound **190** and involved the induction of apoptosis, namely the activation of caspase-8, 9, and 3, release of cytochrome c from mitochondria, decrease of the Bcl-2 level, and increase of the Bax level etc., where caspase was a key mediator of the apoptotic pathway induced by grifolin [124].

Metastasis is reported to be responsible for over 90% of cancer deaths. Targeting cell motility has attracted attention as one of the alternative strategies for the development of anticancer therapies in recent years [180]. Cell motility (migration) is a critical cause of tissue invasion allowing primary tumors to disseminate and metastasize, and cell migration in vitro are thought to be related to many in vivo cellular behaviors such as tumor angiogenesis, cancer cell invasion, and metastasis [181]. Cell motility and angiogenesis inhibitors therefore are considered as potential anticancer candidates in the chemical investigation of natural products. Previous studies showed that cyclohexadepsipeptide **203** possessed various bioactivities such as insecticidal activity, induction of tumor cell apoptosis [132], regulating intracellular ion and cytohomeostasis [134], etc. More interestingly, Zhan et al. observed that compound **203**, from the endophytic fungal strain *F. oxysporum* $EPH2R_{AA}$ of *Ephedra fasciculate*, inhibited migration of the metastatic prostate cancer (PC-3M) and breast cancer (MDA-MB-231) cells with IC_{50} values of 3.0 and 5.0 μM, respectively, and showed potent antiangiogenic activity in HUVEC-2 cells with IC_{50} value of 3.0 μM [182].

3.5 Miscellaneous Activity

Hosoe and coworkers isolated a vasodilator 4-benzyl-3-phenyl-5*H*-furan-2-one belonging to a furanone derivative that could inhibit Ca^{2+}-induced vasocintraction in aortic rings pretreated with high K^+ (60 mM) or norepinephrine (NE) from *Malbranchea filamentosa* IFM 41300 [183]. Generally vasoconstrictor-induced contraction is mediated by Ca^{2+} influx so that inhibitors of Ca^{2+} influx cause vasodilatation. 4-Benzyl-3-phenyl-5*H*-furan-2-one relaxed high K^+-induced vasoconstriction and indicated that its vasodilatory effect may be due to the inhibition of Ca^{2+} influx through voltage-dependent Ca^{2+} channels. In addition, 4-benzyl-3-phenyl-5*H*-furan-2-one weakly inhibited the contractions induced by NE in the presence of a voltage-dependent Ca^{2+} channel blocker nicardipine that could potently inhibit high K^+-induced vasoconstriction [183]. The authors therefore thought that other mechanisms may be partially involved in its vasodilatory activity.

Lipid metabolism normally keeps a delicate balance between synthesis and degradation. When the balance of lipid metabolism is disorder, hyperlipidemia may occur, which in turn can cause atherosclerosis, hypertension, diabetes, etc. Inhibition of pancreatic lipase that is recognized as a key enzyme of dietary triglyceride absorption may inhibit fat absorption and prevent obesity and hyperlipidemia. Liu and colleagues reported that vibralactone from the basidiomycete *Boreostereum vibrans* showed inhibition of pancreatic lipase with an IC_{50} of 0.4 μg mL^{-1}, and its mechanism of inhibition probably involved covalent but reversible modification of the active site serine via acylation by *β*-lactone [184].

The indolizidine alkaloid swainsonine produced by the entomogenous fungus *Metarhizium anisopliae* is a potent inhibitor of various alpha-mannosidases [185]. Because changes in glycosylation have been found to be associated with certain disease processes, the potential of swainsonine as a chemotherapeutic agent is of interest, while destruxins are the main metabolites produced by the same fungus and possess insecticidal effects [186]. From another entomopathogenic fungus *Paecilomyces fumosoroseus*, Asaff et al. isolated a dipicolinic acid which possessed insecticidal potential [187].

Cyclosporin A was originally discovered as a narrow spectrum antifungal peptide produced by the fungus *Tolypocladium inflatum*, nevertheless discovery of its immunosuppressive activity led to its wide use in the organ transplant field such as heart, liver, and kidney transplants and was the only product on the market for many years [9, 179]. Additionally, as is well known, diabetes is a chronic metabolic disorder affecting approximately 5% of the population in industrialized nations. In an extensive search for natural product-based antidiabetic agents capable of mimicking insulin activity, Salituro and coworkers screened the first orally active insulin-mimetic agent demethylasterriquinone B-1 (**241**) that resulted in significant lowering of the glucose levels in two mouse models of diabetes from a tropical endophytic ascomycete *Pseudomassaria* sp. by bioactivity-guided fractionation [188].

4 Bioproduction of Secondary Metabolites from Medicinal Mushrooms

Medicinal mushrooms are abundant sources of a wide range of useful compounds with interesting biological activities. At present, commercial products from medicinal mushrooms are mostly obtained through their field-cultivation. However, in this production system it is difficult to control the product quality and the productivity of desired metabolites is also low. Submerged fermentation of mushrooms is viewed as a promising technology for the efficient production of their valuable compounds, and several interesting review outlines the major factors that affect the submerged cultivation of mushroom mycelia, including media, oxygen supply, shear, mixing, and cultivation strategies [189–191].

Truffles are thought to be a 'miracle of nature' with superior nutritional attributes, distinctive taste, and thrilling smell. Known as 'black diamonds', they are of high commercial value at \$1,000 kg^{-1} and \$48,300 kg^{-1} for white truffles. Submerged fermentation of truffle for production of dry cell weight and polysaccharides (extracellular and intracellular) was recently carried out successfully by Tang et al. [192–194], as a typical

Table 3 Effect of initial sucrose concentration on the cell growth and production of EPS, IPS during submerged fermentation of *T. sinense* in shake flasks (modified from [192])

	Initial sucrose concentration			
	20 g L^{-1}	50 g L^{-1}	80 g L^{-1}	125 g L^{-1}
Maximum DW (g L^{-1})	12.50 (day 4)[a]	18.75 (day 8)	22.81 (day 11)	24.07 (day 16)
Cell yield on sugar (g DW/g sugar)	0.60	0.38	0.32	0.28
Maximum EPS production (g L^{-1})	0.79	1.99	2.33	2.85
EPS productivity (g $L-^{\times 1}$ per day)	0.07	0.22	0.13	0.17
Maximum IPS production (g L^{-1})	1.21	2.26	2.92	2.69
IPS productivity (g/L per day)	0.30	0.28	0.26	0.17

[a] Culture time when the maximum cell mass was reached

example shown in Table 3. They also confirmed the existence of androstenol in the truffle fermentation broth [195], which suggested that truffle fermentation could be a promising alternative for androstenol production on the large scale.

G. lucidum (Fr.) Krast (Polyporaceae) is a famous traditional Chinese medicinal mushroom. Ganoderic acids and *Ganoderma* polysaccharides are two types of useful bioactive components. Interestingly, recent studies show that ganoderic acids have attractive biological activities including suppressing the growth of human solid tumor and the proliferation of a highly metastatic lung cancer cell line [100, 196, 197] as well as anti-HIV-1 [198]. In *G. lucidum* fermentation, Zhong and his colleagues demonstrated the response of the cell growth and metabolite biosynthesis to process parameters including pH, dissolved oxygen and substrate feeding. In shake flask fermentation of *G. lucidum*, Fang and Zhong [199] found an initial pH value of 6.5 was the best for mycelial growth and GA biosynthesis, while its level at 3.5 was beneficial to the accumulation of *Ganoderma* polysaccharides. In stirred bioreactor fermentation, Tang and Zhong [200] observed that 25% of dissolved oxygen was beneficial for *G. lucidum* growth, while 10% of dissolved oxygen was favorable for the specific production (i.e., content) of ganoderic acid. Based on the favorable effect of oxygen limitation on the ganoderic acid biosynthesis, Fang and Zhong [201] developed an interesting two-stage fermentation process by combining conventional shake-flask fermentation (i.e., the first-stage culture) with static culture (i.e., the second-stage culture) in order to enhance the metabolite production, and the ganoderic acid production was greatly enhanced in this novel culture process. Furthermore, the first-stage culture was successful in a conventional stirred-tank bioreactor [202], and the second-stage culture was successfully scaled up in a new self-designed multi-layer static reactor [203]. During submerged fermentation of medicinal mushroom *C. militaris* in fermenters for production of valuable cordycepin, an interesting two-stage dissolved oxygen control approach was also developed [204]. Those cultivation strategies may also be helpful to other higher fungi fermentation for enhancing value-added metabolite production.

Separation, purification and chemical structure identification of natural compounds is very important to various applications. Ganoderic acid T (GA-T) and ganoderic acid Me (GA-Me) were found to have significant antitumor activities [100, 196, 197]. GA mixtures can be produced in quantity through mushroom fermentation. However, to obtain a large amount of pure GAs for biological activity tests, efficient separation and purification methods still need to be developed.

Chromatographic purification is widely used for various bio-products such as plasmid DNA, enzymes and natural products [205–207]. In the analysis and isolation of triterpenes from *G. lucidum*, the crude triterpene extracts are usually subjected to qualitative analysis and semipreparative separation using silica gel TLC plates or silica gel column chromatography [198, 208]. RP-HPLC methods were also used for the complete separation of triterpenes isolated from the *Ganoderma* mycelia [209, 210]. However, modern hyphenated techniques, such as GC–MS, HPLC–MS, HPLC–MS–MS and HPLC–NMR, have not been applied, even though these techniques may provide useful structural information online on these triterpene metabolites and allow the rapid structural determination of known plant constituents with only a minute amount of materials [211, 212]. It is worthwhile to apply modern hyphenated chromatographic techniques to the characterization and determination of a variety of components in *Ganoderma* extracts. We developed an effective HPLC–UV–ESI–MS hyphenated method to determine the presence and distribution of targeted triterpenes in the methanol extract of *Ganoderma* mycelia and then to purify them efficiently.

The semipreparative HPLC separation of the GAs was accomplished after the establishment of the HPLC–UV–ESI–MS hyphenated method. Sample injection volume as a key operating parameter in preparative- and large-scale HPLC was examined [99]. Figure 3 shows HPLC chromatograms for the same sample mixture with two

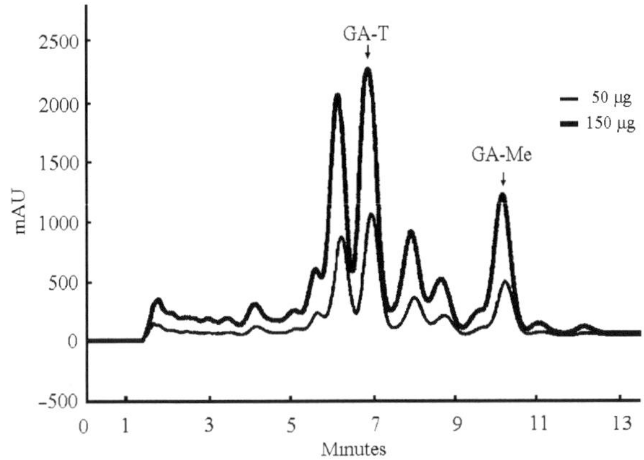

Fig. 3 Semi-preparative HPLC chromatograms showing GA-T and GA-Me peaks. A semi-preparative column (250 × 10.0 mm^2 I.D.) packed with 5 µm Hypersil ODS2 C_{18} (Elite Co., Dalian, China) was used. The *light curve* corresponds to the 50 µg sample injection and the *heavy curve* 150 µg sample injection

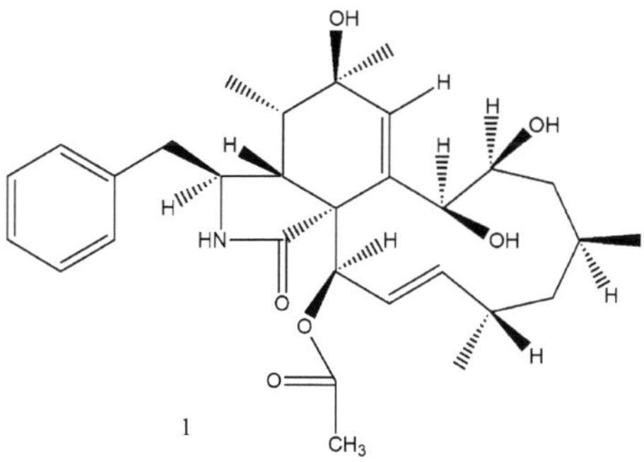

1

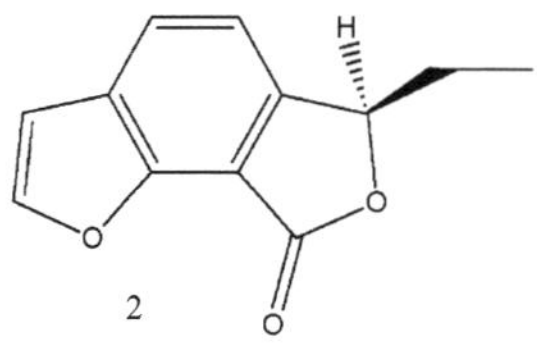

2

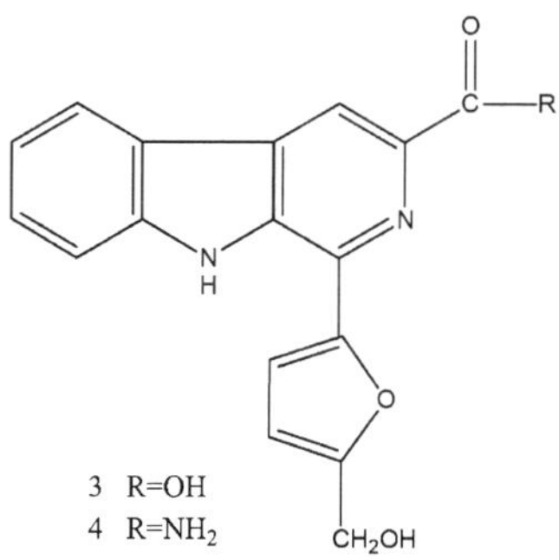

3 R=OH
4 R=NH_2

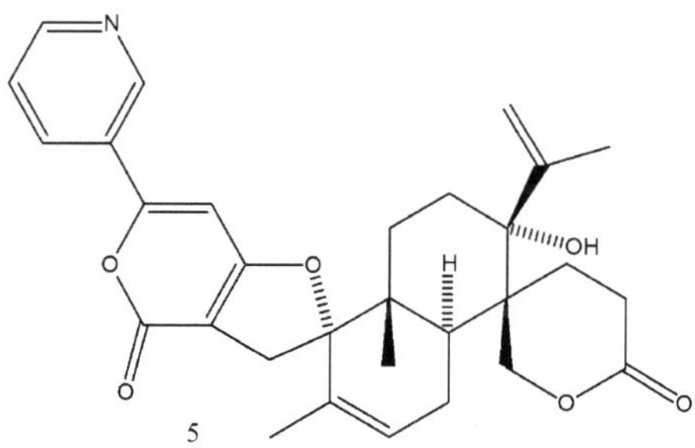

5

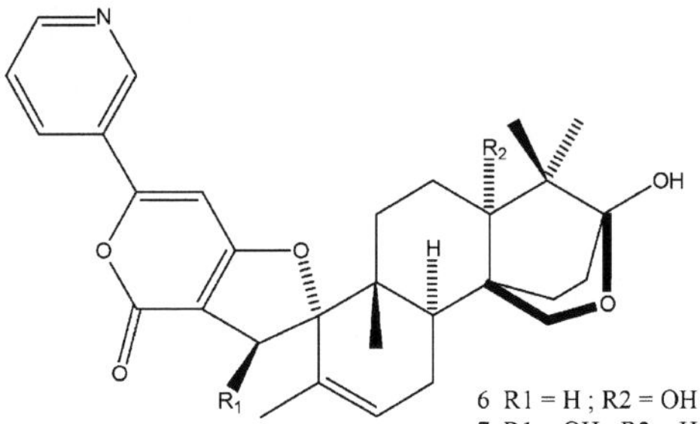

6 R1 = H ; R2 = OH
7 R1 = OH ; R2 = H

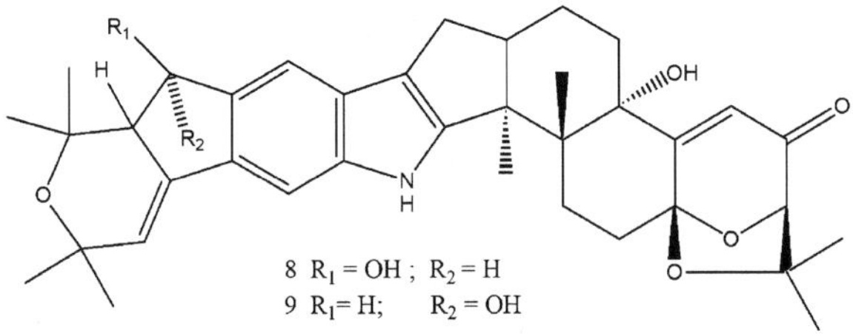

8 R_1 = OH ; R_2 = H
9 R_1= H; R_2 = OH

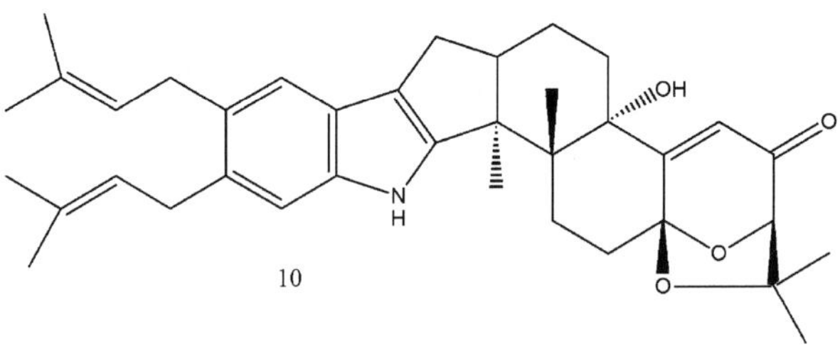

10

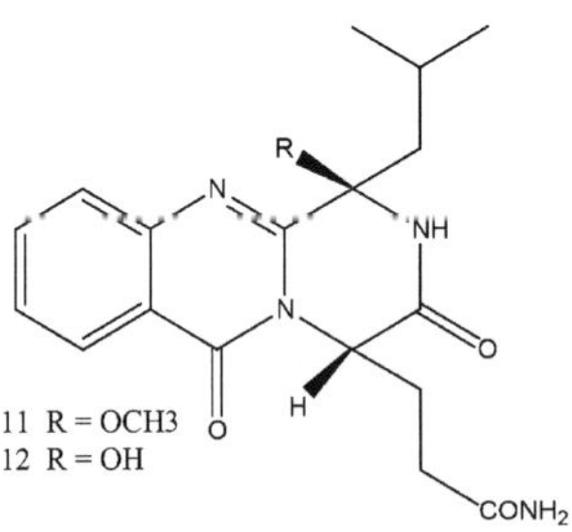

11 R = OCH3
12 R = OH

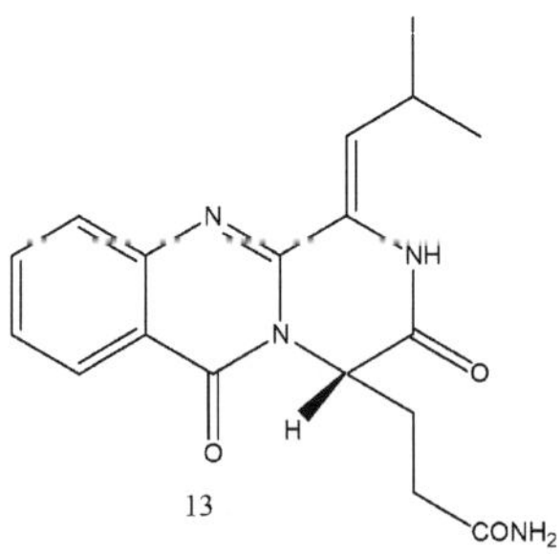

13

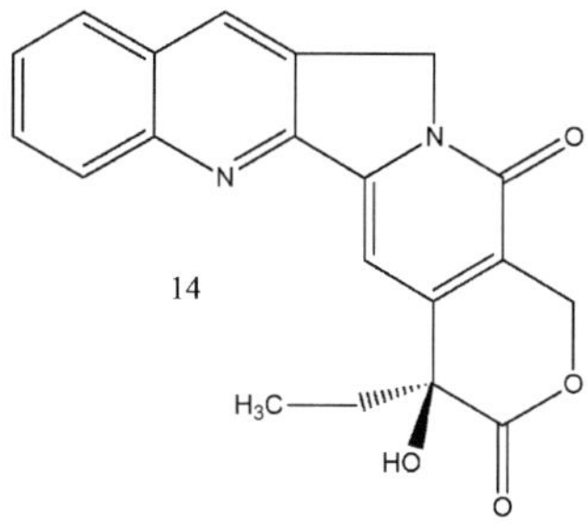

14

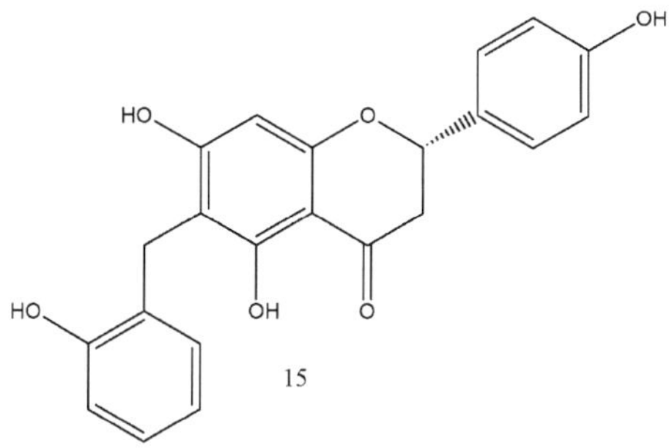

15

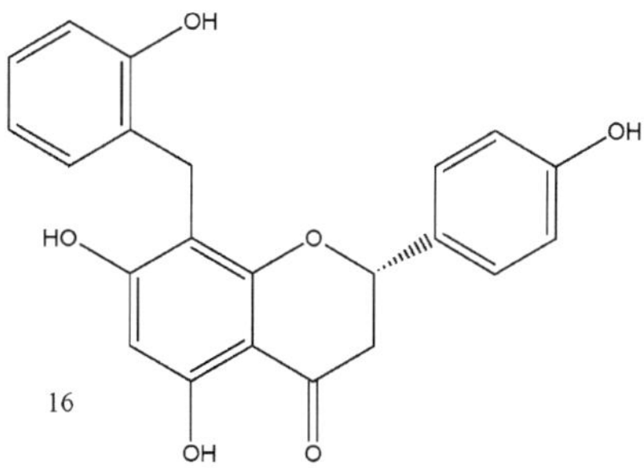

16

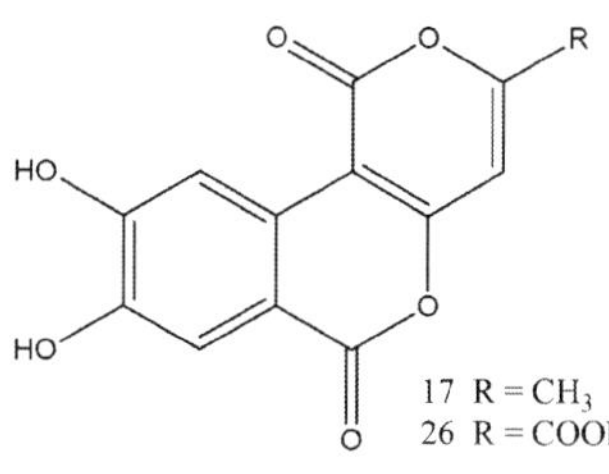

17 R = CH_3
26 R = COOH

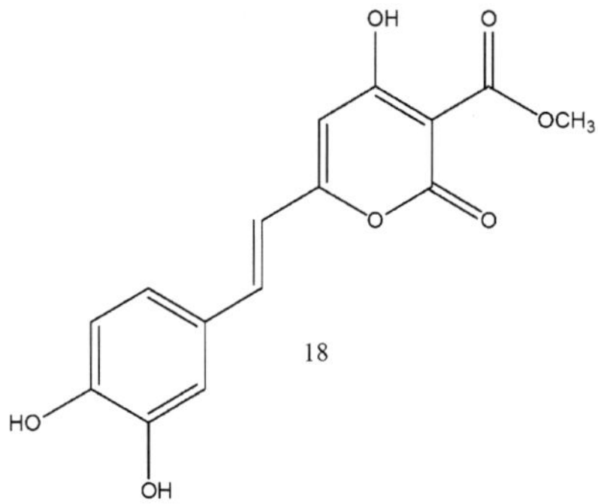

18

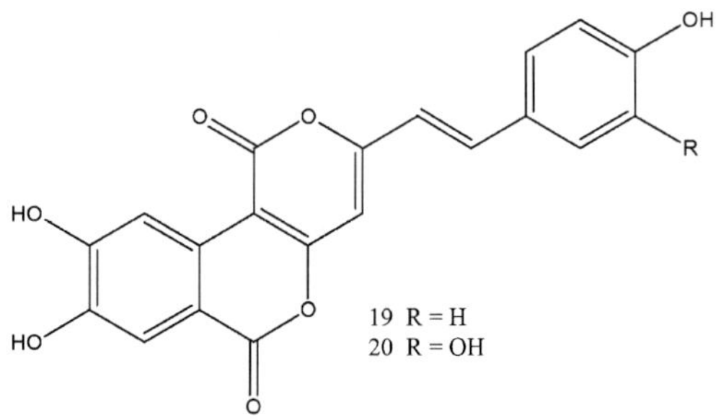

19 R = H
20 R = OH

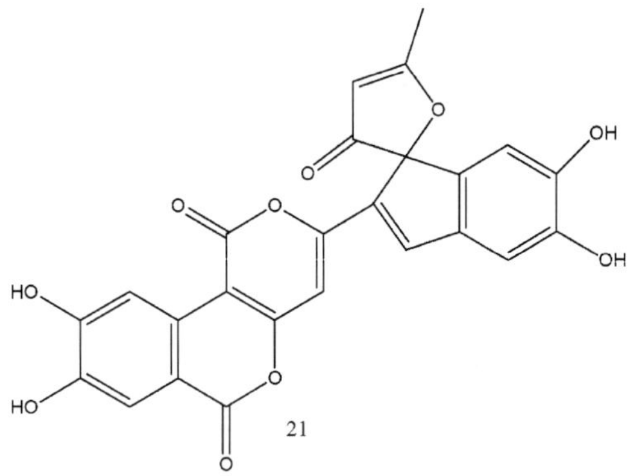

21

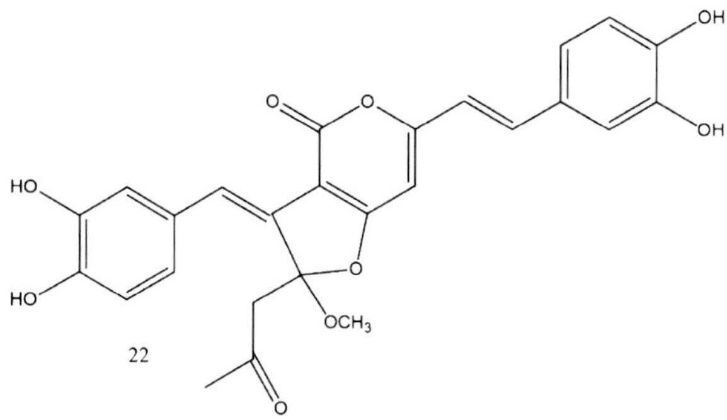

22

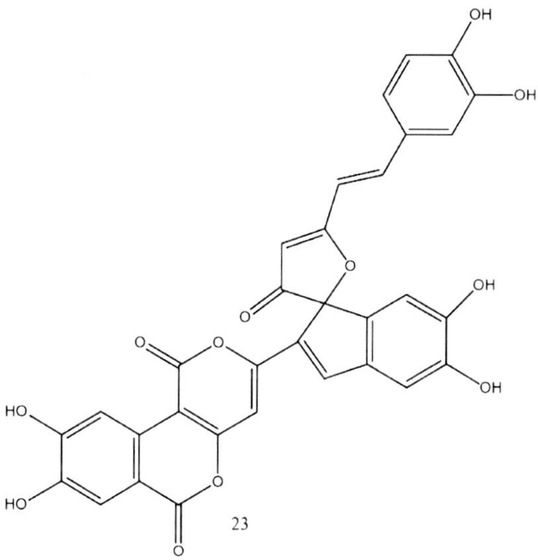

23

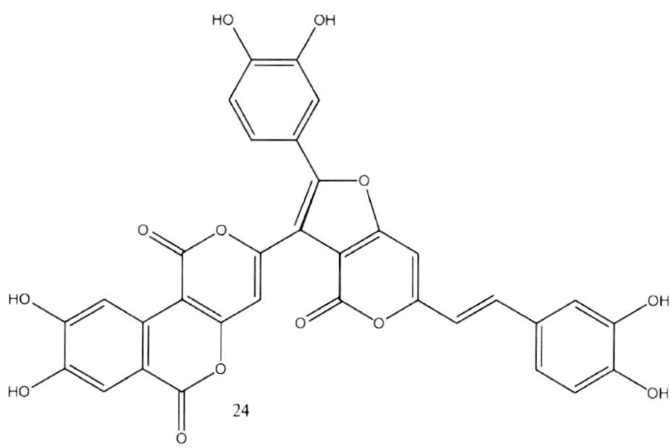

24

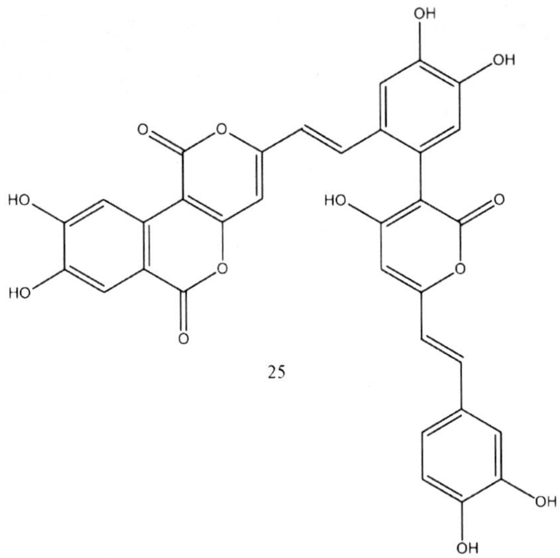
OH
OH
O
O
HO
HO
HO
O
O
O
O
O
OH
OH

25

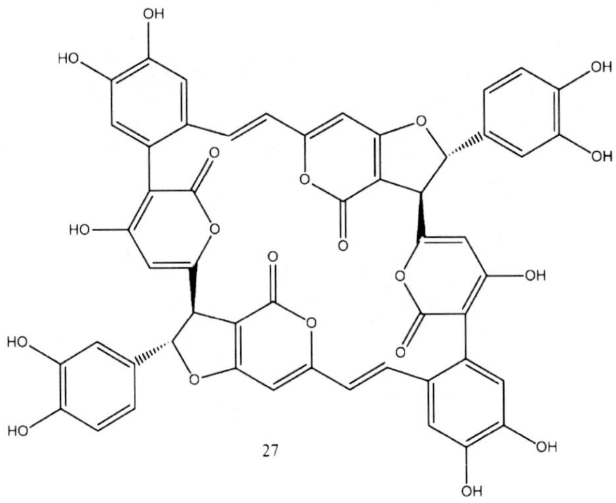
OH
HO
OH
OH
O
O
O
HO
O
O
O
O
OH
HO
O
O
O
HO
O
HO
OH
OH

27

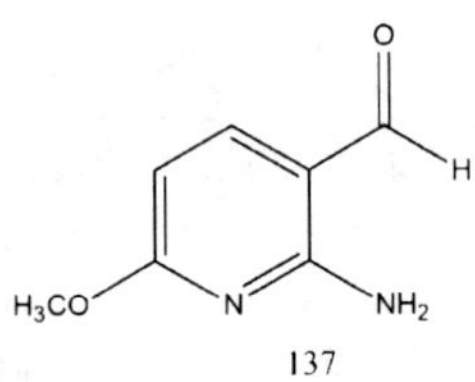
O
H
H3CO
N
NH2

137

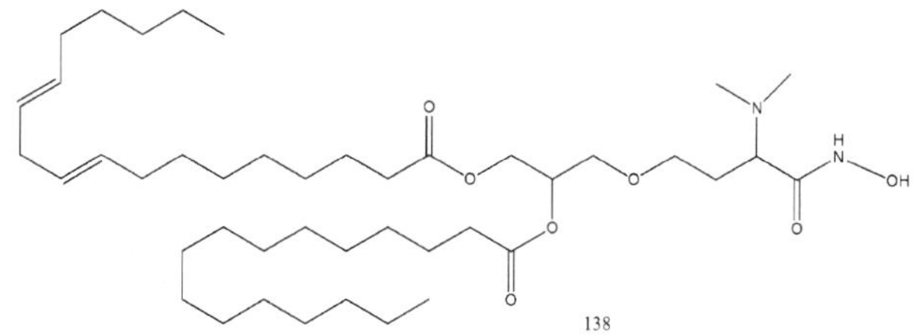

138

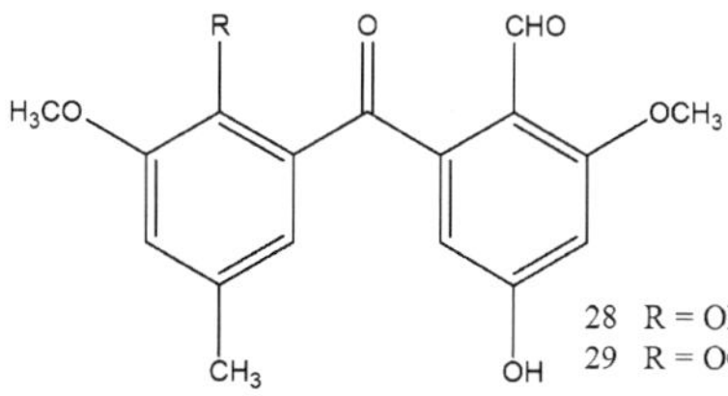

28 R = OH
29 R = OCH_3

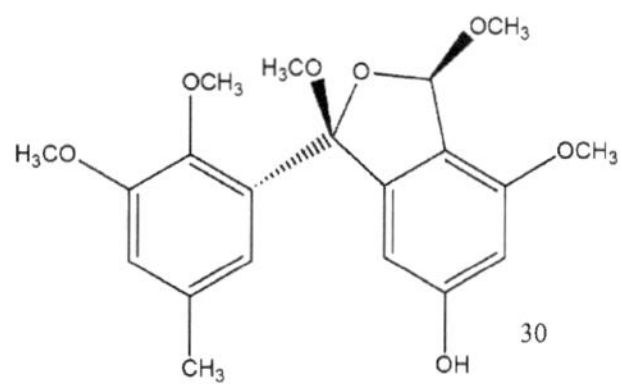

30

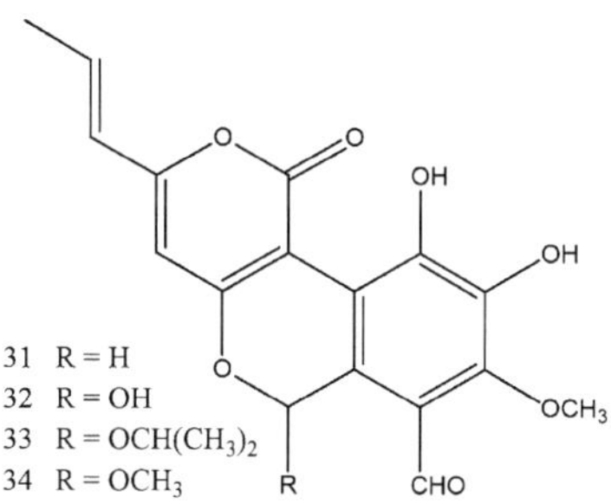

31 R = H
32 R = OH
33 R = $OCH(CH_3)_2$
34 R = OCH_3

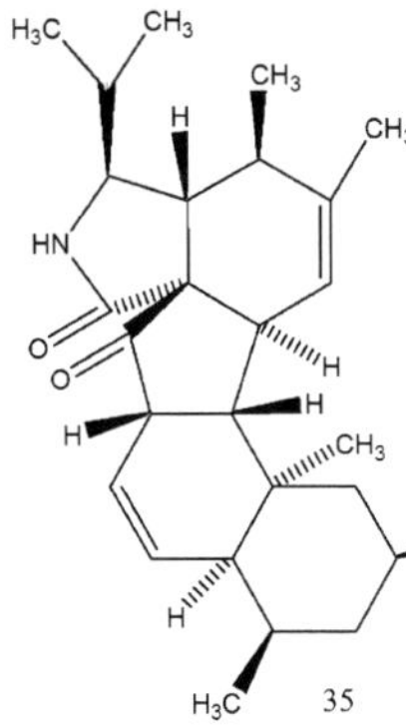

35

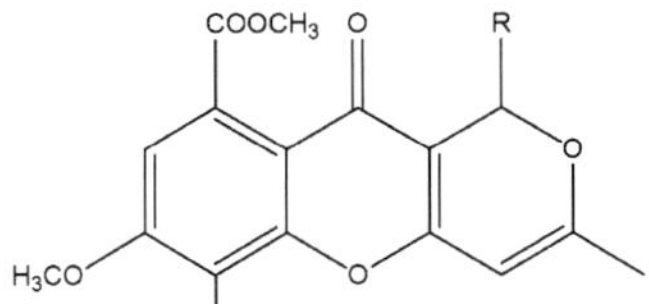

36 R = H
37 R = OCH_3

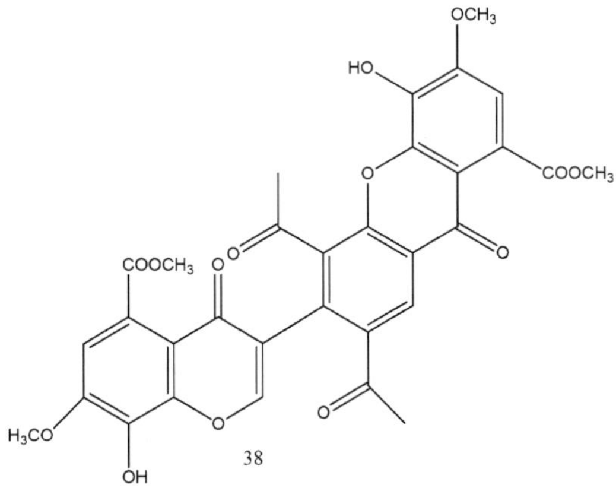

38

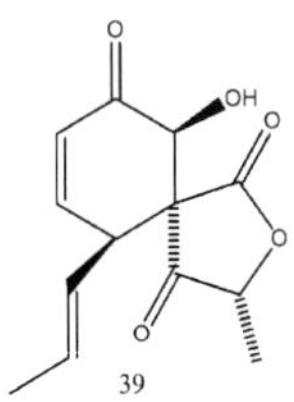

39

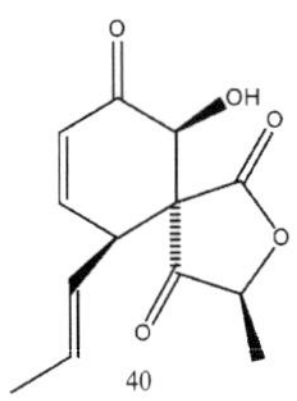

40

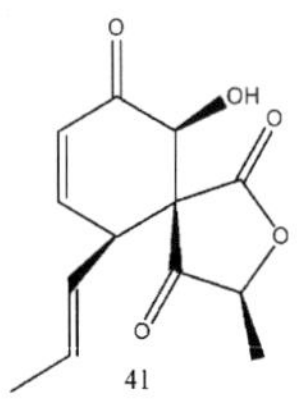

41

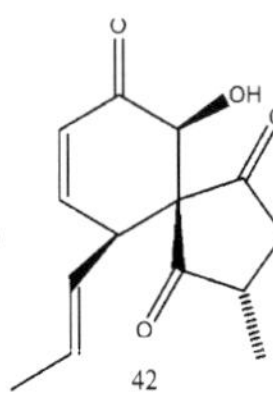

42

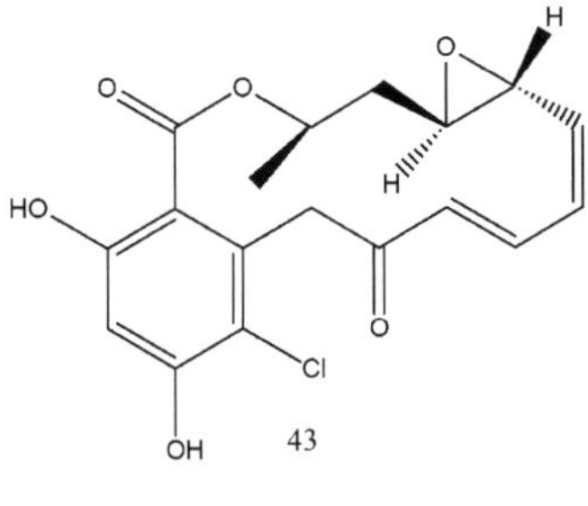

43

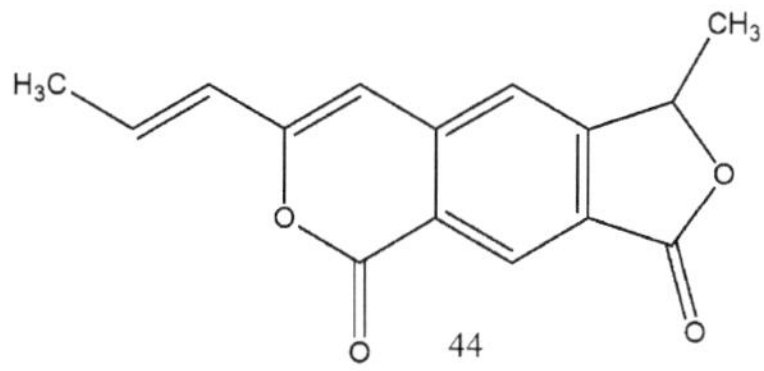

44

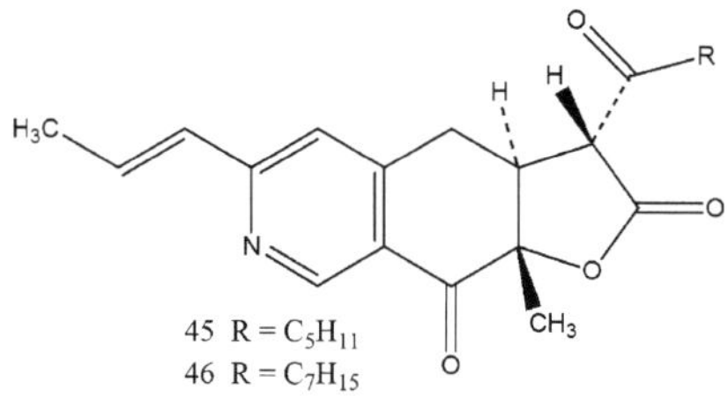

45 R = C_5H_{11}
46 R = C_7H_{15}

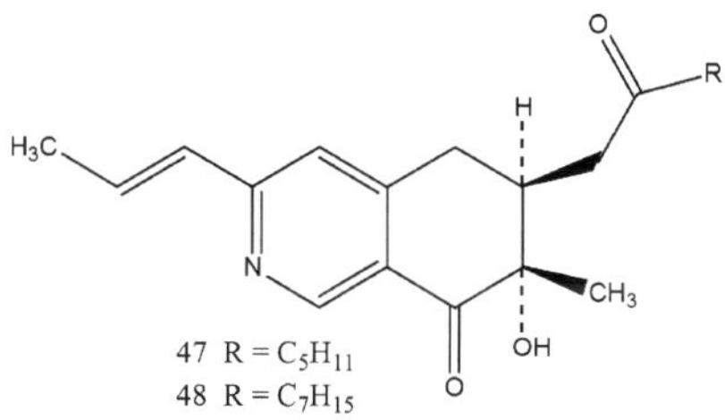

47 R = C_5H_{11}
48 R = C_7H_{15}

49

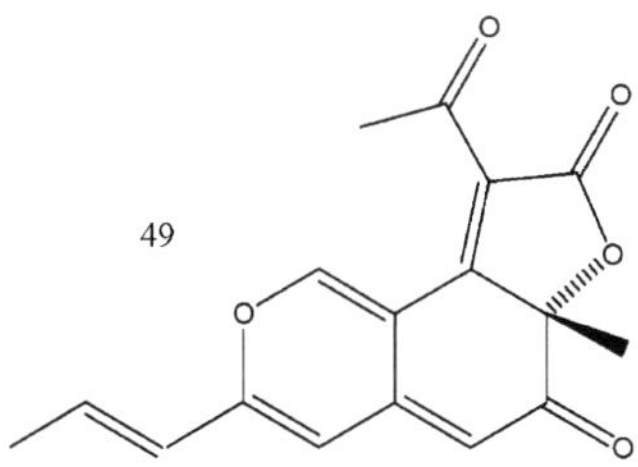

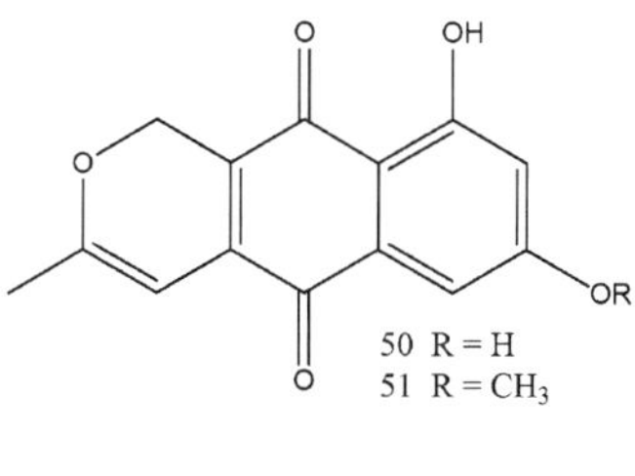

50 R = H
51 R = CH_3

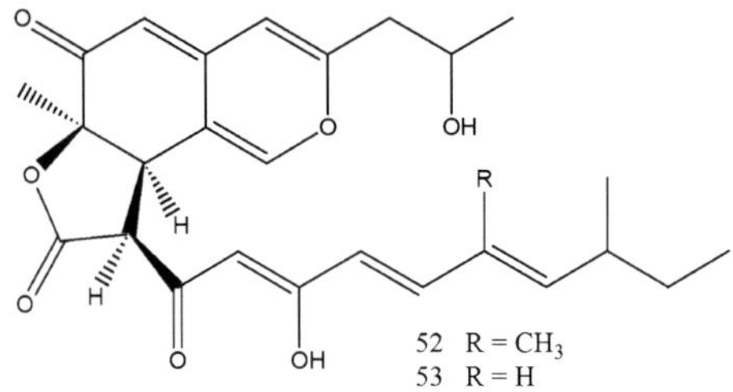

52 R = CH_3
53 R = H

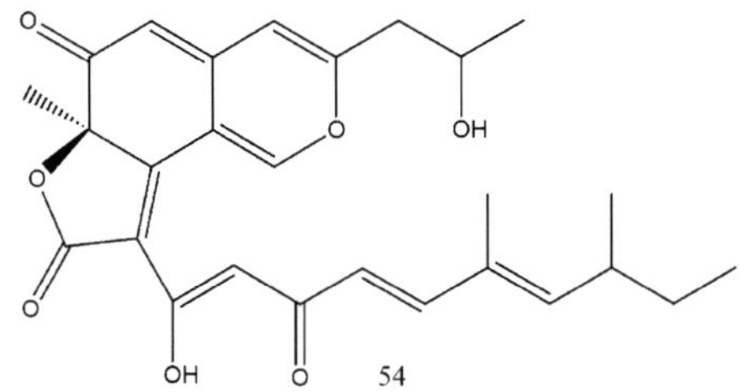

54

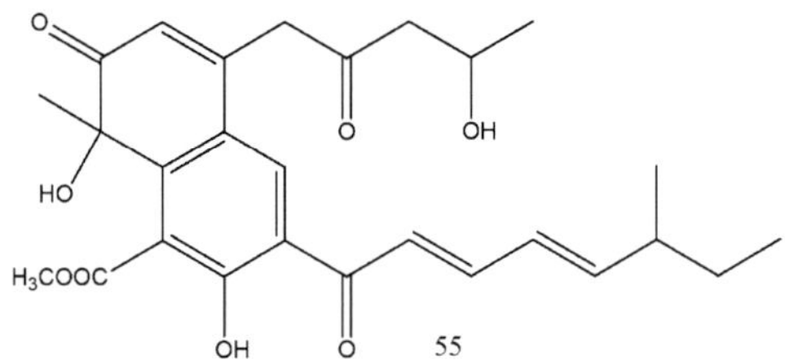

55

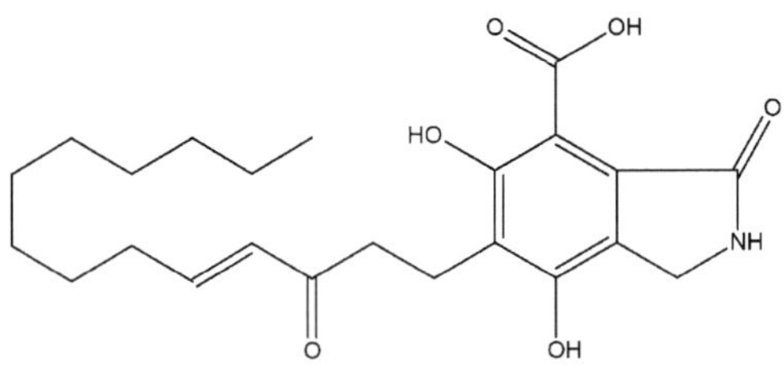

134

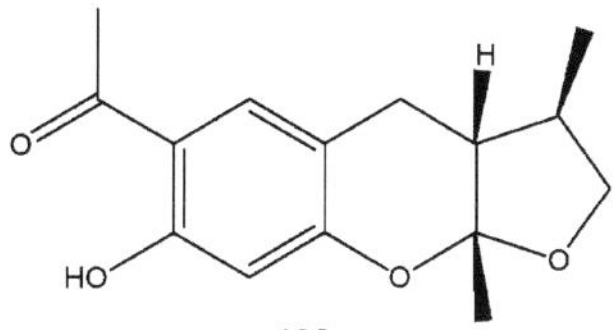

135

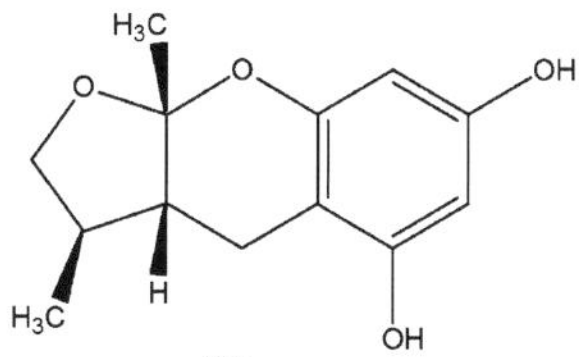

136

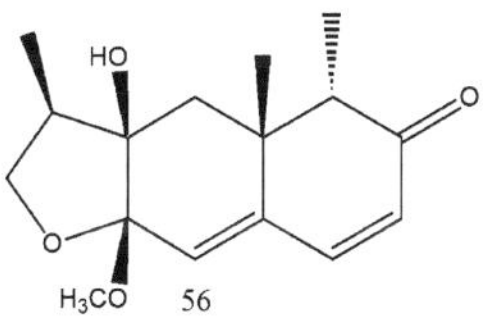

56

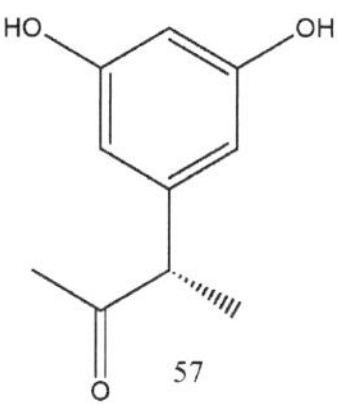

57

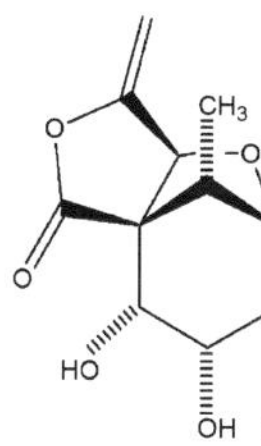

58

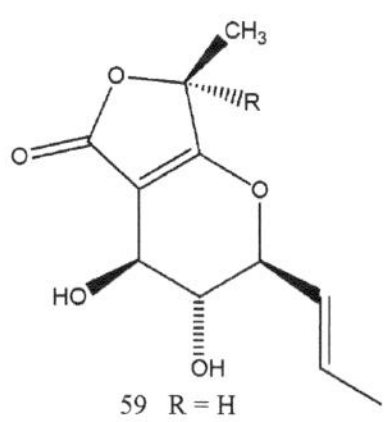

59 R = H
83 R = OH

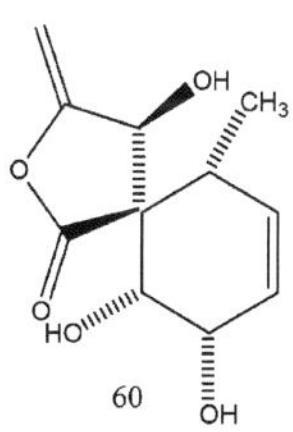

60

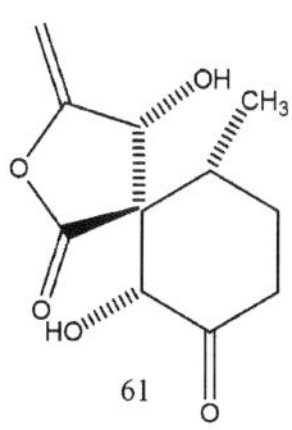

61

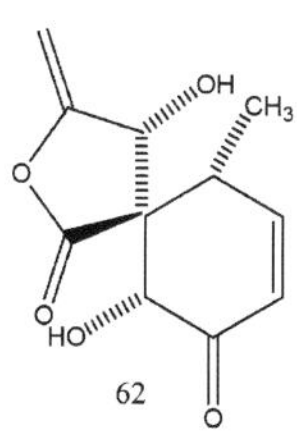

62

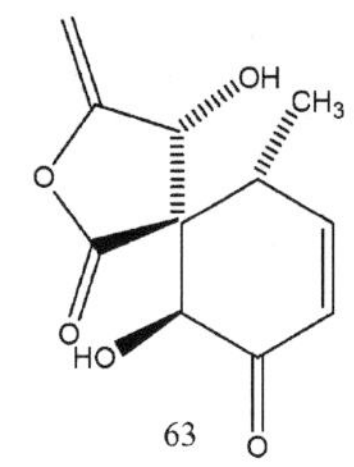

63

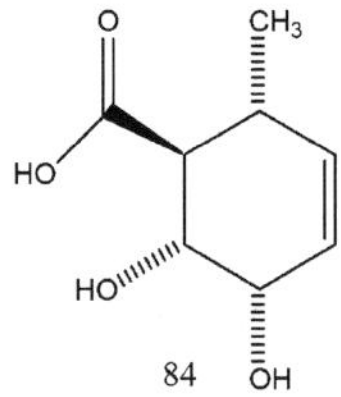

84

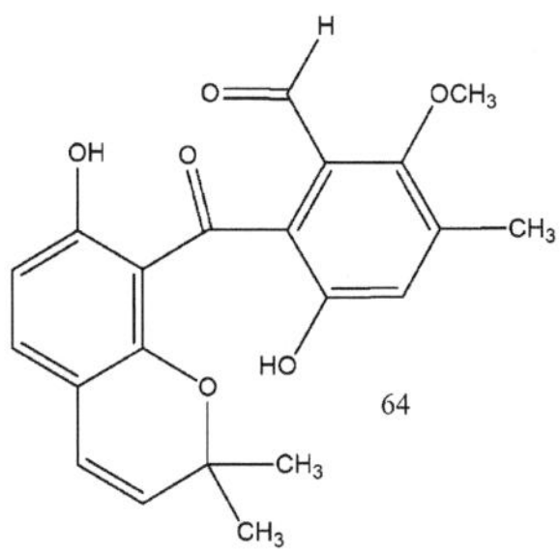

64

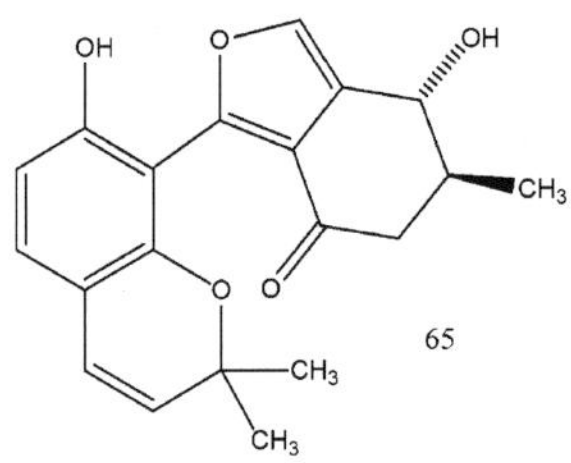

65

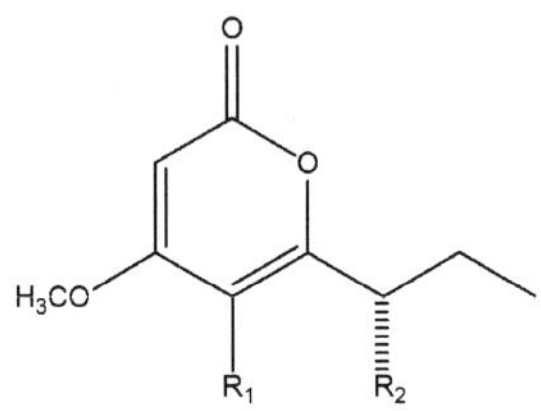

66 $R_1 = CH_3$ $R_2 = OH$
67 $R_1 = CH_2OH$ $R_2 = H$
68 $R_1 = CH_2OH$ $R_2 = OH$
69 $R_1 = CH_3$ $R_2 = H$
70 $R_1 = H$ $R_2 = H$

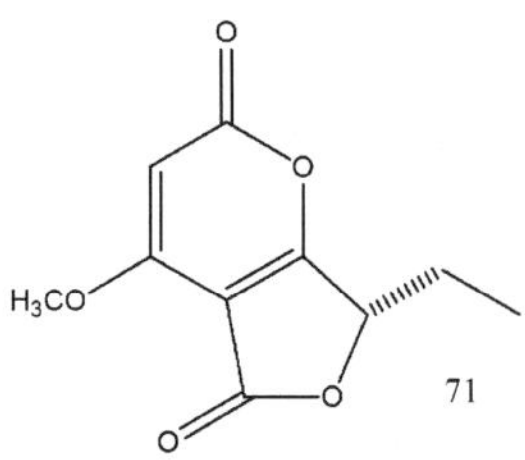

71

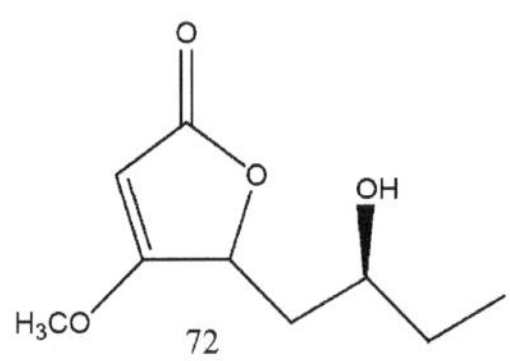

72

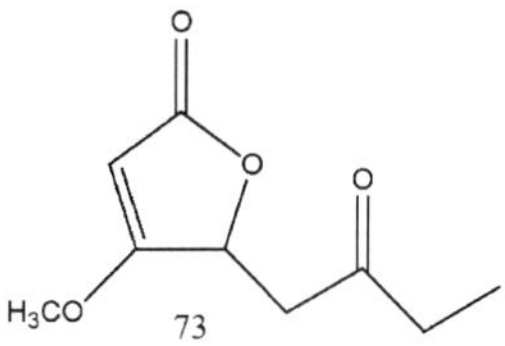

73

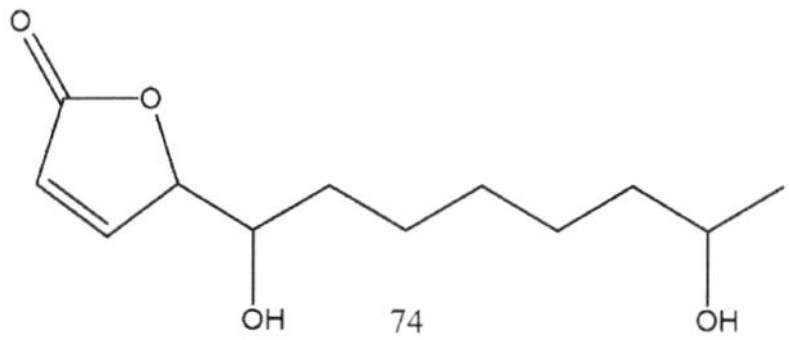

74

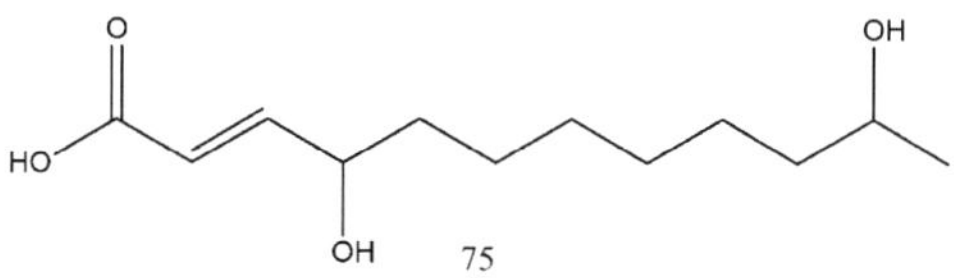

75

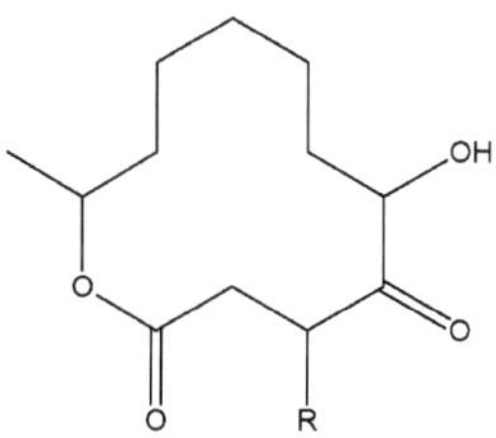

76 R = OH
77 R = S(CH2COOH)

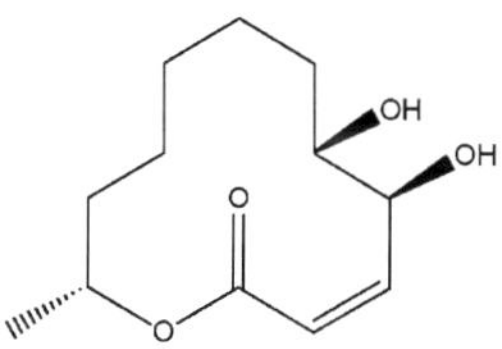

78

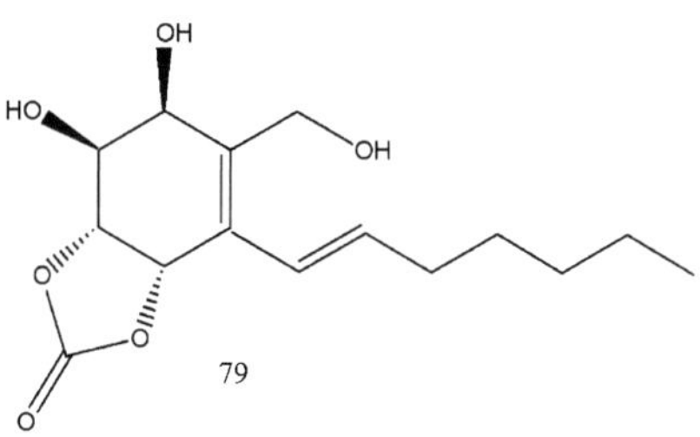

79

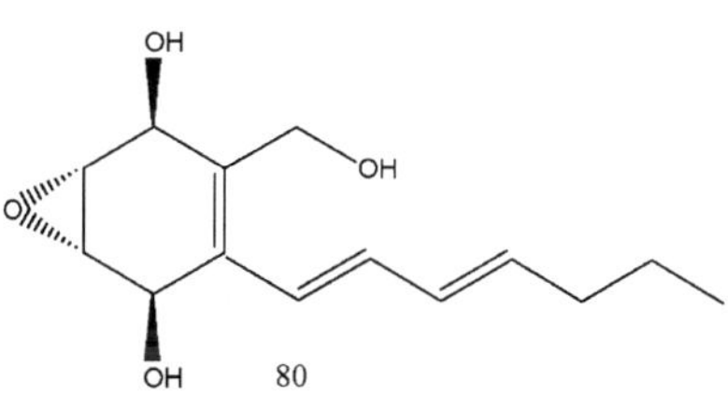

80

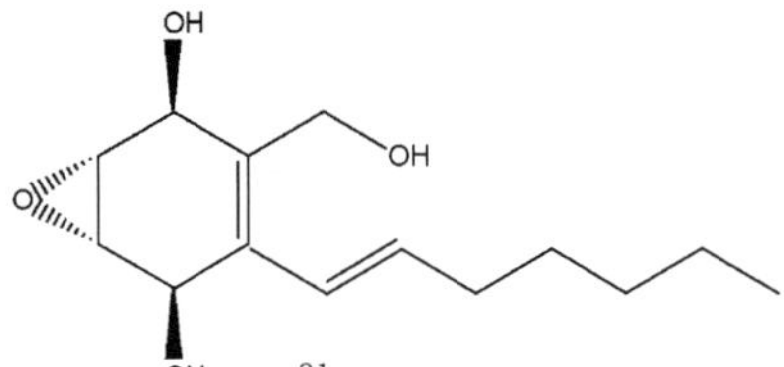
81

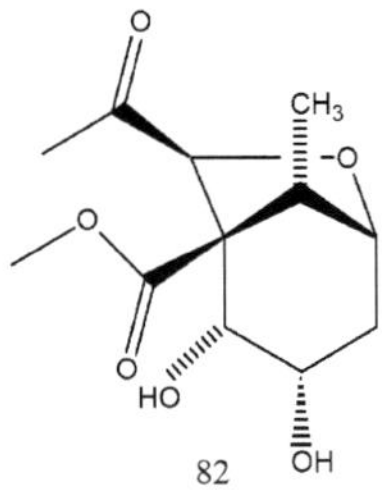
82

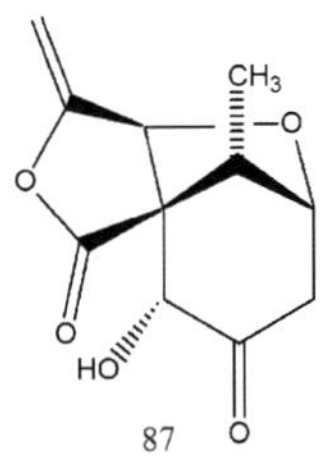
87

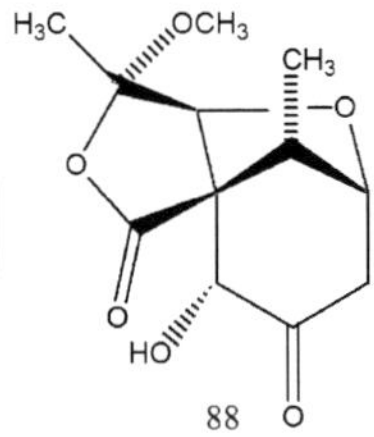
88

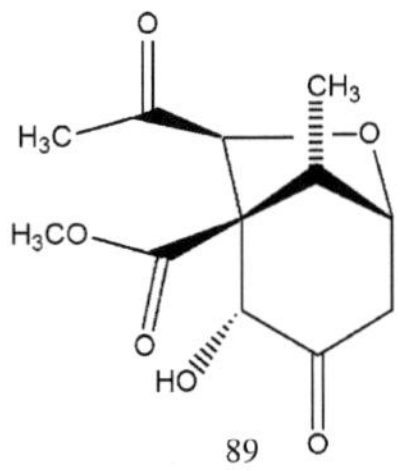
89

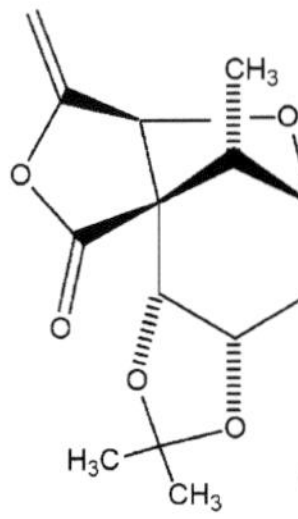
90

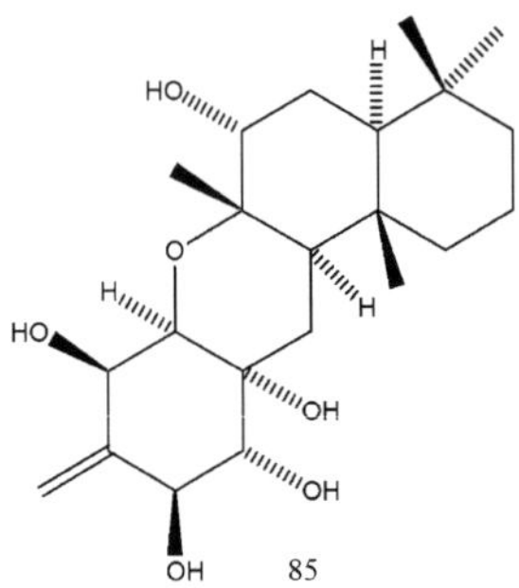
85

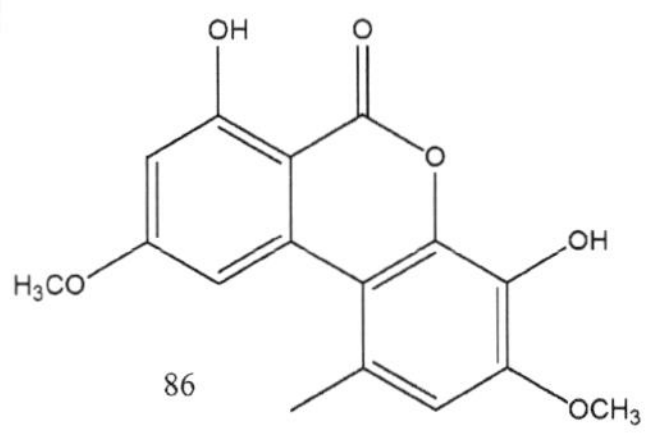
86

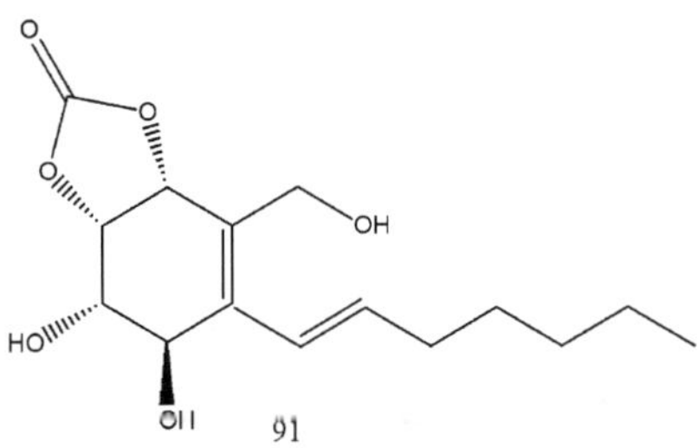
91

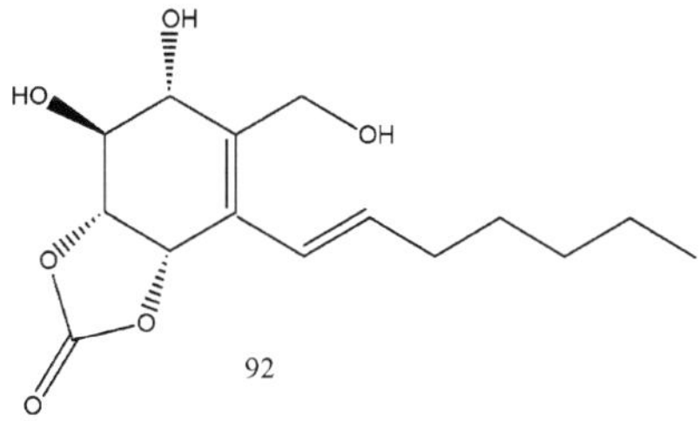

92

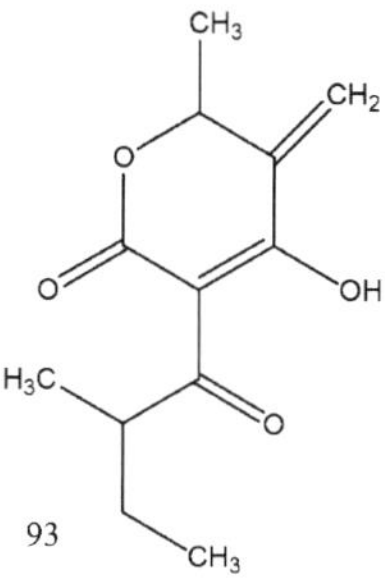

93

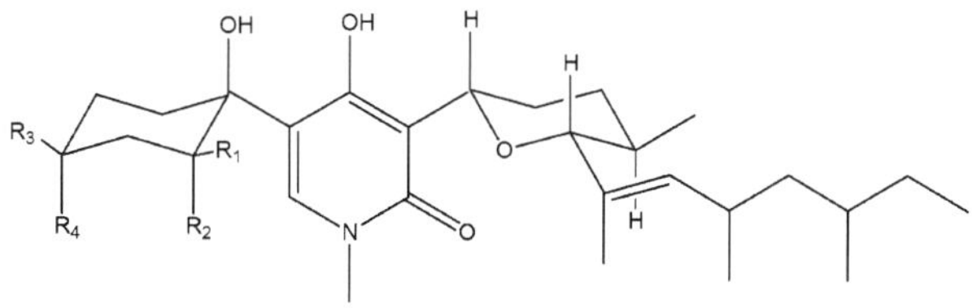

94 $R_1 = H$, $R_2 = OH$, $R_3, R_4 = {=}O$
95 $R_1 = OH$, $R_2 = H$, $R_3 = R_4 = OCH_3$
97 $R_1 = OH$, $R_2 = H$, $R_3, R_4 = {=}O$

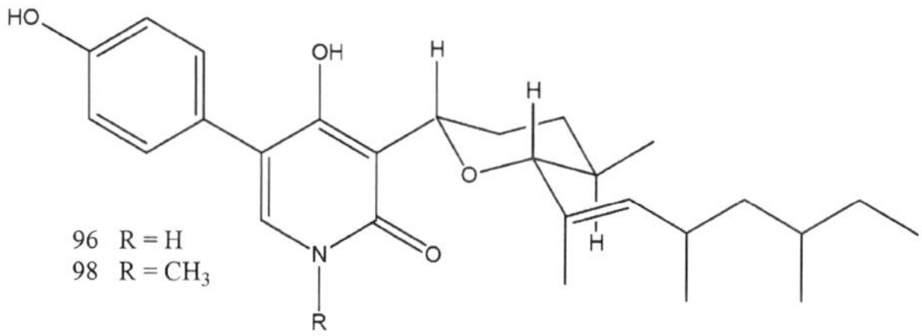

96 R = H
98 R = CH_3

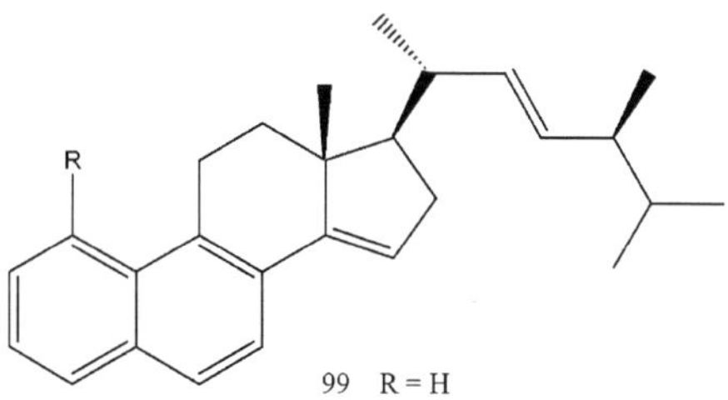

99 R = H
100 R = CH_3

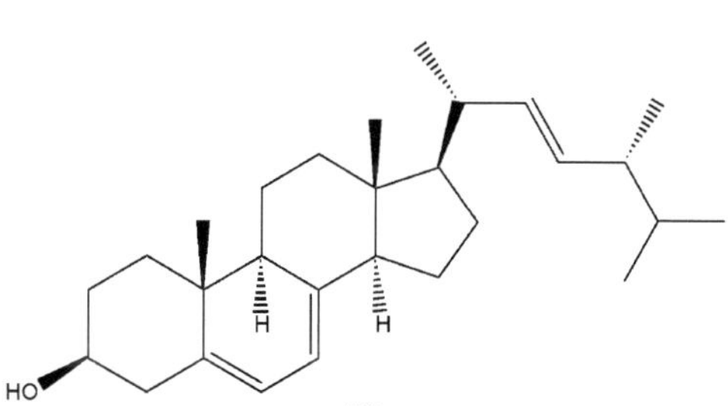

101

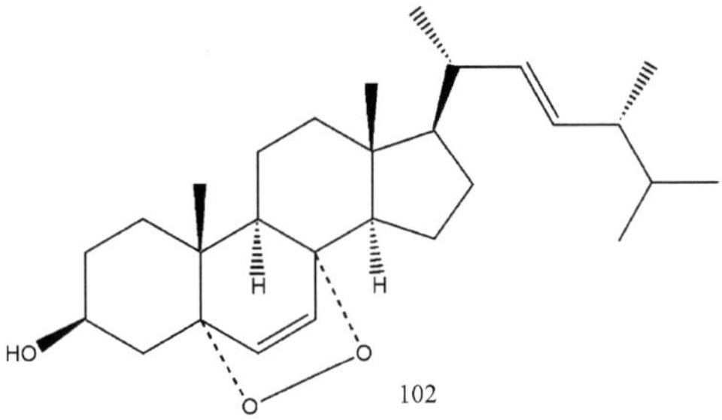

102

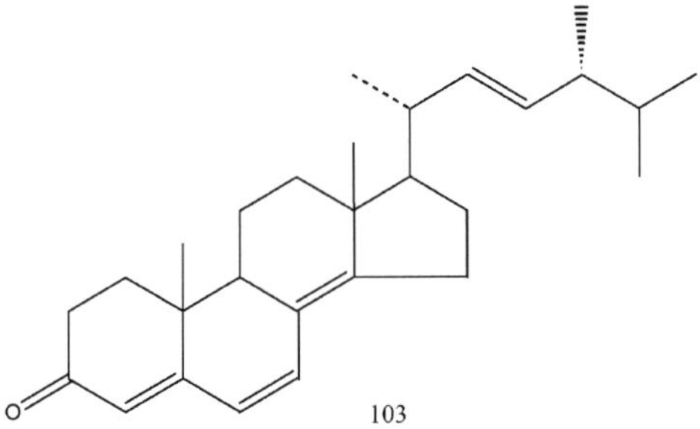

103

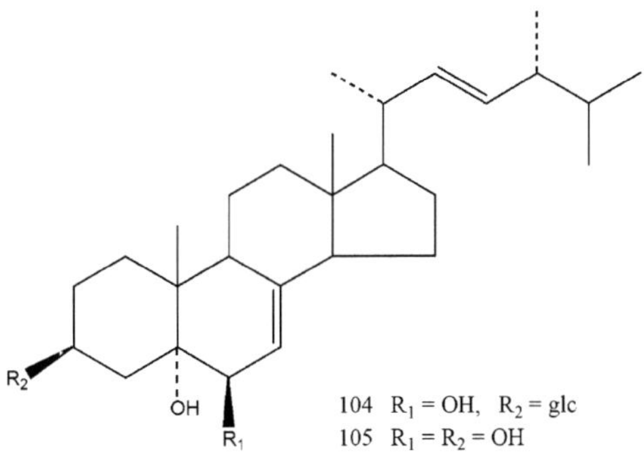

104 R_1 = OH, R_2 = glc
105 R_1 = R_2 = OH

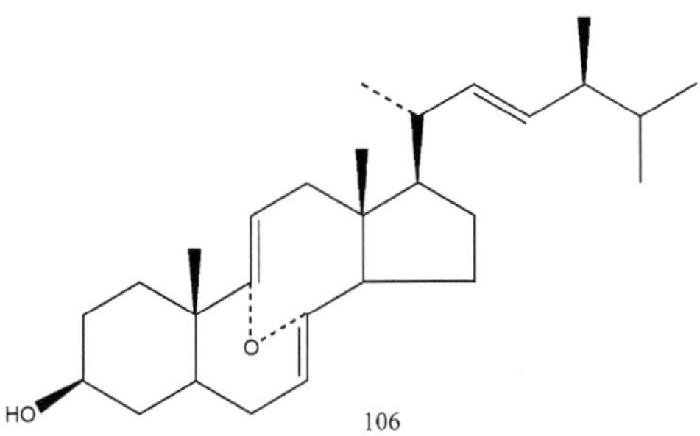

106

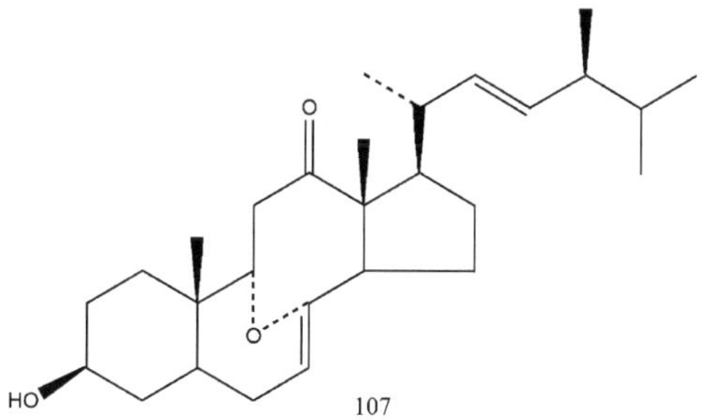

107

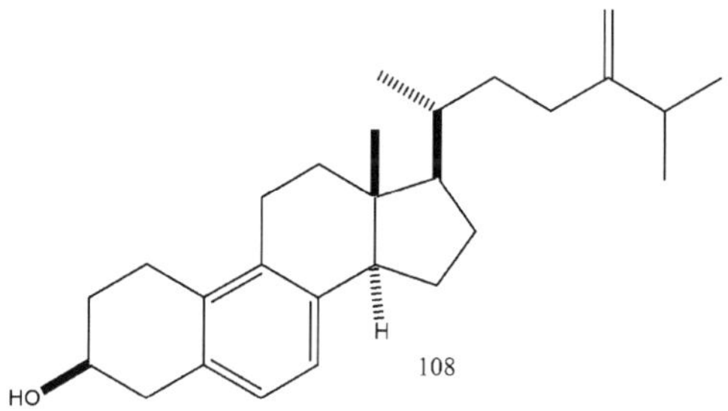

108

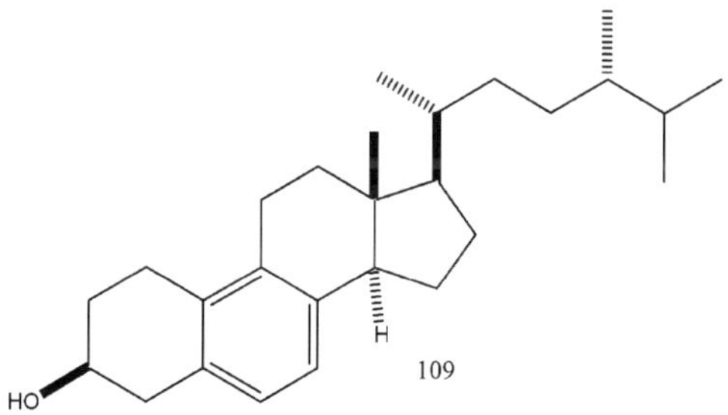

109

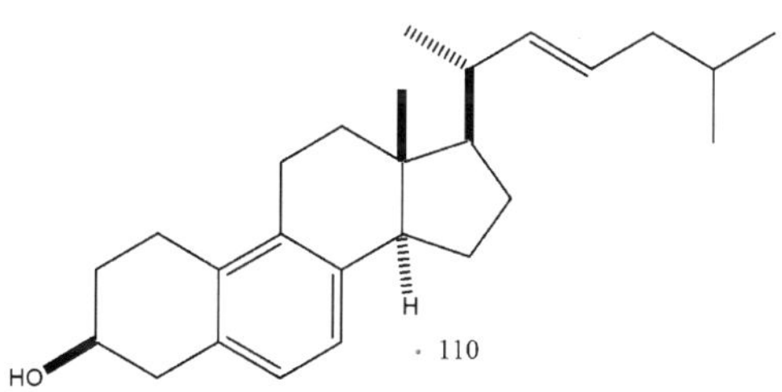

110

Structures of Secondary Metabolites

different sample sizes (50 and 150 μl) using the semipreparative C_{18} column with the aforementioned mobile phase at 4 mL min^{-1} flow rate. The two chromatograms in Fig. 3 are very similar, and this indicates that some operating conditions for the analytical runs could be successfully adopted in the semipreparative separation in this case.

5 Concluding Remarks

It is evident that higher fungi are abundant sources of a wide range of useful natural products with diversified chemical structures and various interesting bioactivities. Nowadays, their commercial products are yet very limited and mainly from field cultivation, which is obviously a time-consuming and labor-intensive process. Fermentation has significant commercial potential, especially for producing value-added secondary metabolites as biopharmaceuticals. Increasing product yields and development of highly efficient production and purification systems will make the technology more attractive and feasible for real application.

More and more studies on secondary metabolites biosynthesis including related genes, enzymes and their biochemical pathways will greatly contribute to powerful manipulation of their biosynthetic routes for directed production. On the one hand, traditional physiological and morphological researches are still necessary and important to many untapped or poorly investigated species including many medicinal mushrooms. On the other hand, modern genomic, proteomic and metabolomic investigations should be conducted on a large scale for significant accumulation of such basic knowledge, which is crucial to the future development of an advanced bioprocessing platform for industrial production of related bioactive secondary metabolites.

Acknowledgments We appreciate the financial support from the National Natural Science Foundation of China (NSFC project Nos. 20762017, 30821005 and 20776084), Program for Excellent Young Talents of Science and Technology of Guizhou Province (No.QKT200786), the National High Technology R&D Program (863 Program project # 2007AA021506) of the Ministry of Science and Technology of China (MOST), and the Shanghai Leading Academic Discipline Project (project nos. B203 and B505).

References

1. Hawksworth DL, Kirk PM, Sutton BC, Pegler DN (1995) Ainsworth and Bisby's Dictionary of the Fungi, 8th edn. CABI, Wallingford, UK, pp 1–616
2. Hawksworth DL (2004) Fungal diversity and its implication for genetic resource collections. Stud Mycol 50:9–18
3. Hawksworth DL (2001) The magnitude of fungal diversity: the 1.5 million species estimate revisited. Mycol Res 105:1422–1432
4. Schmit JP, Mueller GM (2007) An estimate of the lower limit of global fungal diversity. Biodivers Conserv 16:99–111
5. Manoharachary C, Sridhar K, Singh R, Adholeya A, Suryanarayanan TS, Rawat S, Johri BN (2005) Fungal biodiversity: distribution, conservation and prospecting of fungi from India. Curr Sci 89:58–71

6. Hawksworth DL, Rossman AY (1997) Where are all the undescribed fungi? Phytopathology 87:888–891
7. Kirk PM, Cannon PF, David JC, Stalpers JA (2001) Ainsworth and Bisby's Dictionary of the Fungi, 9th edn. CABI, Wallingford, UK, pp 1–655
8. Yang ZL, Zang M (2003) Tropical affinities of higher fungi in southern China. Acta Botanica Yunnanica 25:129–144
9. Stähelin HF (1996) The history of cyclosporin A (Sandimmune®) revisited: another point of view. Experientia 52:5–13
10. Wei J, Si SY, Chen XP, Wang L, Zhuang XL, Zou J, Chen HS, Cao LX, Feng H, Chen L (2006) Development and application of microbial natural products database. J Chin Antibiotics 31:119–121 (in Chinese)
11. Muegge I (2001) Heald SL and Brittelli D. Simple selection criteria for drug-like chemical matter. J Med Chem 44:1841–1846
12. Bérdy J (2005) Bioactive microbial metabolites: a personal view. J Antibiotics 58:1–26
13. Zhong J-J (ed.) (2004) Advances in biochemical engineering/biotechnology: biomanufacturing, vol 87. Springer-Verlag, Heidelberg
14. Strobel GA (2003) Endophytes as sources of bioactive products. Microbe Infect 5:535–544
15. König GM, Kehraus S, Seibert SF, Abdel-Lateff A, Müller D (2006) Natural products from marine organisms and their associated microbes. Chembiochem 7:229–238
16. Jiang W, Si SY, Chen XP,Wang L, Zhuang XL, Zou J, Chen HS, Cao LX, Feng H, Chen L (2006) Development and application of microbial natural products database. Chin J Antibiotics 31:119–121 (in Chinese)
17. Eicher T, Hauptmann S (2003) The chemistry of heterocycles: struture, reactions, syntheses and applications, 2nd edn. Wiley, Weinheim, pp 1–16
18. Liu JK (2005) N-Containing compounds of macromycetes. Chem Rev 105:2723–2742
19. Qin XD, Dong ZJ, Liu JK (2006) Two new compounds from the ascomycete *Daldinia concentrica*. Helv Chim Acta 89:450–455
20. Qin XD, Dong ZJ, Liu JK, Yang LM, Wang RR, Zheng YT, Lu Y, Wu YS, Zheng QT (2006) Concentricolide, an anti-HIV agent from the ascomycete *Daldinia concentrica*. Helv Chim Acta 89:127–133
21. Wang F, Liu JK (2004) A pair of novel heptentriol stereoisomers from the ascomycete *Daldinia concentrica*. Helv Chim Acta 87:2131–2134
22. Dong ZJ, Wang F, Wang RR, Yang LM, Zheng YT, Liu JK (2007) Chemical constituents of the fruiting bodies from the basidiomycete Suillus granulatus. Chin Tradit Herb Drugs 38:17–19 (in Chinese)
23. Wang YH, Tang JG, Wang RR, Yang LM, Dong ZJ, Du L, Shen X, Liu JK, Zheng YT (2007) Flazinamide, a novel b-carboline compound with anti-HIV actions. Biochem Biophys Res Commun 355:1091–1095
24. Kozlovsky AG, Zhelifonova VP, Antipova TV (2005) The fungus *Penicillium citrinum*, isolated from permafrost sediments, as a producer of ergot alkaloids and new quinoline alkaloids quinocitrinines. Appl Biochem Microbiol 41:499–502
25. Zhang Y, Li C, Swenson DC, Gloer JB, Wicklow DT, Dowd PF (2003) Novel antiinsectan oxalicine alkaloids from two undescribed fungicolous *Penicillium* spp. Org Lett 5:773–776
26. Smetanina OF, Kalinovsky AI, Khudyakova YV, Pivkin MV, Dmitrenok PS, Fedorov SN, Ji H, Kwak JY, Kuznetsova TA (2007) Indole alkaloids produced by a marine fungus isolate of *Penicillium janthinellum* Biourge. J Nat Prod 70:906–909
27. Xin ZH, Fang Y, Du L, Zhu T, Duan L, Chen J, Gu QQ, Zhu WM (2007) Aurantiomides A-C, quinazoline alkaloids from the sponge-derived fungus *Penicillium aurantiogriseum* SP0-19. J Nat Prod 70:853–855
28. Puri SC, Verma V, Amna T, Qazi GN, Spiteller M (2005) An endophytic fungus from *Nothapodytes foetida* that produces camptothecin. J Nat Prod 68:1717–1719
29. Kantola J, Kunnari T, Mäntsälä P, Ylihonkoa K (2003) Expanding the scope of aromatic polyketides by combinatorial biosynthesis. Comb Chem High Throughput Screen 6:501–512

30. Hopwood DA (2004) Cracking the polyketide code. PLOS Biol 2:166–169
31. Staunton J, Weissman KJ (2001) Polyketide biosynthesis: a millennium review. Nat Prod Rep 18:380–416
32. Mo SY, Yany YC, Shi JG (2003) Isolation and synthesis of phenigrins A and B. Acta Chim Sin 61:1161–1163 (in Chinese)
33. Mo SY, Wen YH, Yang YC, Shi JG (2003) Two benzyl dihydroflavones from *Phellinus igniarius*. Chin Chem Lett 14:810–813
34. Mo SY, Yang YC, He WY, Shi JG (2003) Two pyrone derivatives from fungus *Phellinus Igniarius*. Chin Chem Lett 14:704–706
35. Mo S, Wang S, Zhou G, Yang Y, Li Y, Chen X, Shi J (2004) Phelligridins C-F: cytotoxic pyrano[4,3-*c*][2]benzopyran-1,6-dione and furo[3,2-*c*]pyran-4-one derivatives from the fungus *Phellinus igniari*us. J Nat Prod 67:823–828
36. Wang Y, Mo SY, Wang SJ, Li S, Yang YC, Shi JG (2005) A unique highly oxygenated pyrano[4,3-*c*][2]benzopyran-1,6-dione derivative with antioxidant and cytotoxic activities from the fungus *Phellinus igniarius*. Org Lett 7:1675–1678
37. Wang Y, Shang XY, Wang SJ, Mo SY, Li S, Yang YC, Ye F, Shi JG, He L (2007) Structures, biogenesis, and biological activities of pyrano[4,3-*c*]isochromen-4-one derivatives from the fungus *Phellinus igniarius*. J Nat Prod 70:296–299
38. Wang Y, Wang SJ, Mo SY, Li S, Yang YC, Shi JG (2005) Phelligridimer A, a highly oxygenated and unsaturated 26-membered macrocyclic metabolite with antioxidant activity from the fungus *Phellinus igniarius*. Org Lett 7:4733–4736
39. Mao XL Economic fungi of China, 1st edn. Sciences Press, Beijing, 1998, pp 246
40. Wang F, Tan JW, Liu JK (2004) Vibratilicin: a novel compound from the Basidiomycete *Cortinarius vibratilis*. Helv Chim Acta 87:1912–1915
41. Hu L, Tan JW, Liu JK (2003) Chemical constituents of the basidiomycete *Cortinarius umidicola*. Z Naturforsch 58c:659–662
42. Quang DN, Harinantenaina L, Nishizawa T, Hashimoto T, Kohchi C, Soma GI, Asakawa Y (2006) Inhibitory activity of nitric oxide production in RAW 264.7 cells of daldinals A–C from the fungus *Daldinia childiae* and other metabolites isolated from inedible mushrooms. J Nat Med 60:303–307
43. Kang HS, Jun EM, Park SH, Heo SJ, Lee TS, Yoo ID, Kim JP (2007) Cyathusals A, B, and C, antioxidants from the fermented mushroom *Cyathus stercoreus*. J Nat Prod 70:1043–1045
44. Lösgen S, Schlörke O, Meindl K, Herbst-Irmer R, Zeeck A (2007) Structure and biosynthesis of chaetocyclinones, new polyketides produced by an endosymbiotic fungus. Eur J Org Chem 2191–2196
45. Oh H, Swenson DC, Gloer JB (1998) Chaetochalasin A: a new bioactive metabolite from *Chaetomium brasiliense*. Tetrahedron Lett 39:7633–7636
46. Rether J, Erkel G, Anke T, Sterner O (2004) Oxaspirodion, a new inhibitor of inducible TNF-a expression from the ascomycete *Chaetomium subspirale* production, isolation and structure elucidation. J Antibiotics 57:493–495
47. Turbyville TJ, Wijeratne EMK, Liu MX, Burns AM, Seliga CJ, Luevano LA, David CL, Faeth SH, Whitesell L, Gunatilaka AAL (2006) Search for Hsp90 inhibitors with potential anticancer activity: isolation and SAR studies of radicicol and monocillin I from two plant-associated fungi of the sonoran desert. J Nat Prod 69:178–184
48. Ma J, Li Y, Ye Q, Li J, Hua Y, Ju D, Zhang D, Cooper R, Chang M (2000) Constituents of red yeast rice, a traditional food and medicine. J Agric Food Chem 48:5220–5225
49. Heber D, Yip I, Ashley JM, Elashoff DA, Elashoff RM, Go VL (1999) Cholesterol-lowering effects of a proprietary Chinese red-yeast-rice dietary supplement. Am J Clin Nutr 69: 231–236
50. Juzlová P, Martínková L, Kren V (1996) Secondary metabolites of the fungus *Monascus*: a review. J Ind Microbiol Biotechnol 16:163–170
51. Wild D, Tóth G, Humpf HU (2002) New *Monascus* Metabolite Isolated from Red Yeast Rice (Angkak, Red Koji). J Agric Food Chem 50:3999–4002

52. Wild D, Tóth G, Humpf HU (2003) New *Monascus* metabolites with a pyridine structure in red fermented rice. J Agric Food Chem 51:5493–5496
53. Knecht A, Cramer B, Humpf HU (2006) New Monascus metabolites: structure elucidation and toxicological properties studied with immortalized human kidney epithelial cells. Mol Nutr Food Res 50:314–321
54. Thines E, Anke H, Sterner O (1998) Trichoflectin, a bioactive azaphilone from the Ascomycete *Trichopezizella nidulus*. J Nat Prod 61:306–308
55. Quang DN, Hashimoto T, Fournier J, Stadler M, Radulovic N, Asakawa Y (2005) Sassafrins A–D, new antimicrobial azaphilones from the fungus *Creosphaeria sassafras*. Tetrahedron 61:1743–1748
56. Wang XN, Tan RX, Liu JK (2005) Xylactam, a new nitrogen-containing compound from the fruiting bodies of ascomycete *Xylaria euglossa*. J Antibiotics 58:268–270
57. Holler U, Konig GM, Wright AD (1999) Three new metabolites from marine-derived fungi of the genera coniothyrium and microsphaeropsis. J Nat Prod 62:114–118
58. Oh H, Swenson DC, Gloer JB, Shearer CA (2001) Massarilactones A and B: novel secondary metabolites from the freshwater aquatic fungus *Massarina tunicate*. Tetrahedron Lett 42:975–977
59. Oh H, Swenson DC, Gloer JB, Shearer CA (2003) New bioactive rosigenin analogues and aromatic polyketide metabolites from the freshwater aquatic fungus *Massarina tunicate*. J Nat Prod 66:73–79
60. Li C, Nitka MV, Gloer JB (2003) Annularins A–H: new polyketide metabolites from the freshwater aquatic fungus *Annulatascus triseptatus*. J Nat Prod 66:1302–1306
61. Smith CJ, Abbanat D, Bernan VS, Maiese WM, Greenstein M, Jompa J, Tahir A, Ireland CM (2000) Novel polyketide metabolites from a species of marine fungi. J Nat Prod 63:142–145
62. Liu Z, Jensen PR, Fenical W (2003) A cyclic carbonate and related polyketides from a marine-derived fungus of the genus *Phoma*. Phytochemistry 64:571–574
63. Osterhage C, Schwibbe M, König GM, Wright AD (2000) Differences between marine and terrestrial *Phoma* species as determined by HPLC-DAD and HPLC-MS. Phytochem Anal 11:288–294
64. Kock I, Krohn K, Egold H, Draeger S, Schulz B, Rheinheimer J (2007) New massarilactones, Massarigenin E, and coniothyrenol, isolated from the endophytic fungus *Coniothyrium* sp. from *Carpobrotus edulis*. Eur J Org Chem 2186:2186–2190
65. Krohn K, Ullah Z, Hussain H, Fláorke U, Schulz B, Draeger S, Pescitelli G, Salvadori P, Antus S, Kurtán T (2007) Massarilactones E–G, new metabolites from the endophytic fungus *Coniothyrium* sp., associated with the plant *Artimisia maritime*. Chirality 19:464–470
66. Davis RA, Andjic V, Kotiw M, Shivas RG (2005) Phomoxins B and C: polyketides from an endophytic fungus of the genus *Eupenicillium*. Phytochemistry 66:2771–2775
67. Brady SF, Clardy J (2000) CR377, a new pentaketide antifungal agent isolated from an endophytic fungus. J Nat Prod 63:1447–1448
68. Jayasinghe L, Abbas HK, Jacob MR, Herath WHMW, Nanayakkara NPD (2006) *N*-Methyl-4-hydroxy-2-pyridinone analogues from *Fusarium oxysporum*. J Nat Prod 69:439–442
69. Wu X, Liu X, Jiang G, Lin Y, Chan W, Vrijmoed LLP (2005) Xyloketal G, a novel metabolite from the mangrove fungus *Xylaria sp.* 2508. Chem Nat Compd 41:27–29
70. Liu X, Xu F, Zhang Y, Liu L, Huang H, Cai X, Lin Y, Chan W (2006) Xyloketal H from the mangrove endophytic fungus *Xylaria* sp. 2508. Russ Chem Bull 55:1091–1092
71. Pettigrew JD, Bexrud JA, Freeman RP, Wilson PD (2004) Total synthesis of (±)-xyloketal D and model studies towards the total synthesis of (–)-xyloketal A. Heterocycles 62:445–452
72. Pettigrew JD, Wilson PD (2006) Synthesis of xyloketal A, B, C, D, and G analogues. J Org Chem 71:1620–1625
73. Qin XD, Liu JK (2004) Natural aromatic steroids as potential molecular fossils from the fruiting bodies of the ascomycete *Daldinia concentrica*. J Nat Prod 67:2133–2135
74. Wang F, Liu JK (2005) Two new steryl esters from the basidiomycete *Tricholomopsis rutilans*. Steroids 70:127–130
75. Zhang Y, Mill GL, Nair MG (2003) Cyclooxygenase inhibitory and antioxidant compounds from the fruiting body of an edible mushroom, *Agrocybe aegerita*. Phytomedicine 10:386–390

76. Gao JM, Zhang AL,Wang CY,Liu JK (2002) Constituents from the basidiomycetes *Paxillus panuoides*. Acta Botanica Boreali-Occidentalia Sin 22:391–395
77. Gao J, Hu L, Liu J (2001) A novel sterol from Chinese truffles *Tuber indicum*. Steroids 66:771–775
78. Wu SH, Luo XD, Ma YB, Liu JK, Wu DG, Zhao B, Lu Y, Zheng QT (2000) Two Novel Secoergosterols from the Fungus *Tylopilus plumbeoviolaceus*. J Nat Prod 63:534–536
79. Barrero AF, Oltra JE, Poyatos JA, Jiménez D, Oliver E (1998) Phycomysterols and other sterols from the fungus *Phycomyces blakesleeanus*. J Nat Prod 61:1491–1496
80. Lu H, Zou WX, Meng JC, Hu J, Tan RX (2000) New bioactive metabolites produced by *Colletotrichum* sp., an endophytic fungus in *Artemisia annua*. Plant Sci 151:67–73
81. Bok JW, Lermer L, Chilton J, Klingeman HG, Towers GH (1999) Antitumor sterols from the mycelia of *Cordyceps sinensis*. Phytochemistry 51:891–898
82. Nam KS, Jo YS, Kim YH, Hyun JW, Kim HW (2001) Cytotoxic activities of acetoxyscirpenediol and ergosterol peroxide from *Paecilomyces tenuipes*. Life Sci 69:229–237
83. Oh GS, Hong KH, Oh H, Pae HO, Kim IK, Kim NY, Kwon TO, Shin MK, Chung HT (2001) 4-Acetyl-12,13-epoxyl-9-trichothecene-3,15-diol isolated from the fruiting bodies of *Isaria japonica* YASUDA induces apoptosis of human leukemia cells (HL-60). Biol Pharm Bull 24:785–789
84. Stierle A, Strobel G, Stierle D (1993) Taxol and taxane production by Taxomyces andreanae, an endophytic fungus of Pacific yew. Science 260:214–216
85. Strobel G, Yang X, Sears J, Kramer R, Sidhu RS, Hess WM (1996) Taxol from *Pestalotiopsis microspora*, an endophytic fungus of *Taxus wallachiana*. Microbiology 142:435–440
86. Shrestha K, Strobel GA, Shrivastava SP, Gewali M (2001) Evidence for paclitaxel from three new endophytic fungi of Himalayan yew of Nepal. Planta Med 67:374–376
87. Wang J, Li G, Lu H, Zheng Z, Huang Y, Su W (2000) Taxol from Tubercularia sp. strain TF5, an endophytic fungus of Taxus mairei. FEMS Microbiol Lett 193:249–253
88. Zhou DP, Ping WX, Sun JQ, Zhou XH, Liu XL, Yang DZ, Zhang JP, Zheng XQ (2001) Isolation of taxol producing fungi. J Microbiol 21:18–20 (in Chinese)
89. Zhou DP, Sun JQ, Yu HY, Ping WX, Zheng XQ (2001) Nodulisporium,a genus new to China. Mycosystema 20:277–278
90. Li JY, Strobel G, Sidhu R, Hess WM, Ford EJ (1996) Endophytic taxol-producing fungi from bald cypress, Taxodium distichum. Microbiology 142:2223–2226
91. Strobel GA, Ford E, Li JY, Sears J, Sidhu RS, Hess WM (1999) Seimatoantlerium tepuiense gen. nov., a unique epiphytic fungus producing taxol from the Venezuelan Guyana. Syst Appl Microbiol 22:426–433
92. Strobel G, Daisy B (2003) Bioprospecting for microbial endophytes and their natural products. Microbiol Mol Biol Rev 67:491–502
93. Shibata H, Tokunaga T, Karaswa D, Hirota A, Nakayama M, Nozaki H, Tada T (1989) Isolation and characterization of new bitter diterpenoids from the fungus Sarcodon scabrosus. Agric Biol Chem 53:3373–3375
94. Ohta T, Kita T, Kobayashi N, Obara Y, Nakahata N, Ohizumi Y, Takaya Y, Oshima Y (1998) Scabronine A, a novel diterpenoid having potent inductive activity of the nerve growth factor synthesis, isolated from the mushroom, *Sarcodon scabrosus*. Tetrahedron Lett 39: 6229–6232
95. Kita T, Takaya Y, Oshima Y (1998) Scabronines B, C, D, E and F, novel diterpenoids showing stimulating activity of nerve growth factor-synthesis, from the mushroom Sarcodon scabrosus. Tetrahedron 54:11877–11886
96. Ma BJ, Zhu HJ, Liu JK (2004) Isolation and characterization of new bitter diterpenoids from the basidiomycete Sarcodon scabrosus. Helv Chim Acta 87:2877–2881
97. Ma BJ, Liu JK (2005) A new bitter diterpenoid from *Sarcodon scabrosus*. J Basic Microbiol 45:328–330
98. Bills GF, Polishook JD, Goetz MA, Sullivan RF, Jr JFW (2002) *Chaunopycnis pustulata* sp. nov., a new clavicipitalean anamorph producing metabolites that modulate potassium ion channels. Mycol Prog 1:3–17

99. Tang W, Gu T, Zhong JJ (2006) Separation of targeted ganoderic acids from *Ganoderma lucidum* by reversed phase liquid chromatography with ultraviolet and mass spectrometry detections. Biochem Eng J 32:205–210
100. Tang W, Liu JW, Zhao WM, Wei DZ, Zhong JJ (2006) Ganoderic acid T from Ganoderma lucidum mycelia induces mitochondria mediated apoptosis in lung cancer cells. Life Sci 80:205–211
101. Wu TS, Shi LS, Kuo SC (2001) Cytotoxicity of *Ganoderma lucidum* triterpenes. J Nat Prod 64:1121–1122
102. Gao JJ, Min BS, Ahn EM, Nakamura N, Lee HK, Hattori M (2002) New triterpene aldehydes, lucialdehydes A–C, from *Ganoderma lucidum* and their cytotoxicity against murine and human tumor cells. Chem Pharm Bull 50:837–840
103. Akihisa T, Nakamura Y, Tokuda H, Uchiyama E, Suzuki T, Kimura Y, Uchikura K, Nishino H (2007) Triterpene acids from *Poria cocos* and their anti-tumor-promoting effects. J Nat Prod 70:948–953
104. Quang DN, Hashimoto T, Tanaka M, Takaoka S, Asakawa Y (2004) Tyromycic acids B–E, new lanostane triterpenoids from the mushroom *Tyromyces fissilis*. J Nat Prod 67:148–151
105. Quang DN, Hashimoto T, Tanaka M, Asakawa Y (2003) Tyromycic acids F and G: two new triterpenoids from the mushroom *Tyromyces fissilis*. Chem Pharm Bull 51:1441–1443
106. Shao HJ, Qing C, Wang F, Zhang YL, Luo DQ, Liu JK (2005) A new cytotoxic lanostane triterpenoid from the basidiomycete *Hebeloma versipelle*. J Antibiotics 58:828–831
107. Clericuzio M, Mella M, Vita-Finzi P, Zema M, Vidari G (2004) Cucurbitane triterpenoids from *Leucopaxillus gentianeus*. J Nat Prod 67:1823–1828
108. Clericuzio M, Tabasso S, Bianco MA, Pratesi G, Beretta G, Tinelli S, Zunino F, Vidari G (2006) Cucurbitane triterpenes from the fruiting bodies and cultivated mycelia of *Leucopaxillus gentianeus*. J Nat Prod 69:1796–1799
109. Hashimoto T, Asakawa Y (1998) Biologically active substances of Japanese inedible mushrooms. Heterocycles 47:1110–1121
110. Stadler M, Baumgartner M, Grothe T, Mühlbauer A, Seip S, Wollweber H (2001) Concentricol, a taxonomically significant triterpenoid from *Daldinia concentrica*. Phytochemistry 56:787–793
111. Quang DN, Hashimoto T, Tanaka M, Baumgartnerc M, Stadler M, Asakawa Y (2002) Concentriols B, C and D, three squalene-type triterpenoids from the ascomycete *Daldinia concentrica*. Phytochemistry 61:345–353
112. Stadler M, Wollweber H, Mühlbauer A, Henkel T, Asakawa Y, Hashimoto T, Rogers JD, Ju YM, Wetzstein HG, Tichy HV (2001) Secondary metabolite profiles, genetic fingerprints and taxonomy of *Daldinia* and allies. Mycotaxon 77:379–429
113. Deyrup ST, Gloer JB, O'Donnell K, Wicklow DT (2007) Kolokosides A–D: triterpenoid glycosides from a Hawaiian isolate of *Xylaria sp*. J Nat Prod 70:378–382
114. Tan JW, Dong ZJ, Liu JK (2000) New terpenoids from basidiomycetes *Russula lepida*. Helv Chim Acta 83:3191–3197
115. Tan JW, Dong ZJ, Liu JK (2001) A new sesquiterpenoid from *Russula lepida*. Acta Botanica Sin 43:329–330 (in Chinese)
116. Tan J, Dong Z, Hu L, Liu J (2003) Lepidamine, the first aristolane-type sesquiterpene alkaloid from the basidiomycete *Russula lepida*. Helv Chim Acta 86:307–309
117. Tan JW, Dong ZJ, Ding ZH, Liu JK (2002) Lepidolide, a novel *seco*-ring-A cucurbitane triterpenoid from *Russula lepida* (basidiomycetes). Z Naturforsch 57c:963–965
118. Gao JM, Dong ZJ, Liu JK (2001) A new ceramide from the basidiomycete *Russula cyanoxantha*. Lipids 36:175–180
119. Tan JW, Xu JB, Dong ZJ, Luo DQ, Liu JK (2004) Nigricanin, the first ellagic acid derived metabolite from the basidiomycete *Russula nigricans*. Helv Chim Acta 87:1025–1028
120. Luo DQ, Wang F, Bian XY, Liu JK (2005) Rufuslactone, a new antifungal sesquiterpene from the fruiting bodies of the basidiomycete *Lactarius rufus*. J Antibiotics 58:456–459
121. Hu L, Liu JK (2002) The first humulene type sesquiterpene from *Lactarius hirtipes*. Z Naturforsch 57c:571–574

122. Luo DQ, Gao Y, Gao JM, Wang F, Yang XL, Liu JK (2006) Humulane-type sesquiterpenoids from the mushroom *Lactarius mitissimus*. J Nat Prod 69:1354–1357
123. Luo DQ, Shao HJ, Zhu HJ, Liu JK (2005) Activity in vitro and in vivo against plant pathogenic fungi of grifolin isolated from the basidiomycete *Albatrellus dispansus*. Z Naturforsch 60c:50–56
124. Ye M, Liu JK, Lu ZX, Zhao Y, Liu SF, Li LL, Tan M,Weng XX, Li W, Cao Y (2005) Grifolin, a potential antitumor natural product from the mushroom *Albatrellus confluens*, inhibits tumor cell growth by inducing apoptosis in vitro. FEBS Lett 579:3437–3443
125. Nukata M, Hashimoto T, Yamamoto I, Iwasaki N, Tanaka M, Asakawa Y (2002) Neogrifolin derivatives possessing anti-oxidative activity from the mushroom *Albatrellus ovinus*. Phytochemistry 59:731–737
126. Quang DN, Hashimoto T, Arakawa Y, Kohchi C, Nishizawa T, Soma G, Asakawa Y (2006) Grifolin derivatives from *Albatrellus caeruleoporus*, new inhibitors of nitric oxide production in RAW 264.7 cells. Bioorg Med Chem. 14:164–168
127. Ding ZH, Dong ZJ, Liu JK (2001) Albaconol, A novel prenylated resorcinol (benzene-1,3-diol) from basidiomycetes *Albatrellus confluens*. Helv Chim Acta 84:259–262
128. Liu MH, Qing C, Liu JK, Zhang YL, Wang L, Ji SY (2004) Study on the antitumor activity of albaconol and its effect on DNA topoisomerase II. Chin Pharmacol Bull 20:1224–1228 (in Chinese)
129. Smetanina OF, Kuznetsova TA, Gerasimenko AV, Kalinovsky AI, Pivkin MV, Dmitrenok PC, Elyakov GB (2004) Metabolites of the marine fungus *Humicola fuscoatra* KMM 4629. Russ Chem Bull 53:2643–2646
130. Oh H, Gloer JB, Shearer CA (1999) Massarinolins A-C: new bioactive sesquiterpenoids from the aquatic fungus *Massarina tunicate*. J Nat Prod 62:497–501.
131. Xiao JH, Zhong JJ (2007) Secondary metabolites from *Cordyceps* species and their antitumor activity studies. Recent Patents Biotechnol 1:123–137
132. Jow GM, Chou CJ, Chen BF, Tsai JH (2004) Beauvericin induces cytotoxic effects in human acute lymphoblastic leukemia cells through cytochrome *c* release, caspase 3 activation: the causative role of calcium. Cancer Lett 216:165–173
133. Nilanonta C, Isaka M, Kittakoop P, Trakulnaleamsai S, Tanticharoen M, Thebtaranonth Y (2002) Precursor-directed biosynthesis of beauvericin analogs by the insect pathogenic fungus *Paecilomyces tenuipes* BBC 1614. Tetrahedron 58:3355–3360
134. Wu SN, Chen H, Liu YC, Chiang HT (2002) Block of L-type Ca2+ current by bauvericin, a toxic cyclopeptide, in the NG108-15 neuronal cell line. Chem Res Toxicol 15:854–860
135. Jia JM, Ma XC, Wu CF, Wu LJ, Hu GS (2005) Cordycedipeptide A, a new cyclodipeptide from the culture liquid. Chem Pharm Bull 53:582–583
136. Rukachaisirikul V, Chantaruk S, Tansakul C, et al. (2006) A cyclopeptide from the insect pathogenic fungus *Cordyceps* sp. BCC 1788. J Nat Prod 69:305–307
137. Krasnoff SB, Reátegui RF, Wagenaar MM, Gloer JB, Gibson DM (2005) Cicadapeptins I and II: new aib-containing peptides from the entomopathogenic fungus *Cordyceps heteropoda*. J Nat Prod 68:50–55
138. Ravindra G, Ranganayaki RS, Raghothama S, Srinivasan MC, Gilardi RD, Karle IL, Balaram P (2004) Two novel hexadepsipeptides with several modified amino acid residues isolated from the fungus Isaria. Chem Biodivers 1:489–504
139. Sabareesh V, Ranganayaki RS, Raghothama S, Bopanna MP, Balaram H, Srinivasan MC, Balaram P (2007) Identification and characterization of a library of microheterogeneous cyclohexadepsipeptides from the fungus *Isaria*. J Nat Prod 70:715–729
140. Degenkolb T, Berg A, Gams W, Schlegel B, Gräfe U (2003) The occurrence of peptaibols and structurally related peptaibiotics in fungi and their mass spectrometric identification via diagnostic fragment ions. J Pept Sci 9:666–678
141. Chugh JK, Wallace BA (2001) Peptaibols: models for ion channels. Biochem Soc Trans 29:565–570
142. Summers MY, Kong F, Feng X, Siegel MM, Janso JE, Graziani EI, Carter GT (2007) Septocylindrins A and B: peptaibols produced by the terrestrial fungus *Septocylindrium* sp. LL Z1518. J Nat Prod 70:391 396

143. Xie ST, Song XY, Shi M, Chen XL, Sun CY, Zhang YZ (2006) Antimicrobial activities of trichokonins: peptaibols-like antimicrobial peptides produced by *Trichoderma koningii* SMF2. J Shandong Univ 41(6):140–144 (in Chinese)
144. Liu JK (2006) Natural terphenyls: developments since 1877. Chem Rev 106:2209–2223
145. Quang DN, Hashimoto T, Asakawai Y (2006) Inedible mushrooms: a good source of biologically active substances. Chem Rec 6:79–99
146. Hu L, Liu JK (2001) Two novel phenylacetoxylated *p*-terphenyls from *Thelephora ganbajun*. Zang Z Naturforsch 56c:983–987
147. Hu L, Gao JM, Liu JK (2001) Unusual poly(phenylacetyloxy)-substituted 1,1′:4′,1″-terphenyl derivatives from fruiting bodies of the basidiomycete *Thelephora ganbajun*. Helv Chim Acta 84:3342–3349
148. Yun BS, Lee IK, Kim JP, Yoo ID (2000) Curtisians A and D, new free radical scavengers from the mushroom *Paxillus curtisii*. J Antibiotics 53:114–122
149. Quang DN, Hashimoto T, Nukada M, Yamamoto I, Tanaka M, Asakawa Y (2003) Curtisians E–H: four *p*-terphenyl derivatives from the inedible mushroom *Paxillus curtisii*. Phytochemistry 64:649–654
150. Quang DN, Hashimoto T, Nukada M, Yamamoto I, Tanaka M, Asakawa Y (2003) Antioxidant activity of custisians I–L from the inedible mushroom *Paxillus curtisii*. Planta Medica 69:1063–1066
151. Quang DN, Hashimoto T, Nukada M, Yamamoto I, Tanaka M, Asakawa Y (2003) Curtisians M—Q: five novel *p*-terphenyl derivatives from the mushroom. Paxillus curtisii Chem Pharm Bull 51:1064–1067
152. Ma BJ, Hu Q, Liu JK (2006) A new *p*-terphenyl derivative from fruiting bodies of the basidiomycete *Sarcodon laevigatum*. J Basic Microbiol 46:239–242
153. Gao JM, Zhang AL, Zhang CL, Liu JK (2004) Paxillamide: a novel phytosphingosine derivative from the fruiting bodies of *Paxillus panuoide*s. Helv Chim Acta 87:1483–1487
154. Gao JM, Zhang AL, Wang CY, Wang XH, Liu JK (2002) A new ceramide from the Ascomycete *Tuber indicum*. Chin Chem Lett 13:325–326
155. Gao JM, Zhang AL, Chen H, Liu JK (2004) Molecular species of ceramides from the ascomycete *Tuber indicum*. Chem Phys Lipids 131:205–213
156. Gao JM, Wang CY, Zhang AL, Liu JK (2001) A new trihydroxy fatty acid from the ascomycete, Chinese truffle *Tuber indicum*. Lipids 36:1365–1370
157. Nicholas GM, Blunt JW, Munro MHG (2001) Cortamidine oxide, a novel disulfide metabolite from the New Zealand basidiomycete (mushroom) *Cortinarius* species. J Nat Prod 64:341–344
158. Wang F, Luo DQ, Liu JK (2005) Aurovertin E, a new polyene pyrone from the basidiomycete *Albatrellus confluens*. J Antibiotics 58:412–415
159. Qin XD, Liu JK (2004) Three new homologous 3-alkyl-1,4-benzoquinones from the fruiting bodies of *Daldinia concentrica*. Helv Chim Acta 87:2022–2024
160. Eyberger AL, Dondapati R, Porter JR (2006) Endophyte fungal isolates from *Podophyllum peltatum* produce Podophyllotoxin. J Nat Prod 69:1121–1124
161. Puri SC, Nazir A, Chawla R, Arora R, Riyaz-ul-Hasan S, Amna T, Ahmed B, Verma V, Singh S, Sagar R, Sharma A, Kumar R, Sharma RK, Qazi GN (2006) The endophytic fungus *Trametes hirsuta* as a novel alternative source of podophyllotoxin and related aryl tetralin lignans. J Biotechnol 122:494–510
162. Soman AG, Gloer JB, Koster B, Malloch D (1999) Sporovexins A–C and a new preussomerin analog: antibacterial and antifungal metabolites from the Coprophilous fungus *Sporormiella vexans*. J Nat Prod 62:659–661
163. Chadwick DJ, Whelan J (eds) (1992) Secondary metabolites: their function and evolution. Ciba Foundation Symposium 171. Wiley, Chichester, UK, pp 1–328
164. Demain AL (2006) From natural products discovery to commercialization: a success story. J Ind Microbiol Biotechnol 33:486–495
165. Zjawiony JK (2004) Biologically active compounds from aphyllophorales (Polypore) fungi. J Nat Prod 67:300–310

166. Song XY, Shen QT, Xie ST, Chen XL, Sun CY, Zhang YZ (2006) Broad-spectrum antimicrobial activityand high stability of trichokonins from *Trichoderma koningii* SMF2 against plant pathogens. FEMS Microbiol Lett 260:119–125
167. Vicente MF, Basilio A, Cabello A, Peláez F (2003) Microbial natural products as a source of antifungals. Clin Microbiol Infect 9:15–32
168. Neamati N (2001) Structure-based HIV-1 integrase-inhibitor design: a future perspective. Exp Opin Invest Drugs 10:281–296
169. Singh SB, Jayasuriy H, Dewey R, Polishook JD, Dombrowski AW, Zink DL, Guan Z, Collado J, Platas G, Pelaez F, Felock P, Hazuda DJ (2003) Isolation, structure, and HIV-1-integrase inhibitory activity of structurally diverse fungal metabolites. J Ind Microbiol Biotechnol 30:721–731
170. Kim KM, Kwon YG, Chung HT, Yun YG, Pae HO, Han JA, Ha KS, Kim TW, Kim YM (2003) Methanol extract of *Cordyceps pruinosa* inhibits in vitro and in vivo inflammatory mediators by suppressing NF-kappa B activation. Toxicol Appl Pharm 190:1–8
171. Lipsky LP, Abramson SB, Crofford L, Dubois RN, Simon LS, Van de Putte LB (1998) The classification of cyclooxygenases inhibitors. J Rheumatol 25:2298–2303
172. Smith CJ, Zhang Y, Koboldt CM, Muhammad J, Zweifel BS, Shaffer A, Talley JJ, Masferrer JL, Seibert K, Isakson PC (1998) Pharmacological analysis of cyclooxygenase-1 in inflammation. Proc Natl Acad Sci U S A 95:13313–13318
173. Sun B, Fukuhara M (1997) Effects of co-administration of butylated hydroxytoluene, butylated hydroxyanisole and flavonoids on the activation of mutagens and drug-metabolizing enzymes in mice. Toxicology 122:61–72
174. Yen GC, Wu JY (1999) Antioxidant and radical scavenging properties of extracts from *Ganoderma tsugae*. Food Chem 65:375–379
175. Sun J, He H, Xie BJ (2004) Novel antioxidant peptides from fermented mushroom *Ganoderma lucidum*. J Agric Food Chem 52:6646–6652
176. Fang LZ, Qing C, Shao HJ, Yang YD, Dong ZJ, Wang F, Zhao W, Yang WQ, Liu JK (2006) Hypocrellin D, a cytotoxic fungal pigment from fruiting bodies of the ascomycete *Shiraia bambusicola*. J Antibiot 59:351–354
177. Whitesell L, Lindquist SL (2005) Hsp90 and the chaperoning of cancer. Nat Rev Cancer 5:761–772
178. Chen Y, Ding J (2004) Heat shock protein 90: novel target for cancer therapy. Chin J Cancer 23:968–974 (in Chinese)
179. Demain AL (1999) Pharmaceutically active secondary metabolites of microorganisms. Appl Microbiol Biotechnol 52:455–463
180. Fenteany G, Zhu S (2003) Small-molecule inhibitors of actin dynamics and cell motility. Curr Top Med Chem 3:593–616
181. Carmeliet P (2003) Angiogenesis in health and disease. Nat Med 9:653–660
182. Zhan J, Burns AM, Liu MX, Faeth SH, Gunatilaka AAL (2007) Search for cell motility and angiogenesis inhibitors with potential anticancer activity: beauvericin and other constituents of two endophytic strains of *Fusarium oxysporum*. J Nat Prod 70:227–232
183. Hosoe T, Iizuka T, Komai SI, Wakana D, Itabashi T, Nozawa K, Fukushima K, Kawai KI (2005) 4-Benzyl-3-phenyl-5*H*-furan-2-one, a vasodilator isolated from *Malbranchea filamentosa* IFM 41300. Phytochemistry 66:2776–2779
184. Liu DZ, Wang F, Liao TG, Tang JG, Steglich W, Zhu HJ, Liu JK (2006) Vibralactone: a lipase inhibitor with an unusual fused β-lactone produced by cultures of the basidiomycete *Boreostereum vibrans*. Org Lett 8:5749–5752
185. Sim KL, Perry D (1997) Analysis of swainsonine and its early metabolic precursors in cultures of *Metarhizium anisopliae*. Glycoconjugate J 14:661–668
186. Skrobek A, Shah FA, Butt TM (2008) Destruxin production by the entomogenous fungus *Metarhizium anisopliae* in insects and factors influencing their degradation. Biocontrol 53:361–373
187. Asaff A, Cerda-García-Rojas C, Torre M de la (2005) Isolation of dipicolinic acid as an insecticidal toxin from *Paecilomyces fumosoroseus*. Appl Microbiol Biotechnol 68:542–547

188. Salituro GM, Pelaez F, Zhang BB (2001) Discovery of a small molecule insulin receptor activator.Recent Prog Horm Res 56:107–126
189. Wagner R, Mitchell DA, Sassaki GL, de Almeida Amazonas MAL, Berovic M (2003) Current techniques for the cultivation of *Ganoderma lucidum* for the production of biomass, ganoderic acid and polysaccharides. Food Technol Biotechnol 41:371–382
190. Zhong JJ, Tang YJ (2004) Submerged cultivation of medicinal mushrooms for production of valuable bioactive metabolites. Adv Biochem Eng Biotechnol 87:25–59
191. Tang YJ, Zhu LW, Li HM, Li DS (2007) Submerged culture of mushrooms in bioreactors – challenges, current state-of-the-art, and future prospects. Food Technol Biotechnol 45:221–229
192. Tang YJ, Zhu LL, Li DS, Mi ZY, Li HM (2008) Significance of inoculation density and carbon source on the mycelial growth and tuber polysaccharides production by submerged fermentation of Chinese truffle *Tuber sinense*. Process Biochem 43:576–586
193. Liu RS, Li DS, Li HM, Tang YJ (2008) Response surface modeling the significance of nitrogen source on the cell growth and tuber polysaccharides production by submerged cultivation of Chinese truffle *Tuber sinense*. Process Biochem 43:868–876
194. Tang YJ, Zhu LL, Liu RS, Li HM, Li DS, Mi ZY (2008) Quantitative response of cell growth and tuber polysaccharides biosynthesis by medicinal mushroom Chinese truffle *Tuber sinense* to metal ion in culture medium. Bioresour Technol 99:7606–7615
195. Wang G, Li YY, Li DS, Tang YJ (2008) Determination of 5α-androst-16-en-3α-ol in truffle fermentation broth by solid-phase extraction coupled with gas chromatography-flame ionization detector/electron impact mass spectrometry. J Chromatogr B 870:209–215
196. Wang G, Zhao J, Liu JW, Huang YP, Tang W, Zhong JJ (2007) Enhancement of IL-2 and IFN-γ expression and NK cells activity involved in the anti-tumor effect of ganoderic acid Me in vivo. Int Immunopharmacol 7:864–870
197. Chen NH, Liu JW, Zhong JJ (2008) Ganoderic acid Me inhibits tumor invasion through downregulating matrix metalloproteinases 2/9 gene expression. J Pharmacol Sci 108:212–216
198. El-Mekkawy S, Meselhy MR, Nakamura N, Tezuka Y, Hattori M, Kakiuchi N, Shimotohno K, Kawahata T, Otake T (1998) Anti-HIV-1 and anti-HIV-1-protease substances from *Ganoderma lucidum*. Phytochemistry 49:1651–1657
199. Fang QH, Zhong JJ (2002) Effect of initial pH on production of ganoderic acid and polysaccharide by submerged fermentation of *Ganoderma lucidum*. Process Biochem 37:769–774
200. Tang YJ, Zhong JJ (2003) Role of oxygen supply in submerged fermentation of *Ganoderma lucidum* for production of *Ganoderma* polysaccharide and ganoderic acid. Enzyme Microb Technol 32:478–484
201. Fang QH, Zhong JJ (2002) Two-stage culture process for improved production of ganoderic acid by liquid fermentation of higher fungus *Ganoderma lucidum*. Biotechnol Prog 18:51–54
202. Tang YJ, Zhong JJ (2002) Fed-batch fermentation of *Ganoderma lucidum* for hyperproduction of polysaccharide and ganoderic acid. Enzyme Microb Technol 31:20–28
203. Tang YJ, Zhong JJ (2003) Scale-up of a liquid static culture process for hyperproduction of ganoderic acid by the medicinal mushroom *Ganoderma lucidum*. Biotechnol Prog 19:1842–1846
204. Mao XB, Zhong JJ (2004) Hyperproduction of cordycepin by two-stage dissolved oxygen control in submerged cultivation of medicinal mushroom *Cordyceps militaris* in bioreactors. Biotechnol Prog 20:1408–1413
205. Li Y, Dong XY, Sun Y (2005) High-speed chromatographic purification of plasmid DNA with a customized porous hydrophobic adsorbent. Biochem Eng J 27:33–39
206. Harsa S, Furusaki S (2001) Chromatographic separation of amyloglucosidase from the mixtures of enzymes. Biochem Eng J 8:257–261
207 Schwarz M, Hillebrand S, Habben S, Degenhardt A, Winterhalter P (2003) Application of high-speed countercurrent chromatography to the large-scale isolation of anthocyanins. Biochem Eng J 14:179–189
208. Kikuchi T, Kanomi S, Kadota S, Murai Y, Tsubono K, Ogita Z (1986) Constituents of the fungus Ganoderma lucidum (F.R.) Karst. 1. Structures of ganoderic acid-C2, acid-E, acid-I, AND acid-K, lucidenic acid-F AND related-compounds. Chem Pharm Bull 34:3695–3712

209. Lin L, Shiao S (1987) Separation of oxygenated triterpenoids from *Ganoderma lucidum* by high performance liquid chromatography. J Chromatogr A 410:195–200
210. Shiao M, Lin L, Chen C (1989) Determination of stereo and positional isomers of oxygenated triterpenoids by reversed phase high performance liquid chromatography. J Lipid Res 30:287–291
211. Mutlib A, Strupczewski J, Chesson S (1995) Application of. hyphenated LC/NMR and LC/MS techniques in rapid identification in vitro and in vivo metabolites of iloperidone. Drug Metab Dispos 23:951–964
212. Mutlib A, Klein J (1998) Application of liquid chromatography/mass spectrometry in accelerating the identification of human liver cytochrome P450 isoforms involved in the metabolism of iloperidone. J Pharmacol Exp Ther 286:1285–1293

Adv Biochem Engin/Biotechnol (2009) 113: 151-176
DOI: 10.1007/10_2008_31

Published online: 17 April 2009

Large Scale Culture of Ginseng Adventitious Roots for Production of Ginsenosides

Kee-Yoeup Paek, Hosakatte Niranjana Murthy, Eun-Joo Hahn, and Jian-Jiang Zhong

Abstract Ginseng (*Panax ginseng* C. A. Meyer) is one of the most famous oriental medicinal plants used as crude drugs in Asian countries, and now it is being used worldwide for preventive and therapeutic purposes. Among diverse constituents of ginseng, saponins (ginsenosides) have been found to be major components responsible for their biological and pharmacological actions. On the other hand, difficulties in the supply of pure ginsenosides in quantity prevent the development of ginseng for clinical medicines. Cultivation of ginseng in fields takes a long time, generally 5–7 years, and needs extensive effort regarding quality control since growth is susceptible to many environmental factors including soil, shade, climate, pathogens and pests. To solve the problems, cell and tissue cultures have been widely explored for more rapid and efficient production of ginseng biomass and ginsenosides. Recently, cell and adventitious root cultures of *P. ginseng* have been established in large scale bioreactors with a view to commercial application. Various physiological and engineering parameters affecting the biomass production and ginsenoside accumulation have been investigated. Advances in adventitious root cultures including factors for process scale-up are reviewed in this chapter. In addition, biosafety analyses of ginseng adventitious roots are also discussed for real application.

K.-Y. Paek (✉), H.N. Murthy, and E.-J. Hahn
Research Center for the Development of Advanced Horticultural Technology, Chungbuk National University Cheongju 361-763, Republic of Korea
e-mail: paekky@cbnu.ac.kr

H.N. Murthy
Department of Botany, Karnatak University, Dharwad – 580 003 India

J.-J. Zhong (✉)
School of Life Sciences and Biotechnology, Key Laboratory of Microbial Metabolism (Ministry of Education), Shanghai Jiao Tong University, 800 Dong-Chuan Road, Shanghai 200240, China
e-mail: jjzhong@sjtu.edu.cn

Keywords Adventitious roots, Bioreactor cultures, Ginseng, Ginsenosides, *Panax ginseng*, Suspension cultures

Contents

Abbreviations

2, 4-D	2, 4-Dichloraophenoxy acetic acid
DW	Dry weight
FW	Fresh weight
IAA	Indole-3-acetic acidM
IBA	Indole-3-butyric acid
MJ	Methyl jasmonate
MS medium	Murashige and Skoog medium
NAA	α-Naphthalene acetic acid
vvm	Air volume per medium volume per minute

1 Introduction

Ginseng (*Panax ginseng* C. A. Meyer; family Araliaceae), a valuable medicinal plant of oriental region (China, Japan and Korea), has been used as a healing drug and health tonic since ancient time [1]. Ginseng has various bioactive effects on human health such as antitumor, antineoplastic, antimitogenic and antistress, antiaging, and immunomodulation activities [2]. Recent years have witnessed a tremendous resurgence in the interest and use of medicinal plant products and nutraceuticals (functional foods), especially in North America [3]. *P. ginseng* figures prominently in sales of botanical products in the United States, which totals about $3 billion annually [4]. The major secondary metabolites in ginseng roots are an array of triterpene saponins, collectively called ginsenosides [5, 6], which are glycosylated derivatives of two major aglycones, panaxadiol and panaxatriol. At present, 30 ginsenosides have been identified of which the ginsenosides Rb1, Rb2, Rc, Rd, Re, Rf, Rg1, and Rg2 (Fig. 1) are considered to be most relevant for pharmacological activities [5, 7].

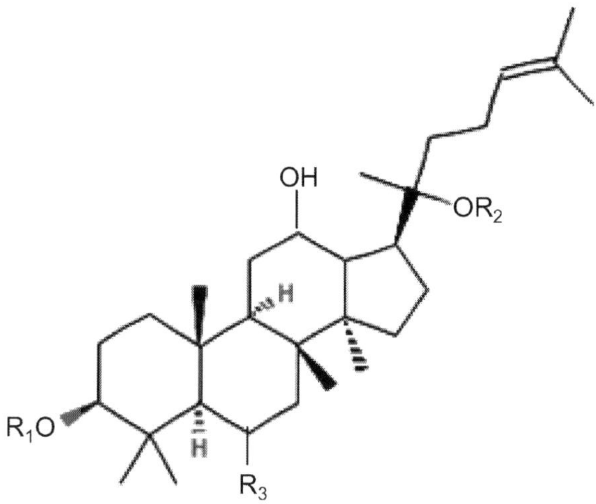

Fig. 1 Structure of ginsenosides in the roots of ginseng

Nowadays, wild ginseng has become extremely scarce and the ginseng supply depends almost extensively on field cultivation which is a time-consuming and labor-intensive process [8]. It takes 5–7 years from sowing to harvest, during which much care is needed since plant growth is susceptible to many environmental factors including soil, shade, climate, pathogens and pests. There has been extensive work on cell and organ culture for more rapid, stable and efficient production of ginseng biomass and ginsenosides, which has resulted in remarkable progress [9]. However, large scale and commercial production of ginsenosides through cell cultures were hampered by low yield and productivity [9]. In recent years new approaches have been developed, e.g., culturing differentiated cells, especially adventitious roots [10, 11]. Adventitious root cultures demonstrated the ability of higher accumulation of biomass and desired compounds in levels compared to natural roots. Furthermore, adventitious roots have a high stability against physical and chemical conditions during large-scale cultures. We have been working on suspension cultures of adventitious roots of Korean ginseng (*P. ginseng* C. A. Meyer) in bioreactors during the past decade. As a result, various bioprocessing methodologies were successfully developed for the production of root biomass and ginsenosides. This review highlights advances on adventitious root cultures, ginsenoside production and commercialization of *P. ginseng* C.A. Meyer.

1.1 Induction of Adventitious Roots and Establishment of Suspension Cultures

Calli were induced from *P. ginseng* C. A. Meyer roots, collected from the wild, on Murashige and Skoog (MS) semi-solid medium supplemented with 4.52 μM 2,4-dichlorophenxyacetic acid (2,4-d), 0.46 μM kinetin and 3% sucrose in the dark at 25 °C [12]. Adventitious roots were induced from the calli on MS medium supplemented with 24.6 µM indole butyric acid (IBA) and maintained in MS liquid medium supplemented with 24.6 µM IBA.

1.2 Adventitious Root Cultures in Airlift Bioreactors

1.2.1 Effect of Growth Regulators on Proliferation of Adventitious Roots

Proliferation of adventitious roots in bioreactors (5 and 20 L capacity airlift bioreactors; Fig. 2) depends on the growth regulators supplemented to the culture medium. Among four growth regulators tested (2, 4-D, IBA, IAA and NAA with 4.9, 9.8, 14.8, 19.7, 24.6, 34.4, 44.3 and 49.3 µM, respectively), 24.6 µM IBA and 9.8 µM NAA were found best for lateral root induction and proliferation. Further experiments revealed that IBA (24.6 µM; Fig. 3a) was more suitable for the induction and proliferation of adventitious roots compared to NAA (9.8 µM) since the roots on NAA containing medium were thick and developed transversely, which resulted in a lower proliferation rate (Fig. 4; [12]). The roots on IBA supplemented medium were higher in their fresh and dry biomass (Fig. 3b, c). Higher content of total ginsenosides (8.09 ± 0.30 mg g^{-1} DW) was also obtained in IBA containing medium compared to NAA containing medium (6.26 ± 0.45 mg g^{-1} DW; Table 1).

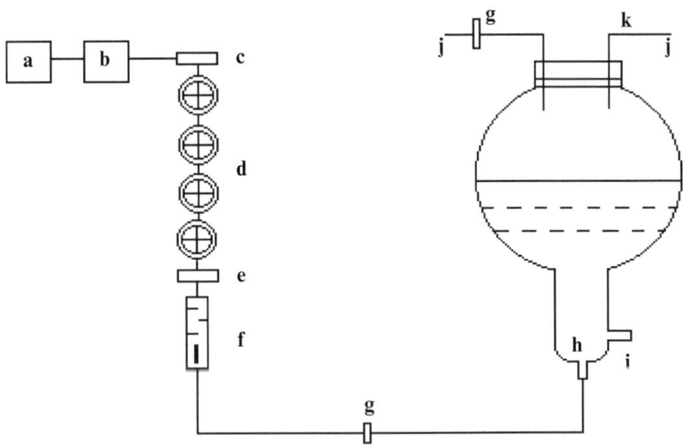

Fig. 2 Schematic diagram of an air-lift bioreactor: (a) air compressor, (b) air reservoir, (c) air cooling device, (d) air filter system, (e) air dryer, (f) air flow meter, (g) membrane filter, (h) glass sparger, (i) medium sampling port, (j) vent, (k) pre filter

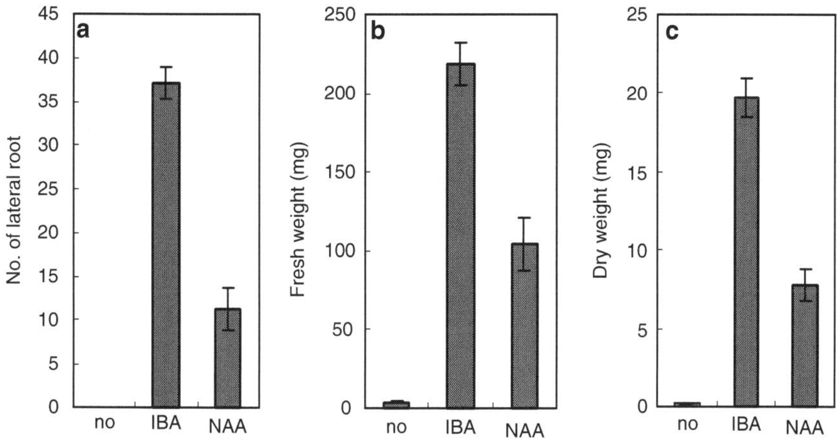

Fig. 3a–c Effects of auxins on the number of lateral roots (**a**), fresh weight of lateral roots (**b**), and dry weight of lateral roots (**c**), in *Panax ginseng* cultured for 28 days on MS semi-solid medium

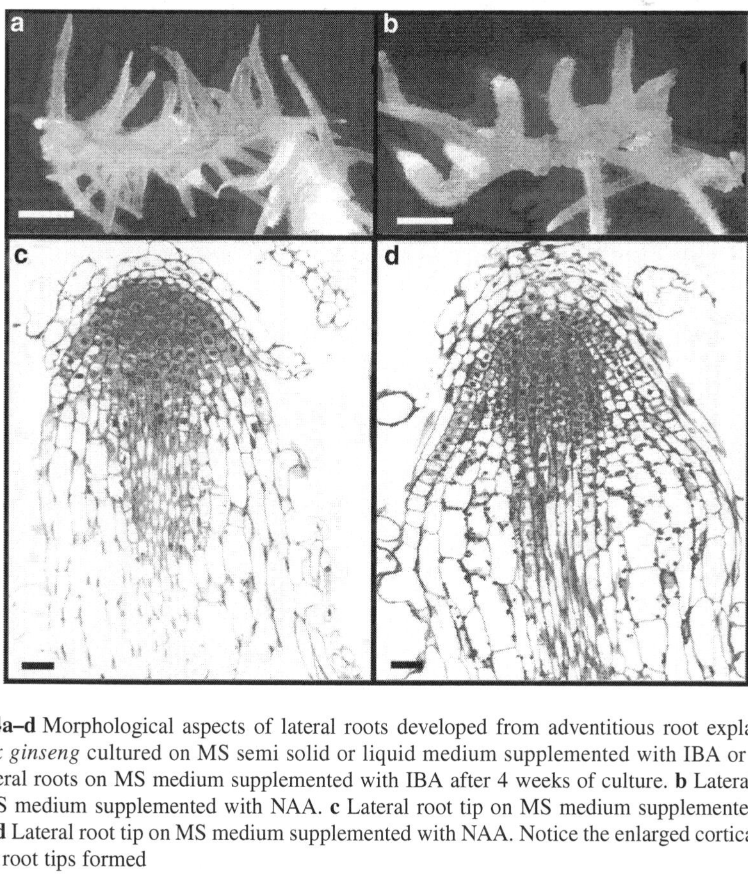

Fig. 4a–d Morphological aspects of lateral roots developed from adventitious root explants of *Panax ginseng* cultured on MS semi solid or liquid medium supplemented with IBA or NAA. **a** Lateral roots on MS medium supplemented with IBA after 4 weeks of culture. **b** Lateral roots on MS medium supplemented with NAA. **c** Lateral root tip on MS medium supplemented with IBA. **d** Lateral root tip on MS medium supplemented with NAA. Notice the enlarged cortical cells in the root tips formed

Table 1 Ginsenoside contents in auxin-induced lateral roots of *Panax ginseng*

Ginsenoside content (mg g^{-1} DW)				
Auxin[a]	Protopanaxadiol-type ginsenoside[b]	Protopanaxtriol-type ginsenoside[c]	Total[d]	Protopanaxdiol/protopanaxatriol
IBA	4.68	3.41	8.09 ± 0.30	1.37
NAA	3.89	2.37	6.26 ± 0.45	1.64

[a]24.6 μM IBA or 9.8 μM NAA was added to culture medium
[b]Protopanaxadiol-type ginsenoside: RB1 +Rb2 + Rc + Rd
[c]Protopanaxtriol-type ginsenoside: Re +Rf + Rg
[d]Total ginsenoside content represents mean ± standard error of three replicates

Table 2 Effect of ammonium/nitrate ratios (NH_4^+/NO_3^-; mM) in medium on growth of ginseng adventitious roots and ginsenoside production. The roots were cultured for 5 weeks in 10-L balloon type bubble bioreactors containing 5 L MS medium with different concentrations of NH_4NO_3/KNO_3 (mM)

Concentration of NH_4^+/NO_3^- (mM)	Biomass (DW g^{-1})	Growth ratio[a]	Residual NO_3^- (mM)	Ginsenosides (mg g^{-1} DW)	Ginsenoside yield (mg L^{-1})
0.0: 18.5	42.3	30.2	1.60	9.89 ± 0.12[b]	83.6
7.19: 18.5	47.5	33.9	0.39	7.94 ± 0.26	75.4
14.38: 18.5	40.4	28.9	7.02	6.32 ± 0.04	51.1
21.57: 18.5	37.7	37.7	7.69	5.36 ± 0.24	40.4
28.75: 18.5	35.9	25.6	25.61	4.56 ± 0.32	32.7
14.38: 0.0	13.6	9.72	0.50	3.22 ± 0.44	8.76
14.38: 9.4	42.4	30.3	4.50	6.26 ± 0.54	59.5
14.38: 18.8	45.6	32.6	12.49	6.52 ± 0.14	53.1
14.38: 26.2	44.8	32.0	29.01	5.25 ± 0.52	47.0
14.38: 37.6	40.1	28.7	35.05	4.58 ± 0.42	36.8

[a]Growth ratio was calculated from increase in DW. Values are quotients of DW after culture and DW of the inoculum
[b]Mean ± standard error of three replicates

Plant growth regulators are one of the key factors influencing biomass increase and secondary metabolite production. For ginseng callus induction and cell growth, 2,4-D has been most commonly used. Choi et al. [13, 14] reported that 2,4-D in the range of 1–5 mg L^{-1} was essential for cell growth, while 0.1 mg L^{-1} was essential for ginsenoside synthesis. In cell suspension cultures of *Panax quinauefolium*, Zhang et al. [15] demonstrated that not only individual growth regulators' level but also their combination have significant influence on cell growth and ginsenoside accumulation. Similar results were also reported from long-term cell cultures of *P. ginseng* [16]. In a large-scale culture process, NAA and kinetin were used as growth regulators instead of 2,4-D, perhaps because of health and safety concerns about this carcinogen [14, 17]. We have tested various collections of mountain ginseng (*P. ginseng* C. A. Meyer) to induce adventitious roots, which suggested that 2,4-D was needed only for callus induction. Induction and proliferation of adventitious roots were conducted on the medium supplemented with 24.6 μM IBA only.

1.2.2 Effect of Ammonium/Nitrate Ratios on Adventitious Root Growth and Ginsenoside Accumulation

Table 2 illustrates the effects of ammonium/nitrate ratios in MS medium on growth of adventitious roots and accumulation of ginsenosides in ginseng bioreactor cultures. Nitrate rather than ammonium was necessary for both root growth and ginsenoside production. A low ammonium concentration combined with a high nitrate concentration was favorable to root growth, showing the largest amount of root biomass at an ammonium/nitrate ratio of 7.19/18.5. Maximum ginsenoside yield was achieved at a ratio of 0 (mM) ammonium to 18.5 (mM) nitrate followed by a ratio of 7.19:18.5. When ammonium was used as sole nitrogen source, root biomass and ginsenoside content decreased by one third.

Generally lower concentrations of ammonium to nitrate ratios are favorable to plant cultures [18]. Similar observations were also reported with cell cultures of *P. ginseng* and *P. notoginseng* [15, 19]. Increased accumulation of ginsenoside content was reported when ammonium nitrogen was reduced and nitrate content increased [15, 20]. Based on the previous results including ours, MS medium containing 7.19 mM: 18.5 mM ammonium and nitrate was found suitable for the production of root biomass as well as accumulation of ginsenosides.

1.2.3 Effects of Carbon Source on Adventitious Root Growth and Ginsenoside Production

Sucrose is a major carbon and energy source for plant cultures. Carbon consumption is directly correlated with biomass accumulation as well as metabolic status of cells and organs. To improve cell growth and ginsenoside production, different types and concentrations of sugar were tested. Choi et al. [13, 14] found that the optimal sucrose concentration for ginseng cell growth was between 40 and 50 g L^{-1}, above which cell growth was inhibited. Furuya et al. [21] were able to improve ginseng cell growth by supplying 5 g L^{-1} glucose and 20 g L^{-1} sucrose at the beginning, and adding another 20 g L^{-1}sucrose in 2 weeks of postinoculation in a batch culture process. In adventitious root cultures of ginseng, root biomass (fresh and dry weight) increased with raising sucrose concentration up to 5% (50 g L^{-1}) (Table 3). Root fresh and dry weights decreased with sucrose concentrations higher than 5%. Thus, 5% initial sucrose concentration is generally used for bioreactor culture of ginseng adventitious roots.

1.2.4 Culture Conditions

Temperature, light and aeration showed profound influences on cell and organ cultures. It was observed that incubation of cultures at 20 °C was most suitable for ginseng root growth and ginsenoside accumulation among 10, 15, 20, 25 and 30 °C temperature regimes tested (Table 4; [22]). Choi et al. [13] and Furuya [23] demonstrated the stimulatory effect of light conditions on ginseng cell growth and

Table 3 Effect of sucrose concentration on growth of ginseng adventitious roots and ginsenoside production. The roots were cultured for 5 weeks in 5-L balloon type bubble bioreactors containing 3 L MS medium

	Biomass					
Sucrose (%)	FW (g)	DW (g)	% DW	Growth ratio[a]	Ginsenoside (mg g^{-1} DW)	Ginsenoside yield (mg L^{-1})
1	194.7 ± 0.4	14.81 ± 0.11	8.12	15.06[a]	6.53 ± 0.05[b]	34.4
2	271.8 ± 5.6	24.19 ± 0.21	8.90	23.04	7.13 ± 0.49	57.5
3	294.7 ± 3.5	26.43 ± 1.01	8.97	15.17	8.01 ± 0.63	70.6
5	346.0 ± 0.7	36.40 ± 0.15	10.52	34.67	10.02 ± 1.33	121.6
7	269.8 ± 8.8	33.19 ± 0.19	12.20	31.67	6.38 ± 1.36	70.6
9	236.7 ± 5.6	32.74 ± 0.62	13.83	31.18	4.98 ± 0.12	54.4

[a]Growth ratio was calculated from increase in DW. Values are quotients of DW after cultivation and DW of the inoculum
[b]Mean ± standard error of three replicates

Table 4 Effect of incubation temperature on growth of ginseng adventitious roots and ginsenoside production. The roots were cultured for 5 weeks in 5-L balloon type bubble bioreactors containing 3 L MS medium

Temperature (°C)	Fresh weight (g L^{-1})	Dry weight (g L^{-1})	Ginsenosides (mg g^{-1} DW)
10	11.24 d[a]	7.12 e	4.77 c
15	3,147 c	5.08 c	4.93 b
20	92.28 a	9.53 a	5.46 a
25	74.57 b	7.72 b	4.34 d
30	7.12 e	0.85 e	026 e

[a]Mean separation by Duncan's multiple range test at *P* 0.05

accumulation of ginsenosides. However, such impacts were not clearly observed in ginseng adventitious root cultures with respect to light intensity and light quality (red, blue and blue plus red range; [10]). The optimal temperature for ginseng adventitious root cultures (in bioreactors) was 20 °C for most root lines and dark conditions were preferred for simplicity and equipment operation. Table 5 shows the influence of aeration rate on ginseng root biomass and ginsenoside production [22]. Relatively higher root biomass and ginsenoside contents were obtained when aeration rate was increased after a certain period of culture initiation. The increment of aeration rate during cultivation increased volumetric mass transfer (K_La), which caused increases in root biomass and ginsenoside content.

1.2.5 Effect of Oxygen, Carbon Dioxide and Ethylene on Adventitious Root Growth and Ginsenoside Production

The impact of gaseous components, namely oxygen, carbon dioxide and ethylene, was tested on growth of adventitious roots and accumulation of secondary metabolites in airlift bioreactor cultures. The composition of air supplied to the bioreactors (0.1 air volume culture volume per min; 400 mL min^{-1}) was the same as atmospheric air (N_2 78%, O_2 20.8%, Ar 0.9%, CO_2 0.03%, Ne, He), which was enriched by supplying pure oxygen, carbon dioxide and ethylene along with incoming air. The gas varied between 30% and 40% O_2, 2.5% and 5% CO_2, 10 and 20 ppm of C_2H_5; among those 40% O_2 enhanced the accumulation of root biomass and the production of ginsenosides. Enhanced levels of carbon dioxide and ethylene were unfavorable to the cultures with low ginsenoside production (Fig. 5; [24]).

1.3 Elicitation

Various attempts have been made to increase ginsenoside yield, among which the use of methyl jasmonate (MJ) was found most beneficial for ginsenoside productivity [25, 26]. Among concentrations of MJ tested (0, 10, 30, 50, 100 and 150 μM), 100 μM MJ was responsible for enhancing ginsenoside accumulation. However, MJ treatments prevented biomass accumulation, inhibiting the root growth (Fig. 6). To achieve accumulation of root biomass as well as ginsenosides, two-stage culture

Table 5 Effect of aeration rate on growth of ginseng adventitious roots and ginsenoside production. The roots were cultured for 5 weeks in 5-L balloon type bubble bioreactors containing 3 L MS medium

Aeration rate (vvm)	Initial K_La (h^{-1})	Dry weight ($g\ L^{-1}$)	Ginsenoside ($mg\ g^{-1}$ DW)
0.05	3.77	12.16 g*	3.98 d
0.1	6.98	13.16 d	4.86 c
0.2	10.84	12.65 e	3.96 d
0.3	12.99	12.40 f	3.13 e
0.05–0.1[a]	3.77–6.98	13.22 c	4.88 c
0.05–0.2[b]	3.77–10.84	13.54 a	5.25 a
0.05–0.3[c]	3.77–12.99	13.50 b	5.20 b

* Mean separation by Duncan's multiple range test at P 0.05
[a]Aeration rate increased after 20 days of culture
[b]Aeration rate increased after 15 days of culture
[c]Aeration rate increased after 10 days of culture

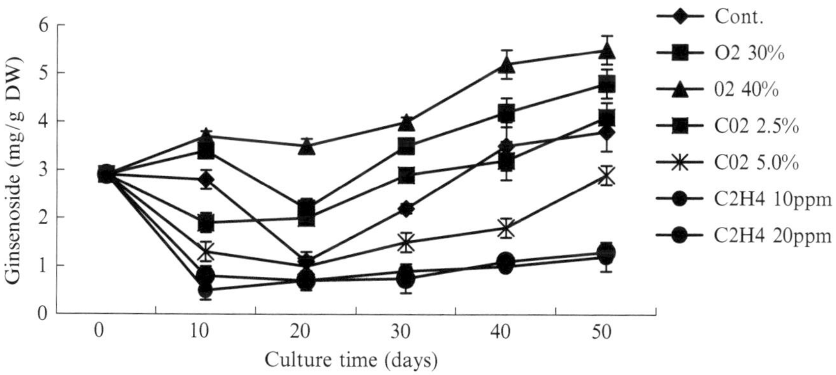

Fig. 5 Changes of ginsenoside content in ginseng adventitious roots as affected by O_2, CO_2 and C_2H_4 supplied throughout the culture cycle

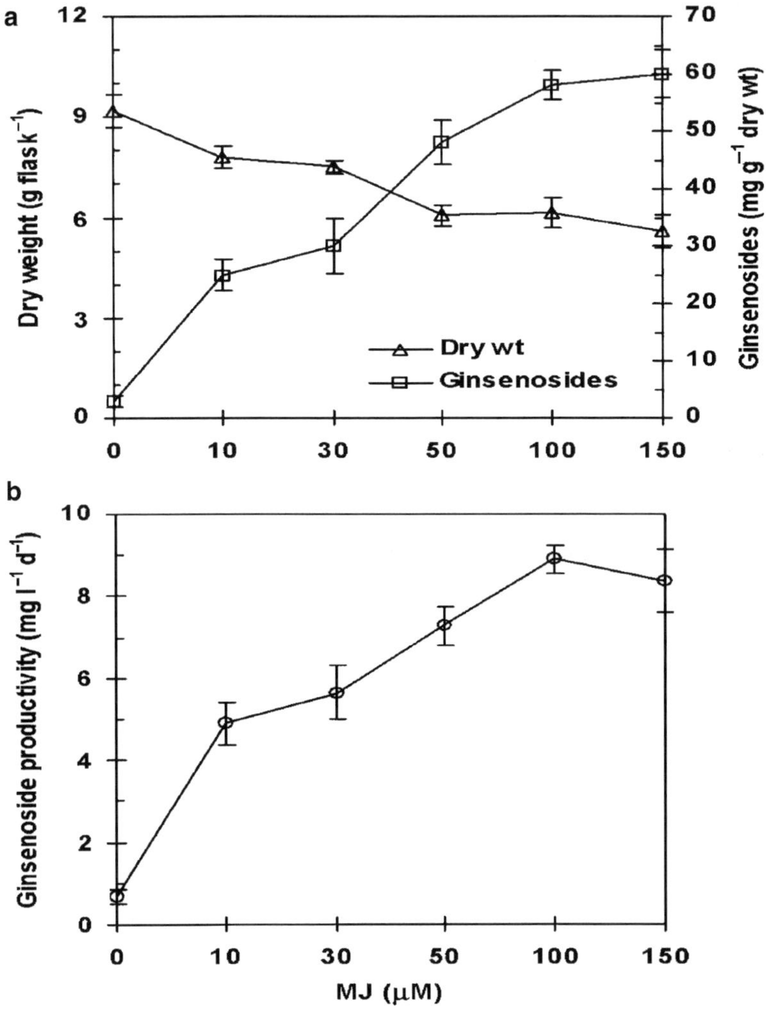

Fig. 6a,b Root dry weight and ginsenoside contents (**a**), and ginsenoside productivity (**b**) as affected by MJ after 40 days of culture in shake flasks containing 300 mL MS medium (without NH_4NO_3). The values are expressed as means of three replicates with standard deviation. The ginsenoside productivity was calculated as: Ginsenoside productivity (mg L^{-1} per day) =total ginsenoside content (mg g^{-1}) × dry weight of harvested roots (g) per volume of culture medium (L) per culture day (d)

was adopted by Kim et al. [26]: the adventitious roots were cultured on elicitor-free medium for the first 40 days of culture and subsequently elicited with 100 μM MJ for another 10 days (Fig. 7). The elicitation led to a sevenfold increase of total ginsenosides and differential accumulation of Rb and Rg group ginsenosides compared with the control (Table 6). The results indicated that MJ elicitation in two-stage culture was useful for enhancing ginsenoside accumulation and manipulating

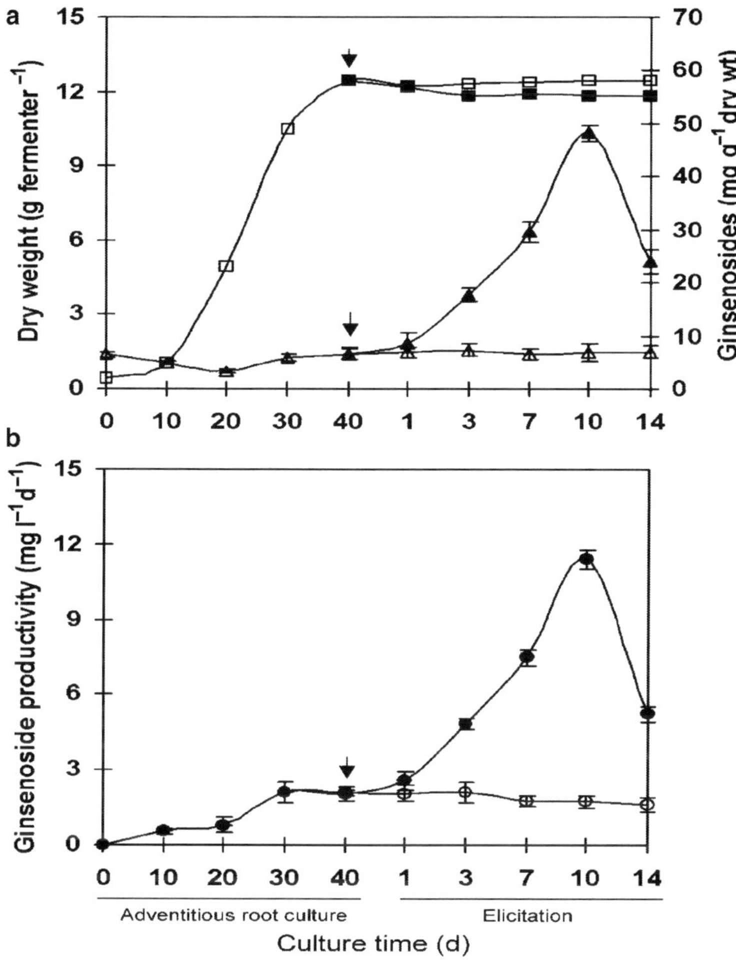

Fig. 7a,b Changes of root dry weight and ginsenoside content (**a**) and ginsenoside productivity (**b**) during 40 days of culture period (without MJ treatment) and during 14 days of elicitation period (with 100 μM MJ treatment) in 5-L bioreactors containing 4 L MS medium (without NH_4NO_3). *Open symbols* (*square*, *triangle*, *circle*) represent root dry weight, ginsenoside content, and ginsenoside productivity without elicitation, respectively. *Filled symbols* (*square*, *triangle*, *circle*) represent root dry weight, ginsenoside content, and ginsenoside productivity with elicitation (100 μM MJ). *Arrows* indicate the time of MJ treatment

Rb and Rg group ginsenosides as well. The results are also very interesting in view of the ginsenosides heterogeneity manipulation, which is a very significant issue for application [27].

Effects of methyl jasmonate (MJ) and salicylic acid (SA) on the antioxidative defense mechanism in ginseng adventitious root cultures were studied by Ali et al. [28]. Single treatment of 200 μM MJ and SA decreased the biomass compared to

Table 6 Effect of methyl jasmonate (MJ) elicitation on ginsenoside contents in ginseng adventitious roots cultured in 5-L bioreactors containing 4 L MS medium. MJ was added to the culture at 100 μM after 40 days of culture. Samples were taken at day 10 after elicitation

	Ginsenosides (mg g^{-1} DW)[a]										
	Rb group				Rg group						
MJ (μM)	Rb1	Rb2	Rc	Rd	Re	Rf	Rg1	TRb[b]	TRg[b]	Total	Rb/Rg
0	1.9 ± 0.2	0.7 ± 0.2	0.9 ± 0.1	0.7 ± 0.3	0.8 ± 0.3	0.7 ± 0.3	1.2 ± 0.2	4.2 ± 0.2	2.7 0± 0.2	6.9 ± 0.3	1.6
100	18.5 ± 0.6	11.4 ± 0.1	12.1 ± 0.3	4.5 ± 0.3	0.7 ± 0.1	0.5 ± 0.2	0.6 ± 0.2	46.5 ± 1.2	1.8 ± 0.2	48.3 ± 1.5	25.8

[a]Mean ± standard error of three replicates

[b]TRb =Rb1 +Rb2 +Rc + Rd; TRg =Re +Rf + Rg1

the control. However, both treatments resulted in higher production of ginsenosides in all culture period compared to the control (Fig. 8). Activity of superoxide dismutase (SOD) was inhibited in MJ treated roots, while the activities of monohydroascorbate reductase (MDHAR), dehyroascorbate reductase (DHAR), SOD, guaiacol peroxidase (G-POD), glutathione perodxidase (GPx) and glutathione reductase (GR) were induced in SA-treated roots. Decrease in the activity of catalase (CAT) was observed in both MJ- and SA-treated roots. Activities of ascorbate peroxidase (ASC), redox state (ASC/(ASC+DHA) and total glutathione were higher in SA- than MJ-treated roots while oxidized ascarbate (DHA) decreased in both cases. The results suggest that the roots are protected against the oxidative stress, thus mitigating MJ and SA stress.

Ali et al. [29] studied effect of copper (Cu^{++}) stress on the growth, oxidative metabolism and ginsenoside production in ginseng root cultures. Cu (0, 0.5, 10.0, 25.0 and 50.0 μM) was used to treat the root cultures for 20 and 40 days of culture. High amounts of Cu were accumulated in a concentration-dependent and duration dependent manner (Table 7). Roots treated with 50 μM Cu resulted in 52 and 89% growth inhibition after 20 and 40 days, respectively. Ginsenoside synthesis was stimulated at a Cu concentration between 5 and 25 μM but decreased at 50 μM Cu (Table 8). Malaondialdehyde (MDA), lipoxigenase activity,

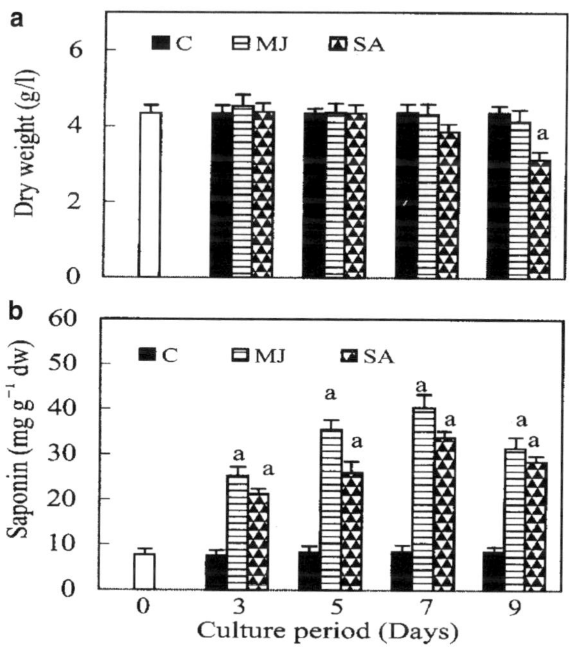

Fig. 8a,b Effects of methyl jasmonate and salicylic acid on root dry weight (**a**) and ginsenoside production (**b**) in *P. ginseng* adventitious roots. Values means ± SE (n=3). *Different letters* in *different bars* differ significantly from the control according to DMRT test, [a]P<0.01, [b]P<0.05

Table 7 Effect of Cu treatment (for 20 and 40 days) on growth of ginseng adventitious roots and metal uptake after 40 days of bioreactor cultures

	20 days		40 days	
Copper (μM)	Dry weight (g^{-1})	Metal uptake (μg g^{-1} DW)	Dry weight (g^{-1})	Metal uptake (μg g^{-1} DW)
Control	2.3 a (0.06)	14.00 a (0.89)	8.0 a (0.14)	21.32 a (0.80)
5	2.2 a (0.19)	53.04 b (3.47)	7.3 a (0.13)	67.97 b (2.36)
10	2.1 a (0.12)	63.20 c (3.80)	7.9 a (0.17)	85.12 c (3.50)
25	1.9 a (0.16)	118.6 d (7.63)	6.8 a (0.16)	128.3 d (3.50)
50	1.1 b (0.14)	473.6 e (9.24)	0.88 b (0.13)	577.8 c (19.1)

The values in parenthesis indicate standard error of three separate experiments. Means followed by the same letter were not significantly different at $P < 0.05$

Table 8 Effect of Cu treatment (for 20 and 40 days) on accumulation of ginsenosides in ginseng adventitious roots after 40 days of bioreactor cultures

	Ginsenoside content (mg g^{-1} DW)				
Copper (μM)	Rg1	Rb	Rc	Rd	Total
20 days					
Control	1.46 c (0.017)	0.69 b (0.003)	0.12 c (0.002)	0.40 b (0.008)	2.68 c (0.023)
5	1.55 c (0.016)	0.61 b (0.001	0.11 c (0.003)	0.56 a (0.004)	2.84 b (0.019)
10	1.83 b (0.016)	0.36 c (0.03)	0.15 b (0.003)	0.60 a (0.013)	2.96 b (0.022)
25	2.25 a (0.016)	1.04 a (0.002)	0.22 a (0.002)	0.59 a (0.008)	4.10 a (0.015)
50	1.24 d (0.013)	0.97 a (0.001)	0.11 c (0.004)	0.25 c (0.008)	2.58 c (0.009)
40 days					
Control	1.41 c (0.005)	0.75 b (0.003)	0.11 a (0.002)	0.63 a (0.005)	2.91 b (0.011)
5	1.88 c (0.005)	0.50 c (0.001)	0.08 a (0.003)	0.21 c (0.005)	2.68 c (0.069)
10	2.01 c (0.010)	0.91 a (0.004)	0.11 a (0.004)	0.37 b (0.003)	3.41 a (0.014)
25	2.13 c (0.007)	0.84 a (0.005)	0.12 a (0.004)	0.80 a (0.003)	3.90 a (0.017)
50	1.17 c (0.008)	0.89 a (0.003)	0.09 a (0.003)	0.21 c (0.005)	2.37 d (0.018)

The values in parenthesis indicate standard error of three separate experiments. Means followed by the same letter were not significantly different at $P < 0.05$

superoxide ion (O_2^-) accumulation, and hydrogen peroxide content did not increase at 5 and 10 μM Cu treated roots but strongly increased at 50 μM Cu, which resulted in the oxidation of ascorbate, glutathione to dehyroascorbate and glutathione disulfide, respectively, indicating oxidative stress. Glutothione metabolism enzymes such as γ-glutamylcytosine synthase, glutathione-*S*-transferase and glutathione peroxidase activities were activated at 5 and 10 μM Cu but were strongly inhibited at 50 μM Cu due to the accumulation of Cu in the root tissue. These results suggest that ginseng roots can grow under Cu stress (5–25 μM) by modulating the antioxidant defense mechanism and Cu stress is useful for enhancing ginsenoside content in ginseng suspension cultures.

2 Scale-Up Cultures

2.1 *Growth Pattern*

The accumulation of root biomass and ginsenosides in 5- and 20-L bioreactors followed a typical growth pattern (Fig. 9). Maximum biomass production was 500 g fresh weight in a 5-L bioreactor with 60 g of inoculum and 2.2 kg fresh weight in a 20-L bioreactor with 240 g inoculum, respectively, in 6 weeks of culture. The yield of ginsenosides increased until root growth reached the exponential growth phase (after 30 days) and decreased thereafter (Fig. 9).

Scale-up cultures of ginseng adventitious roots have been reported [20, 22, 30, 31] and the results demonstrated that the root growth and ginsenoside accumulation were similar in 100-L, 300-L and 500-L airlift bioreactors. Root biomass sharply increased with root-cutting (4 and 6 weeks after culture, i.e., one-time cut after 4 weeks of culture and two-times cut after 4 and 6 weeks, respectively; Fig. 10) during the growth period. Root fresh weight reached 2.8 kg after 8 weeks of culture in a 20-L airlift bioreactor. In large-scale airlift bioreactors root-cutting increased root proliferation, showing 50% increase in fresh weight at the time of harvest (Table 9). The results indicate that the root-cutting is responsible for maximum biomass increment (150-fold) compared to the control.

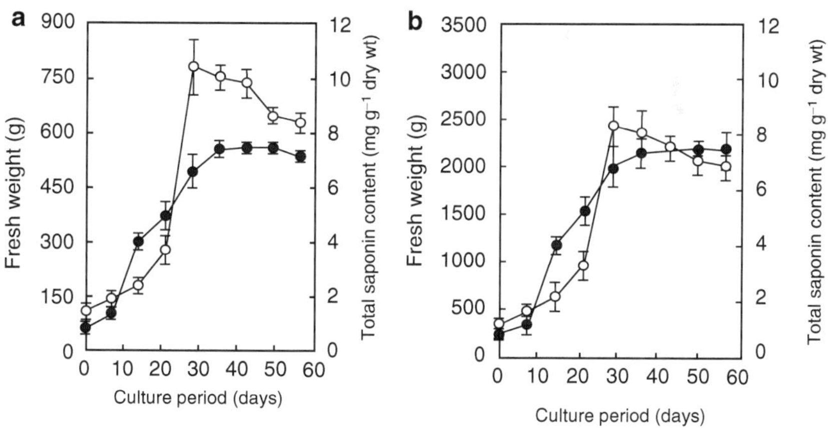

Fig. 9a,b Time course of adventitious root growth (*filled circles*) and total ginsenoside content (*open circles*) in *Panax ginseng* in 5-L (**a**) and 20-L (**b**) bioreactors. *Each point* represents mean of three replicates from two separate trials. *Bars* represent standard deviation

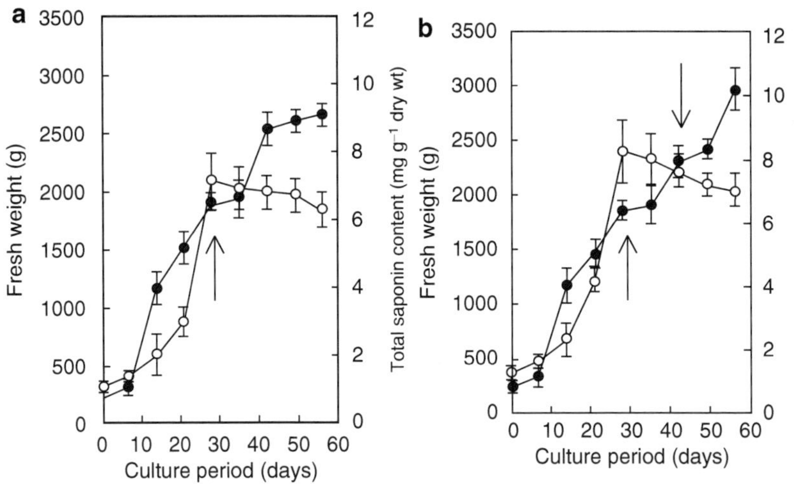

Fig. 10a,b Time course of adventitious root growth (*filled circles*) and total ginsenoside production (*open circles*) in *Panax ginseng* in a 20-L airlift bioreactor as affected by one time root-cutting (**a**) and two times cutting (**b**). *Arrows* indicate the time of root-cutting. *Each point* represents the mean of three replicates from two separate trails. *Bars* represent standard deviation

Table 9 Effect of root-cutting on root growth and ginsenoside content in large-scale ginseng culture[a]

	Without cutting		With cutting	
Working volume (l)	Fresh weight (kg)	Total ginsenoside ($mg\ g^{-1}$ DW)	Fresh weight (kg)	Total ginsenoside ($mg\ g^{-1}$ DW)
100	10.8 ± 0.5	7.5 ± 0.7	14.5 ± 0.7	7.7 ± 0.6
200	31.8 ± 0.9	7.3 ± 0.8	45.2 ± 1.5	7.1 ± 1.2
300	52.7 ± 1.2	7.2 ± 0.6	74.8 ± 3.4	7.0 ± 1.5

[a]Root cutting was performed using a top motor driven blade at 4 weeks after inoculation. Data were collected after 8 weeks of culture and values represent mean ± stranded deviation

2.2 *Effect of Inoculum Density on Growth of Adventitious Roots and Accumulation of Ginsenosides in Large Scale Bioreactors*

Kim [31] investigated the effect of inoculum density (in the range of 0.2–0.5% per culture volume) in small-scale (5- and 20-L bioreactors) and large-scale cultures (500 L and 1,000 L working volume). The results revealed that 0.5% inoculum was suitable for both biomass accumulation and ginsenoside production in 5-L and 20-L bioreactors; whereas 0.4% inoculum was suitable for 500-L and 1,000-L bioreactor cultures (Table 10). Previous results indicated a decrease in productivity during scale up [32, 33], but no decrease was shown in biomass and ginsenoside productivity in our case (Fig. 11), which can be regarded as one of few successful plant cell/tissue culture scale-up examples (e.g., [34]).

Table 10 Effect of inoculum density on biomass accumulation of adventitious roots and ginsenoside accumulation in different working volume of airlift bioreactors

Working volume (L)	Inoculum density (g FW)	Dry weight (g L^{-1})	Ginsenoside accumulation (mg g^{-1} DW)[a]
4	10 (0.25%)[b]	7 ± 0.4	38 ± 0.3
	20 (0.5%)	7 ± 0.2	42 ± 0.9
	40 (1.0%)	6 ± 0.2	43 ± 5.4
15	50 (0.25%)	10 ± 0.1	31 ± 1.6
	75 (0.5%)	13 ± 0.3	33 ± 2.1
	150 (1.0%)	12 ± 0.2	34 ± 2.4
500	1,000 (0.2%)	4 ± 0.3	40 ± 2.1
	2,000 (0.4%)	7 ± 0.2	45 ± 4.5
	3,000 (0.6%)	8 ± 0.2	38 ± 6.3
1,000	2,000 (0.2%)	7 ± 0.4	34 ± 4.7
	3,000 (0.3%)	9 ± 0.6	35 ± 4.5
	4,000 (0.4%)	9 ± 0.5	42 ± 5.5
	5,000 (0.5%)	9 ± 0.3	40 ± 3.8

[a]These values were obtained from roots treated by 100 μM methyl jasmonate for last 8 days of culture
[b]These values in parenthesis present the % of inoculum density per culture volume
All values represent mean ± standard deviation with three trials

To improve the productivity of biomass and secondary metabolites, draw-fill [35] and fed-batch [36] have been suggested in earlier reports. In ginseng adventitious root cultures, a medium replenishment (medium exchange) method was applied [22]: replenishment of 0.75 strength fresh MS medium after 20 days of culture yielded 27.5% increase in dry biomass (28.7 g L^{-1}) and 8.3% increase in ginsenoside content (4.92 mg g^{-1}DW; Table 11), suggesting that the medium replenishment was beneficial for improving root biomass and ginsenoside accumulation.

3 Pilot-Scale Bioreactor Cultures

After the establishment of a large-scale bioreactor system for ginseng adventitious root cultures, a pilot scale bioreactor (10,000 L capacity) was developed by CBN Biotech Company, South Korea (http://www.cbnbiotech.com). Bioreactors and accessory system were constructed by Kihenung Plant Co. Ltd., South Korea. In pilot-scale bioreactors, steam sterilization is required at 125 °C and 1.5 atmospheres for 50 min [22]. The root growth pattern in the pilot-scale bioreactor was the same as that in small scale bioreactors. In addition, a two-stage culture system is suggested: to maximize both root biomass and secondary metabolite contents; in the first stage the adventitious roots are cultured under optimal culture conditions for biomass accumulation followed by physical and chemical elicitations for secondary metabolite accumulation. An average production of ginseng adventitious root biomass is 45 tons (FW) per year on the pilot scale (Fig. 12).

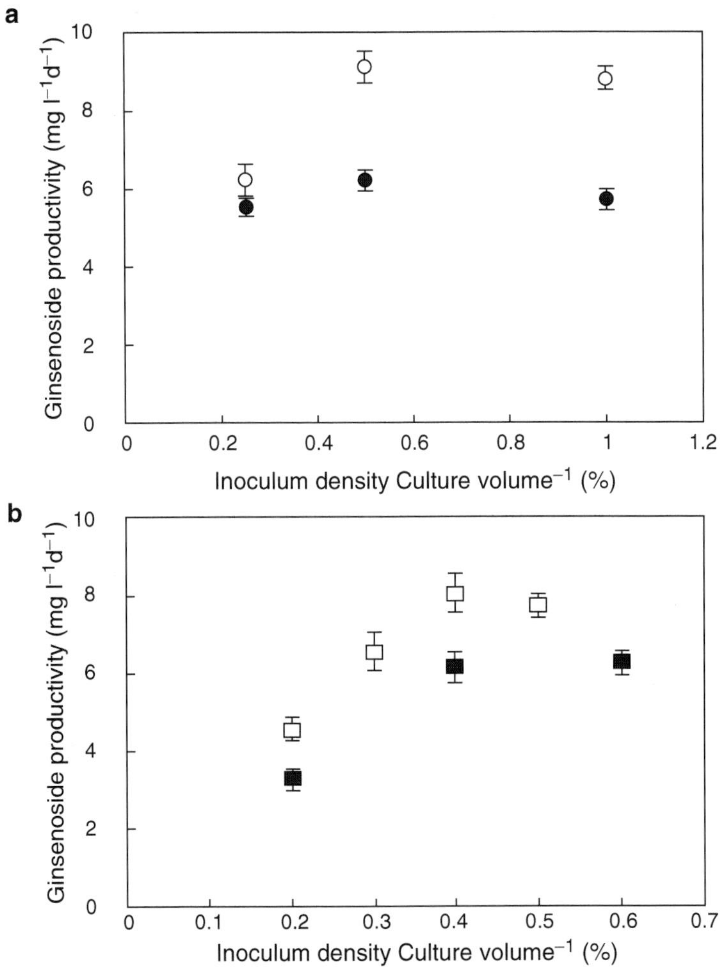

Fig. 11 Effect of inoculum density on ginsenoside productivity in large-scale cultures (**a**) and pilot-scale cultures (**b**) of *P. ginseng* adventitious roots. *Symbols* represent ginsenoside productivity in 4-L cultures (*filled circles*), in 15-L cultures (*open circles*), in 500-L cultures (*filled squares*), and in 1,000-L cultures (*open squares*)

4 Processing of Adventitious Roots and Extraction of Ginsenosides

Processing (drying) is one of the most important steps in the production and commercialization of the adventitious root. The roots have to be dried to low moisture content (5–10%) so that the raw material can be stored and used as various health and pharmaceutical products. A method for the processing was developed by Kim et al. [37] who adopted forced air drying methodology. The effect of various temperature (30, 50 and 70 °C) and duration of treatment (1, 3, 5, 10 and 20 h) on

Table 11 Effect of medium replenishment on root biomass and ginsenoside production after 50 days of ginseng adventitious root culture in 5-L balloon type bubble bioreactors containing 4 L of 1.5 strength MS medium[a]

Strength of medium and replenishment schedule	Dry weight (g^{-1} L)	Ginsenoside content (mg g^{-1} DW)
Control	16.32 ± 0.54[a]	4.17 ± 0.14
Replenishment after 10 days		
0.75 MS	23.49 ± 0.79	4.01 ± 0.16
1.0 MS	24.72 ± 0.51	4.27 ± 0.21
Replenishment after 20 days		
0.75 MS	24.03 ± 0.50	4.94 ± 0.17
0.75 MS	28.66 ± 0.70	4.92 ± 0.15

[a]Values are the mean and standard error of three replicates

Table 12 Effect of drying temperature and duration of treatment on ginsenoside contents in ginseng adventitious roots

Forced air drying		Ginsenosides (mg g^{-1} DW)		
Temp. (°C)	Duration (h)	Triol[a]	Diol[b]	Total
Control		2.0 a[c]	11.2 d	13.2 cd
30	1	1.7 a	12.9 abcd	14.6 abcd
	3	1.6 a	12.7 abcd	14.3 abcd
	5	1.7 a	12.7 abcd	14.4 abcd
	10	1.7 a	11.7 cd	13.3 cd
	20	1.8 a	15.5 ab	17.3 ab
50	1	1.5 a	12.0 bcd	13.5 abcd
	3	1.8 a	11.5 d	13.2 cd
	5	1.9 a	11.7 cd	13.6 abcd
	10	1.5 a	15.9 a	17.4 a
	20	1.6 a	15.2 abc	16.8 abc
70	1	1.8 a	11.9 bcd	13.7 abcd
	3	1.5 a	10.9 d	12.4 e
	5	1.7 a	11.7 cd	13.4 bcd
	10	1.8 a	11.3 bcd	13.1 bcd
	20	1.6 a	11.3 d	12.9 d
Significance[d]				
Drying temperature (A)		NS	*	*
Drying time (B)		NS	NS	NS
A × B		NS	NS	NS

[a]Triol contents=Re + Rf + Rg1 + Rg2 + Rh1 + Rh2
[b]Diol contents=Rb1 + Rb2 + Rb3 + Rc + Rd + Rg3
[c]Mean separation within columns by Duncan multiple range test at 5% level
[d]NS, * are non significant or significant at *P* 0.05, respectively

the concentration of bioactive components was verified. Desirable moisture content (3.13 g water/dry matter, i.e., 10% moisture; Fig. 13) was achieved by drying at 50 °C for 10 h, in which the dried roots also possessed the highest amount of ginsenosides (1.5 mg g^{-1} DW triols, 15.9 mg g^{-1} DW diols and 17.4 mg g^{-1} DW of total ginsenosides; Table 12). Kim et al. [37] also processed the ginseng roots by forced

Fig. 12 Production of *Panax ginseng* adventitious roots in 10-ton scale bioreactors

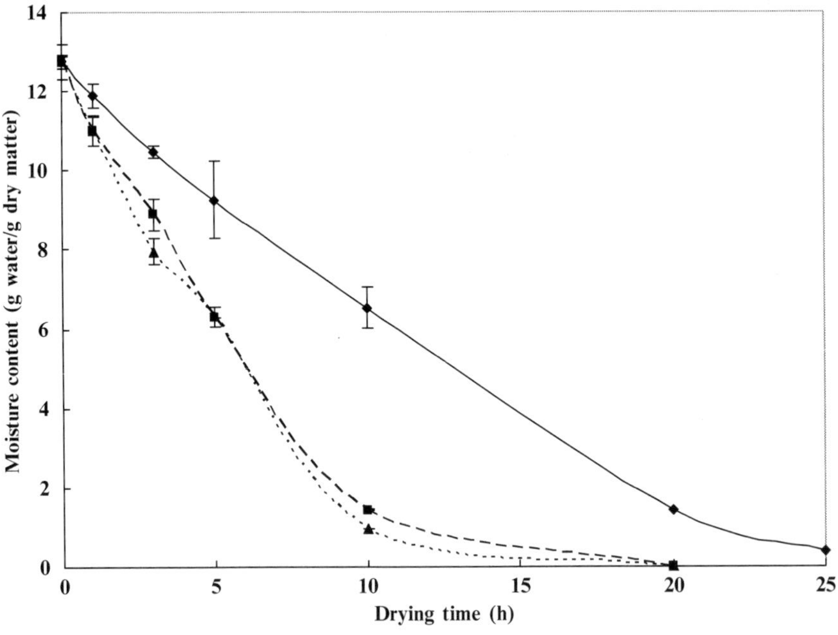

Fig. 13 Changes of the moisture content in *Panax ginseng* adventitious roots as affected by drying temperature (*filled diamonds* – 30 °C; *filled square* – 50 °C; *filled triangle* – 70 °C) and duration of treatment. *Vertical bars* represent standard deviation from the mean values

Table 13 Effect of drying methods and duration of treatment on ginsenoside contents in ginseng adventitious roots

Forced air drying		Ginsenosides (mg g^{-1} DW)		
Temp. (°C)	Duration (h)	Triol[a]	Diol[b]	Total
Forced air drying (50 °C)	10	1.5 a[c]	15.9 a	17.4 a
Far infrared drying (50 °C)	0.5	1.4 a	11.8 bc	13.4 b
	1	1.6 a	11.6 bc	13.2 b
	3	1.5 a	9.3 cd	10.8 bc
	5	1.6 a	10.5 b	12.2 bc
	10	1.4 a	12.4 b	13.8 b
	20	1.4 a	11.2 bc	12.8 b
Freeze drying (−70 °C)	0.5	1.9 a	5.5 ef	7.4 d
	1	1.9 a	4.9 ef	6.8 d
	3	1.7 a	4.6 f	6.3 d
	5	1.9 a	4.7 f	6.6 d
	10	1.8 a	7.6 de	9.4 d
	20	1.9 a	11.8 bc	13.7 b
Significance[d]				
Drying temperature (A)	**	***	***	
Drying time (B)	NS	**	**	
A × B	NS	*	*	

[a]Triol contents=Re + Rf + Rg1 + Rg2 + Rh1 + Rh2
[b]Diol contents=Rb1 + Rb2 + Rb3 + Rc + Rd + Rg3
[c]Mean separation within columns by Duncan multiple range test at 5% level
[d]NS, *, **, *** are non-significant or significant at *P* 0.05, 0.01 and 0.01 respectively

air drying, far infrared drying and freeze drying methods, which revealed that the forced air drying method was superior to the other two methods (Table 13).

Extraction is another important step for the recovery of bioactive compounds from plant raw materials. Extraction technologies must be versatile, relatively simple and safe for operating personnel and consumers. A heat reflux method was applied for the extraction of ginsenosides from powdered roots of ginseng [38]. Four extraction variables, i.e., type, concentration of solvent (water, 10, 30, 50, 70 and 100% ethanol), extraction temperature (40, 60 and 80 °C) and duration (2, 4, 6 and 8 h) were compared. Results revealed that the yield of ginsenosides showed no obvious increase with extraction time (2–8 h) but temperature had a profound influence on extraction of active ingredients from the samples. Treatment of samples with 70% ethanol at 80 °C for 6 h was found suitable for the extraction of ginsenosides, phenolics and polysaccharides. Ultrasonic extraction and microwave extraction were also tested and compared with heat reflux extraction method. Ultrasonic and microwave extraction methods were inferior to conventional heat reflux extraction with less amount of total ginsenosides, while heat reflux extraction yielded higher amount of crude saponins, total ginsenosides (panxatriols plus panaxadiols), and phenolics (Table 14). The extract obtained from the heat reflux also scavenged highest percentage DPPH radicals compared to the extracts from other treatments.

Table 14 Effect of extraction methods on ginsenoside, phenolic and polysaccharide contents and radical scavenging activity on DPPH in the extract of ginseng adventitious roots

Extraction method and duration		Amount (mg g^{-1} DW)					Radical scavenging activity on DPPH (%)	Crude ginsenoside content (%)
		Ginsenosides						
		Triol	Diol	Total	Phenolics	Polysac charides		
Heat reflux extraction 6 h		1.8 b[a]	16.8 a	18.6 a	8.4 a	116.8 b	37.7 a	5.9 a
Ultrasonic extraction	30 min	1.3 c	9.9 cd	11.2 cd	5.9 c	125.5 ab	18.1 c	4.6 bc
	1 h	1.3 c	11.7 bc	13.0 bc	6.1 c	128.2 a	18.6 c	4.6 bc
	2 h	1.4 bc	12.3 b	13.7 b	7.1 b	120.4 abc	23.3 b	5.3 b
	4 h	1.0 c	13.4 b	14.5 b	7.0 b	106.3 c	36.5 a	4.1 c
Microwave	1 min	1.2 c	4.7 fgh	5.9 gh	4.9 d	102.1 ef	12.8 d	2.0 f
	3 min	0.9 c	6.9 ef	7.9 efg	5.2 cd	112.1 b	12.8 d	3.0 de
	5 min	2.8 a	5.9 fg	8.7 ef	5.5 cd	107.5 c	13.0 d	3.6 d
	10 min	1.1 c	8.6 de	9.8 de	5.8 c	95.8 d	12.8 d	4.0 cd
Significance								
Extraction method (A)		**	***	**	***	***	***	***
Extraction time (B)		***	*	NS	***	***	NS	**
A × B		***	**	*	***	NS	*	*

NS, *, **, *** are non-significant as $P<0.05$, 0.01 and 0.01, respectivelyaMean separation within columns by Duncan's multiple range test at 5% level

Table 15 In vitro reverse mutation test of powdered ginseng adventitious roots treated with *S. typhimurium* and *E. coli* without S-9 mix

Dose (μg/mL)	Number of revertans/plate (Mean ± S.D., n=3)				
	Base pair substitution type			Frame-shift	
	TA 100	TA 1535	WP2uvrA$^-$	TA 98	TA1537
0	116 ± 6	9 ± 3	34 ± 4	45 ± 2	8 ± 1
312.5	117 ± 17	10 ± 1	28 ± 3	51 ± 7	6 ± 2
625	120 ± 8	10 ± 2	29 ± 5	49 ± 8	7 ± 1
1,250	107 ± 12	12 ± 2	30 ± 9	50 ± 3	8 ± 2
2,500	123 ± 10	10 ± 1	24 ± 3	48 ± 3	8 ± 1
5,000	106 ± 12	11 ± 2	27 ± 5	42 ± 4	7 ± 1
Positive control	489 ± 25	241 ± 10	265 ± 5	417 ± 15	286 ± 14
Strain		Positive control		Concentration	
TA 100		2-Aminoflurene (AP-2)		0.01 μg per plate	
TA 1535		Sodium azide (SA)		1.0 μg per plate	
WP2uvrA		2-Aminoflurene (AP-2)		0.01 μg per plate	
TA98		2-Aminoflurene (AP-2)		0.1 μg per plate	
TA 1537		*p*-Aminoanthracene (9AA)		80 μg per plate	

5 Biosafety Assessment

Biosafety tests of powdered ginseng adventitious roots were carried out, which is very important for application [39]. The reverse mutation (Table 15), chromosomal aberrations (Table 16) and micronucleus test (Table 17) did not reveal any

Table 16 Chromosome aberration test of powdered ginseng adventitious roots treated to CHL cells

S-9 mix	Test item[a]	Dose (μg/mL)	Chromosome aberration per 100 metaphase cells (Mean ± S.D.)
S-9 mid (–) 6 + 18 h	CMC	0	1.5 ± 0.7
	Powdered adventitious roots	150	1.5 ± 0.0
		300	1.0 ± 0.4
		600	0.5 ± 0.1
	MMC	0.05	26.5 ± 2.1
S-9 mix (+) 6 + 18 h	CMC	0	1.0 ± 0.0
	Powdered adventitious roots	150	1.5 ± 0.7
		300	2.0 ± 0.0
		600	1.0 ± 0.0
	B(a)P	20	28.5 ± 0.7
S-9 mid (–) 24 + 0 h	CMC	0	0.5 ± 0.1
	Powdered adventitious roots	150	1.0 ± 0.0
		300	1.5 ± 0.7
		600	1.0 ± 0.0
	MMC	0.05	36.0 ± 2.8
S-9 mid (–) 48 + 0 h	CMC	0	0.0 ± 0.0
	Powdered adventitious roots	150	1.0 ± 0.0
		300	1.0 ± 0.4
		600	0.5 ± 0.2
	MMC	0.05	27.5 ± 2.1

[a]*CMC* Carobxymethylcellulose sodium salt; *MMC* Mitomycin C; *B(a)P* Benzo(*a*)pyrene

Table 17 Micronulceus test of powdered adventitious roots treated to male ICR mice

Test items	Groups	Dose (mg/kg)	Route	Animal number	Sampling time (h)	PCE/(PCE+NCE)[a] (Mean ± S.D.)	MNPCE[a]/1000PCE (Mean ± S.D.)
Saline	G1	0	PO	6	48	0.483 ± 0.021	0.83 ± 0.41
Powdered adventitious roots	G2	500	PO	6	48	0.492 ± 0.016	0.67 ± 0.26
	G3	1,000	PO	6	48	0.485 ± 0.019	0.75 ± 0.42
	G4	2,000	PO	6	48	0.495 ± 0.014	0.92 ± 0.38
MMC	G5	2	IP	6	24	0.493 ± 0.06	84.50 ± 6.86*

* Significantly different from mice treated with vehicle (p<0.01, by Chi-square test)
[a]*MNPCE* Micronulceated erythrocyte; *PCE* Polychromatic erythrocyte; *NCE* Normochromatic erythrocyte

significant mutagenicity [40]. Furthermore, 13 weeks of repeated dose toxicity of the powdered roots (oral doses from 300 to 900 mg kg^{-1} diet) did not produce mortality or significant changes in the general behavior and gross appearance of internal organs in rats and beagle dogs. These results showed that ginseng adventitious roots are safe and nontoxic at average dietary levels. With the above results, ginseng adventitious roots were approved as a food by the Korean Food Drug Association (KFDA) in April, 2003. Since then, ginseng adventitious roots have been manufactured into various health products, which are available in the market as of 2007.

6 Conclusions and Future Perspectives

Extensive research work has been carried out on ginseng cell and tissue cultures, such as optimization of culture medium, physical conditions and strategies to improve flask and large scale cultures. The work also included maximizing biomass yield and ginsenoside contents. Ginseng adventitious root culture has been evolved as an alternative to cell culture, which has the advantages of higher biomass accumulation and ginsenoside production. For commercialization of ginseng adventitious roots, large-scale cultures have been achieved using airlift bioreactors by determining various factors affecting the production of root biomass and ginsenosides [10, 11, 30]. Elicitation of cultures with methyl jasmonates enhanced total ginsenoside content up to 6–8% compared to conventional cultures. Based on the newly developed technologies, pilot-scale (10-ton) bioreactors were established in CBN Biotech Company in 2002. An average production of ginseng adventitious roots is 45 tons (FW) per year from four 10-ton pilot-scale bioreactors. Ginseng adventitious roots have been manufactured into various health products, which are available in the market.

In addition to the production of ginseng biomass and ginsenosides, other opportunities still exist for the utilization of adventitious root cultures for the production of valuable ingredients such as polysaccharides, biophenols, polyacetylenes, peptidoglycans and various ubiquitous compounds such as fatty acids, alcohols and vitamins. In the light of current success, future research should be focused on development of new bioreactor culture processes such as continuous culture, process monitoring, modeling and control which may be useful for further reducing the production cost.

A current major problem is that the adventitious roots produced in the bioreactor should be processed into products immediately in order to avoid deterioration. The deterioration of fresh ginseng is generally associated with a decline in hardness with browning and mold decay. Post-harvest storage of fresh ginseng adventitious roots has become a major concern for regulation of supply and demand for extending processing period. Therefore, approaches should be developed for extending storage life of fresh ginseng by modifying atmosphere packaging and storage. Future research should also be focused on developing efficient processing methods for extraction of ginsenosides. Furthermore, commercialization of tissue cultured ginseng products needs to remove psychological barriers of consumers. Adventitious roots are as natural as field cultivated roots which are grown in disease-free and pesticide-free environments. Furthermore, our tests indicated that the roots are safe for use by humans.

Acknowledgment This work was financially supported by the Ministry of Education and Human Resources Development, the Ministry of Commerce, Industry, and Energy and the Ministry of Labor. One of the authors (H.N. Murthy) is grateful to Korean Federation of Science and Technology Societies (KOFST) for the award of Brain Pool Fellowship. JJZ thanks the funds from the Shanghai Leading Academic Discipline Project (project # B203).

References

1. Tang W, Eisenbrand G (1992) Panax ginseng C. A. Meyer Chinese drugs of plant origin. Springer, Berlin, pp 710–737
2. Attele AS, Wu JA, Yuan CS (1999) Ginseng pharmacology, multiple constituents and multiple actions. *Biochem Phamacol* 58:1685–1693
3. Briskin DP (2000) Medicinal plants and phytomedicines. Linking plant biochemistry and physiology to human health. *Plant Physiol* 124:507–514
4. Glaser V (1999) Billion-dollar market blossoms as botanicals take root. *Nat Biotechnol* 17:17–18
5. Dewick PM (1997) Medicinal natural products. Wiley, West Sussex, UK
6. Huang KC (1999) Pharmacology of Chinese herbs. CRC, Boca Raton, Florida, USA
7. Bruneton J (1995) Pharmacognosy, phytochemistry, medicinal plants. Lavisior, Paris
8. Proctor JTA (1996) Ginseng: old crops, new directions. In: Janick J (ed) Progress in new crops. ASHS, Arlington, VA, pp 565–577
9. Wu J, Zhong JJ (1999) Production of ginseng and its bioactive components plant cell culture: current technological and applied aspects. *J Biotechnol* 68:89–99
10. Yu KW, Hahn EJ, Paek KY (2000) Production of adventitious ginseng roots using bioreactors. *Korean J Plant Tiss Cult* 27:309–315
11. Hahn EJ, Kim YS, Yu KW, Jeong CS, Paek KY (2003) Adventitious root cultures of *Panax ginseng* C. A. Meyer and ginsenoside production through large-scale bioreactor system. *J Plant Biotechnol* 5:1–6
12. Kim YS, Hahn EJ, Yeung EC, Paek KY (2003) Lateral root development and saponin accumulation as affected by IBA or NAA in adventitious root cultures of *Panax ginseng*. C. A. Meyer. *In Vitro Cell Dev Biol Plant* 39:245–249
13. Choi KT, Ahn IO, Park JC (1994a) Production of ginseng saponin in tissue cultures of ginseng (*Panax ginseng* C.A. *Meyer). Russian J Plant Physiol* 41:784–788
14. Choi KT, Lee CH, Ahn IO, Lee JH, Par JC (1994b) Characteristics of the growth and ginsenosides in the suspension-cultured cells of Korean ginseng (*Panax ginseng* C.A. Meyer). In: Bailey WG, Whitehead C, Proctor JTA, Kyle JT (eds) Proceedings of the International Ginseng Conference, Vancouver, 1994, pp 259–268
15. Zhang YH, Zhong JJ, Yui JT (1996) Effect of nitrogen source on cell growth and production of ginseng saponin and polysaccharide in suspension cultures of *Panax ginseng*. *Biotechnol Prog* 12:567–571
16. Gubar SI, Gul'ko TP, Kunaksh VA (1997) Growth and glycoside accumulation in ginseng callus tissue culture under a long-term action of exogenous phytohormones. *Russian J Plant Physiol* 44:83–89
17. Shamakov NV, Zaitseva GV, Belousova IM, Strogov SV, Smonova GM, Buenko RG, Nosov AM (1991) Large-scale cell cultivation in suspension. *II. Eloboraiton of ginseng cell cultivation on pilot plant. Biotechnologiya* 1:32–34
18. Franklin CI, Dixon RA (1994) Initiation and maintenance of callus and cells suspension cultures. In: Dixon RA, Gonzales RA (eds) Plant cell cultures – a practical approach, 2nd edn. IRL Press, Oxford, pp 1–25
19. Ushiyama K (1991) Large-scale culture of ginseng. In: Komamine A, Misawa M, DiCosmo F (eds) Plant cell culture in Japan: progress in production of useful plant metabolites by Japanese enterprises using cell culture technology. CMA, Hong Kong, China, pp 92–98
20. Yu KW (2000) Production of the useful metabolites through bioreactor culture of Korean ginseng (*Panax ginseng* C.A. Meyer). Ph. D. thesis, Chungbuk National University, Cheongju, South Korea
21. Furuya T, Yoshikawa T, Orihara Y, Oda H (1984) Studies of the culture conditions for *Panax ginseng* cells in jar fermentors. *J Nat Prod* 47:70–75
22. Jeong CS (2007) High density culture of mountain ginseng (*Panax ginseng* C.A. Meyer) adventitious roots in large scale bioreactors for the production of ginseng biomass and ginsenosides. Ph. D. thesis, Chungbuk National University, Cheongju, South Korea

23. Furuya T (1988) Saponins (ginseng saponins). In: Vasil IK (ed) Cell culture and somatic cell genetics of plants, vol. 5. Academic Press, San Diego, CA, pp 213–234
24. Jeong CS, Chakrabarty D, Hahn EJ, Lee HL, Paek KY (2006) Effect of oxygen, carbon dioxide and ethylene on growth and bioactive compound production in bioreactor culture of ginseng adventitious roots. *Biochem Eng J* 27:252–263
25. Yu KW, Gao W, Hahn EJ, Paek KY (2002) Jasmonic acid improves ginsenoside accumulation in adventitious root culture of *Panax ginseng* C.A. *Meyer. Biochem Eng J* 11:211–215
26. Kim YS, Hahn EJ, Murthy HN, Paek KY (2004) Adventitious root growth and ginsenoside accumulation in *Panax ginseng* cultures as affected by methyl jasmonate. *Biotechnol Lett* 26:1619–1622
27. Zhong JJ, Yue CJ (2005) Plant cells: secondary metabolite heterogeneity and its manipulation. *Adv Biochem Eng Biotechnol* 100:53–88
28. Ali MB, Yu KW, Hahn EJ, Paek KY (2006) Methyl jasmonate and salicylic acid elicitation induces ginsenoside accumulation, enzymatic and non-enzymatic antioxidant in suspension culture *Panax ginseng* roots in bioreactors. *Plant Cell Rep* 25:613–620
29. Ali MB, Hahn EJ, Paek KY (2006) Copper-induced changes in the growth, oxidative metabolism, and saponin production in suspension culture roots of *Panax ginseng* in bioreactors. *Plant Cell Rep* 25:1125–1132
30. Choi MS, Son SH, Yun SR, Kwon OW, Seon JH, Paek KY (2000) Pilot scale culture of adventitious roots of ginseng in a bioreactor system. *Plant Cell Tiss Org Cult* 62:187–193
31. Kim YS (2002) Production of ginsenosides through bioreactor cultures of adventitious roots in ginseng (*Panax ginseng* C.A. Meyer). Ph. D. thesis, Chungbuk National University, Cheongju, South Korea
32. Kwok KH, Doran PM (1995) Kinetic and stoichimetric analysis of hairy roots in a segmented bubble column reactor. *Biotechnol Prog* 11:429–435
33. Scragg AH, Morris P, Allan EJ, Bond P, Fowler MW (1987) Effect of scale-up on serpentine formation by *Catharanthus roseus* suspension cultures. *Enzyme Mirob Technol* 9:619–624
34. Zhang ZY, Zhong JJ (2004) Scale-up of centrifugal impeller bioreactor for hyperproduction of ginseng saponin and polysaccharide by high-density cultivation of *Panax notoginseng* cells. *Biotechnol Prog* 20:1076–1081
35. Lipsky AKh (1992) Problems of optimization of plant cell culture process. *J. Biotechnol* 26:83–87
36. Zhang YH, Zhong JJ (1997) Hyperproduction of ginseng saponin and polysaccharide by high density cultivation of *Panax ginseng* cells. *Enzyme Microb Technol* 21:59–63
37. Kim SJ, Murthy HN, Hahn EJ, Lee HL, Paek KY (2007) Effect of processing methods on the concentrations of bioactive components of ginseng (*Panax ginseng* C.A. Meyer) adventitious roots. LWT-Food Sci Technol. doi:10.1016/j.lwt.2007.0.6.012
38. Kim SJ, Murthy HN, Hahn EJ, Lee HL, Paek KY (2007) Parameters affecting the extraction of ginsenosides from the adventitious roots of ginseng (*Panax ginseng* C.A. *Meyer). Sep Purif Technol* 56:401–406
39. Shivakumar G, Yu KW, Paek KY (2005) Biosafe ginseng: a novel source for human well-being. *Eng Life Sci* 5:527–533
40. Shivakumar G, Yu KW, Lee JS, Kang JK, Lee HL, Kim WJ, Paek KY (2006) Tissue cultured mountain ginseng adventitious roots: safety and toxicity evaluation. *Eng Life Sci* 6:372–383

Adv Biochem Engin/Biotechnol (2009) 113: 177-215
DOI: 10.1007/10_2008_19

Published online: 17 April 2009

Approaches to Optimizing Animal Cell Culture Process: Substrate Metabolism Regulation and Protein Expression Improvement

Yuanxing Zhang

Abstract Some high value proteins and vaccines for medical and veterinary applications by animal cell culture have an increasing market in China. In order to meet the demands of large-scale productions of proteins and vaccines, animal cell culture technology has been widely developed. In general, an animal cell culture process can be divided into two stages in a batch culture. In cell growth stage a high specific growth rate is expected to achieve a high cell density. In production stage a high specific production rate is stressed for the expression and secretion of qualified protein or replication of virus. It is always critical to maintain high cell viability in fed-batch and perfusion cultures. More concern has been focused on two points by the researchers in China. First, the cell metabolism of substrates is analyzed and the accumulation of toxic by-products is decreased through regulating cell metabolism in the culture process. Second, some important factors effecting protein expression are understood at the molecular level and the production ability of protein is improved. In pace with the rapid development of large-scale cell culture for the production of vaccines, antibodies and other recombinant proteins in China, the medium design and process optimization based on cell metabolism regulation and protein expression improvement will play an important role. The chapter outlines the main advances in metabolic regulation of cell and expression improvement of protein in animal cell culture in recent years.

Keywords Animal cell culture, Chinese hamster ovary (CHO) cell, Metabolism regulation, Protein expression improvement

---

Y. Zhang
State Key Laboratory of Bioreactor Engineering, East China University of Science and Technology, 130 Meilong Road, Shanghai, 200237, China
e-mail: yxzhang@ecust.edu.cn

Contents

1 Introduction

The earliest commercial application of animal cell culture started more than four decades ago, mainly in the production of human and veterinary vaccines, which accelerated the design of large-scale bioprocesses for mammalian cells [1]. With the advent of recombinant DNA and cell infusion technologies in the 1970s, it became possible to produce recombinant proteins and monoclonal antibodies as well as natural biologicals with genetically-engineered transformed cells or hybridomas which are anchorage-dependent or anchorage-independent. Since then, there has been a dramatic increase in the number of animal cell products and a rapid development of large-scale animal cell culture technology. With the development of new technologies in mammalian cell culture and the increasing demand on protein drugs, recombinant protein drugs produced by mammalian cell culture have evoked worldwide attention .

More and more protein products from animal cell culture have been put on the market in the last two decades. Nowadays, more than 200 kinds of biological products are produced in China ranging from preventive vaccines, as the majority, to therapeutic and diagnostic products. About 800–900 million doses of vaccines are administered annually for the purpose of prophylaxis. Biologicals have played a critical role in the prevention of infectious diseases. Although great progress has been made in the field of biological products, China is adopting vigorous measures to strengthen basic research and improve commercial applications of original new products. Large-scale animal cell culture has become one of the key technologies in the pharmaceutical industry.

The expression of recombinant therapeutics is complex and usually requires the post-translational modification machinery only available in eukaryotic cells. Also, from the perspective of immunogenicity and glycosylation of the therapeutics, animal cell culture becomes a more preferable choice for the production of the recombinant proteins. Mammalian cell lines such as Chinese hamster ovary (CHO), baby hamster kidney (BHK), human embryonic kidney (HEK-293), hybridoma and NS0 (murine myeloma) cells, have been widely used for the production of biological therapeutics.

In recent years, the expression level for heterologous genes in animal cells has been greatly improved, and recombinant antibody production has reached 5 g L^{-1} and more in CHO cell culture [2]. However, further improvement in protein productivity is expected, especially in some non-antibody proteins such as hepatitis B surface antigen, factor VIII, human chorionic gonadotropin, and so on. The researchers in the area continue to pay comprehensive attention to the strategies to improve the heterologous gene expression level in recombinant animal cells. There are many factors which influence the heterologous gene expression in recombinant animal cells. The most crucial one is the employment of metabolism regulation, which can modify the metabolic pathways and the overall cellular physiology to improve cell growth and the heterologous genes expression. Meantime, some additives have been found capable of stimulating the expression of heterologous genes. For biological safety reasons, serum-free or protein-free media have been more preferable in the animal cell culture. Therefore, some new recombinant proteins, plant-derived peptides and synthetic oligopeptides have been used to substitute the serum and animal-derived proteins to develop and optimize the media.

In order to reach a high cell density, many commercial bioreactors operated in fed-batch or draw-and-fill mode have been designed. The physical parameters, such as pH, pO_2, pCO_2, temperature, and redox, have been investigated and controlled for the high cell density and protein production in the bioreactor.

From different view angles, all the factors can be divided into three levels: (1) molecular level regulation, including DNA replication, mRNA transcription, protein translation and the post-translational processing and modification – all these factors affect the protein expression individually or synergistically; (2) cellular regulation, such as glucose and amino acids metabolism, lactate and ammonia accumulation, which will disturb the gene expression and physiology of the cells cultured in the bioreactor; (3) process optimization in bioreactor, namely problems with the aeration-agitation, oxygen and nutrients transferring, agitation and shear.

Some high value proteins and vaccines for medical and veterinary applications by animal cell culture have an increasing market in China. In order to meet the demands of large-scale productions of proteins and vaccines, animal cell culture technology has been widely developed. Although animal cell technology has made great achievements in China, a lot of basic research has to be done around the improvement of specific cell growth and product production. In this chapter, the main advances in metabolic regulation of cell and expression improvement of protein in animal cell culture in recent years in China are introduced.

2 Metabolic Regulation in Animal Cell Culture

Metabolic regulation means the organisms coordinate its metabolic system and pathways to yield metabolic products under normal conditions by it own regulation. The organisms have formed numerous perfect metabolic regulation mechanisms during a long evolution to make sure that all the complicated biochemical reactions occur in an orderly way and give rapid responses to the environmental changes. Under the metabolic regulation, the energy utilization is economic and materials that are essential for cell growth and propagation are just synthesized according to needs. There is no redundancy. Therefore, in order to promote cell growth and product formation and decrease metabolic repressors, changing the inside metabolism pathway by metabolic engineering methods will be an efficient strategy, and regulating cellular metabolism by changing culture conditions is another approach and is used more regularly in industrial practice. In this part, the latter is more focused on. At present, glucose and glutamine are the major energy sources for animal cell culture in vitro. Glucose is the most important material for cell culture as both energy and carbon source. The other one is glutamine which can be utilized as energy source by cell, especially when glucose is deficient. Meanwhile, glucose metabolism and glutamine metabolism are closely interacted with each other. Just for their important status in cell culture, the glucose and glutamine metabolic regulation in cell culture have been extensively investigated, and quantitative and intensive studies on the association of glucose and glutamine metabolic regulation with animal cell growth and product formation were reported. Besides, the influences of some metabolic by-products, including lactate, ammonia and alanine, were thoroughly investigated in animal cell culture.

2.1 Medium Optimization

For the large-scale production of recombinant protein, culture conditions should be optimized to support high density growth of animal cell. The nutritional requirements of cells are very important. Among the nutritional constituents of a cell medium, glucose and glutamine are of central importance.

Generally, both glucose and glutamine are important energy sources in animal cell cultures. As one of the major carbon and energy sources in animal cell culture, glucose is utilized either through the pentose phosphate pathway to provide nucleotides and reducing power for biosynthesis, or through the glycolysis pathway and the subsequent tricarboxylic acids cycle to provide metabolic intermediates and energy for the growth and survival of cell. Glutamine is a major source of energy, carbon, and nitrogen for in vitro culture of animal cells. However, lactate and ammonia, as the major waste products of glucose and glutamine metabolism in animal cell culture, have an inhibitory effect on cell growth and need be controlled at a low level.

Until now, many strategies have been used to decrease the concentrations of lactate and ammonia in cell culture. For example, glutamate could be substituted for glutamine by the co-expression of glutamine synthetase in animal cell to decrease the production of ammonia [3], and the glucose concentration could be controlled at a low level in the bioreactor to decrease the production rate of lactate [4]. Moreover, the intracellular metabolism of animal cell in glutamate-based or low-glucose medium has also been investigated, and it will surely help us to further optimize the medium.

2.1.1 Glucose Metabolism

The control of glucose concentration in medium at optimal level to limit lactate formation is critical for achieving high cell density and product yield. There are two main ways to optimize the medium for reducing byproducts accumulation. One is to lower the available concentration of glucose in the medium to regulate its metabolic flux in cell [5, 6]. This way has been extensively used in the large-scale culture of animal cells, and proved to be very effective to decrease lactate accumulation. The other way is to using the nutrients (such as galactose [7] and fructose [8]) at slow metabolic rates instead of glucose to decrease lactate accumulation. However, it is useful only for some kinds of cells since not all of the cells can use fructose or galactose. Besides, cells may need a very long period to adapt to the new substitution, which makes this way not always the best choice. As to be expected, there are other ways to optimize the medium for regulating glucose metabolism from aspects such as temperature, growth factor, vector, culture method, and so on.

Generally, the optimization of animal cell culture mainly relies on the knowledge of the metabolism of cultured cell line. Kinetic modeling is a good way for a better understanding of cell physiology and metabolism and for process control to optimize cell culture further. Zhou et al. [9] proposed a macrokinetic model to simulate the growth and metabolism of myeloma cells in batch and fed-batch cultures. In this model, the specific glycolysis rate was regarded as the key variable to induce the metabolic shift. During the modeling, they found that the production of lactate from pyruvate was resulted from rapid-glycolysis and the lack of activity of malate–aspartate shuttle to transport NADH into mitochondria. As the most distinctive point, this model is composed of two modes. First, a rapid-glycolysis mode describes the rapid-glycolysis and glutaminolysis rates to produce lactate and ammonia. Second, in the low-glycolysis mode, the specific glucose consumption rate and the glycolysis rate decline and the formation of lactate is terminated. So one can predict the experimental data of glucose uptake and lactate production in both batch and fed-batch cultures by combined macrokinetic and bioreactor models with reasonable accuracy. However, the model is based on the essentially major features of the cell line concerned, and the balance for oxygen consumption and carbon dioxide production as well as the balance of NADPH are not taken into account due to the lack of data. So it needs further optimization or another way to understand the cell metabolism in detail.

Lu et al. [10] developed a method to separate and determine simultaneously the TCA cycle acids and other related substances in cultured mammalian cells by ion-exchange chromatography with suppressed conductivity detection. In this work, it was found that the concentrations of malate, citrate and *cis*-aconitate were distinctively higher than the other acids in the TCA cycle, which showed the large accumulation of these acids within CHO cells. What's more, it strongly proved the hypotheses that a blockage existed in the TCA cycle of abnormal mammalian cells. The fact that no peak of isocitrate was detected indicated a very low isocitrate concentration or no isocitrate within CHO cells, and this further supported the hypotheses that the TCA cycle was blocked at the point of citrate.

Earlier on, some analytical methods were developed to identify extracellular concentrations of nutrients, treating the cell as a "black box". Reitzer et al. [11] reported that glucose played a key role in the anabolism of cultured HeLa cells, rather than in the catabolism. In contrast, Barnabé and Butler [12] considered glucose to be the main energy source in the culture of hybridoma cells, with only 28% ATP derived from glutamine. However, Fitzpatrick et al. [13] found in the culture of a B-lymphocyte hybridoma that glucose and glutamine had an equal effect on the energy supply in the exponential growth phase. However, these conclusions were drawn mainly from experimental results on concentrations of nutrients at the extracellular level. Lu et al. [14] employed their new method to characterize the growth and metabolism of CHO cells at low glucose concentration in glucose pulse culture. It is the first time the metabolic status of CHO cells was analyzed on the basis of the intracellular concentrations of TCA cycle acids. It also makes it possible to gain an insight into the intercellular response to the varying conditions and fully understand the cell metabolism behaviors in order to facilitate the further optimization of animal cell culture. In this work by Lu et al., the metabolism of CHO cells at low glucose concentration was investigated at the intracellular level by the determination of intracellular metabolites, including ATP, organic acids, and amino acids. They found that the growth of CHO cells was limited at below 1.22 mmol L^{-1} glucose. The decrease of intracellular ATP concentration with the glucose concentration showed that the energy shortage might be the reason for the growth limitation at low glucose concentration. The specific consumption rates of the majority of amino acids increased when the glucose concentration was below 0.54 mmol L^{-1}, indicating that these amino acids were mainly utilized in cellular catabolism to replenish the energy deficit. The increased consumption of amino acids could not completely eliminate the energy shortage at low glucose concentration. In their conclusion, glucose was shown to be indispensable as energy supplier for CHO cell growth. Intracellular concentrations of organic acids changed little with the decrease of glucose concentration. High intracellular concentrations of malate and undetectable isocitrate might be the result of the citrate bypass from the TCA cycle for the other anabolisms.

There are two noticeable points in the work. First, the improved methods were used to investigate the growth and metabolism of rCHO cells on the metabolite concentrations at both the intracellular and extracellular levels. Second, the different metabolism shift of rCHO cells was found in glucose pulse culture, in contrast to other animal cell lines. All the results are helpful to optimize and control the culture process.

Most of the applied cell lines in research and production are from mammalian sources. Considering the different nutrition sources, Chen et al. [15] investigated the growth and metabolism of marine fish Chinook salmon embryo (CHSE) cells responding to lack of glucose and glutamine. A peculiar phenomenon that CHSE cells could be passaged into the medium lacking both glucose and glutamine was completely different from mammalian cell culture. Before the work, no cell line had been reported that could grow normally in a medium lacking both glucose and glutamine.

To elucidate the metabolic mechanism, the metabolism parameters, key glycolytic and glutaminolysis enzymes, intracellular ATP and energy charge of CHSE cell were investigated at different glucose and glutamine concentrations. For glucose metabolism in the glutamine-free culture, specific glucose consumption rate, Q_{Glc}, remained unaltered because of the constant hexokinase activity. Meanwhile, the decrease of lactate dehydrogenase (LDH) activity by 20% indicated a shift away from lactate production, offering an explanation for the decreases of specific lactate production rate, Q_{Lac}, and the yield of lactate to glucose, $Y_{Lac/Glc}$, showing that lack of glutamine did not expedite glucose consumption but make it shift to lower lactate production and a more efficient energy metabolism. This result differed from pulse culture of baby hamster kidney (BHK) cells, whose Q_{Glc} reduced with the decrease of glutamine concentration. The results are useful in the design of an optimal medium and the establishment of a process control strategy for mass cultivation of CHSE cells. It provides another view of altering glutamine concentration to influence glucose metabolism for optimizing CHSE cell culture. Every kind of cell lines has its own traits, and an optimal culture method is achieved by a series of research. However, up to now, why this peculiar phenomenon occurred has not been explained, and there probably exists a different set of enzyme systems in CHSE cells.

2.1.2 Glutamine Metabolism

In animal cell culture, glutamine is the most important among all the different amino acids. The amount of glutamine present in commercial mammalian cell media is 2–6 mmol L^{-1}, ten times higher than other amino acids. Glutamine is the major nitrogen source in the synthesis of purine, pyrimidine and aminosaccharide. Actually ammonia in a culture is derived predominantly from the amide group of glutamine. On the other hand, glutamine decomposes to pyrrolidone-carboxylic acid and ammonia spontaneously even at 4 °C, limiting the shelf life of the medium.

In the last 30 years, the effects of glutamine on animal cell culture have been extensively investigated and the mechanisms of glutamine metabolism in various cell lines have gradually become more clearly understood. In animal cell culture, glutamine can provide 30–65% energy for the growth of cells and the synthesis of products. However, the cells often use excessive glutamine to produce ammonia. Ammonia has long been identified as one of the most important inhibitory substances for mammalian cells. High concentrations of ammonia can result in the decrease of specific growth rate and final cell density, the inhibition of virus proliferation in cells, as well as specific alterations of protein glycosylation. Although a number of protocols for ammonia removal have been established [16–22], most reports focus on how to reduce ammonia formation in mammalian cell culture. The most common preventative measure is to perfuse culture medium. Despite its simplicity, perfusion has its drawbacks due to the large requirement for culture medium, nutrients of which are less completely utilized than in batch and fed-batch systems, and the dilution of product which increases costs of downstream processes. Another method is to replace glutamine with non- or less-ammoniagenic compounds, such as glutamine derivatives (dipeptides), alanine, glutamate, asparagine, α-ketoglutarate, and pyruvate [23]. Glutamate seems to be a suitable candidate as it has one less amino group than glutamine and is relatively thermo-stable and low-priced. However, the adaptation of various cell lines to glutamate-based media is always inevitable, with different durations by different cell lines, and, in general, the cell yields in glutamate-based media were less than in glutamine-based control.

Huang et al. [23] presented a less-ammoniagenic medium for the growth of Vero cells by substitution of glutamine with glutamate to reduce the ammonia production and investigated some limitations governing Vero cell growth in a glutamate-based medium. With the glutamine-free medium supplementing asparagine to a glutamate-based medium, cell yield in batch culture was higher than with glutamine-based medium. This result is very different from other research.

In this work, glutamate was supplemented to substitute for glutamine in the medium to formulate DMEM-glu. Cells cultured in DMEM were directly transferred to and subcultured in a DMEM-glu medium. Vero cells preferred to use glutamine. For the cell cultures in two kinds of media there were no remarkable changes in the metabolism patterns of amino acids except for glutamine, asparagine, and alanine. In DMEM-glu, alanine and asparagine accumulation were very low. On the other hand, in DMEM, Vero cells continuously accumulated alanine up to 0.98 mmol L^{-1} at the end of the batch culture. However, in DMEM-glu culture alanine decreased from 0.20 mmol L^{-1} on day 1–0.036 mmol L^{-1} on day 3 and then remained constant at a low level. Similarly, and as expected, in DMEM-glu culture ammonia concentration remained at a low level between 0.30 and 0.44 mmol L^{-1}, in comparison with 3.0 mmol L^{-1} at the end of the culture in DMEM culture.

Alanine and ammonia are the main nitrogenous end products of glutamine metabolism in glutamine-based medium. In this case the sums of net alanine and ammonia productions were 3.29 mmol L^{-1} in DMEM and −0.199 mmol L^{-1} in DMEM-glu, indicating the cellular requirement for aminonitrogen in the DMEM-glu medium. When 4 mmol L^{-1} NH_4Cl was added to DMEM-glu to formulate DMEM-glu-NH_3, the growth was improved and the maximum viable cell density was significantly higher than that obtained without NH_4Cl and comparable to that obtained in DMEM culture. During the cultivation, ammonia levels decreased gradually. In previous research, ammonia was a major inhibitive by-product to cell growth, but the work by Huang et al. found that it depended on the cell lines according to their metabolism.

Zhou et al. [24] researched the effect of glutamine concentration on the growth, metabolism and endostatin production of microencapsulated rCHO cells. In the initial glutamine concentrations from 2.69 to 9.05 mmol L^{-1}, the maximum density of viable cells and multiplication ratios almost kept constant. This indicated that when the glutamine concentration was above 2.69 mmol L^{-1}, the growth of microencapsulated cells was independent of glutamine. This will be the basis for medium design. However, glutamine plays a significant role in the endostatin production. The endostatin concentration reached its peak of 546 ng mL^{-1} with the initial glutamine concentration of 4.97 mmol L^{-1}. There might be an optimal glutamine concentration for endostatin production.

To reduce the harmful effect of ammonia, glutamine synthetase gene was recombinated into the CHO cell to synthesize glutamine with glutamate and ammonia. Zhang et al. [25] elucidated the metabolic characteristics of CHO-GS cell (recombinant CHO cell harboring glutamine synthetase gene) in the medium with or without glutamine. In the medium without glutamine, the cells grew normally as those in the medium with glutamine because of the existence of glutamine synthetase (Table 1). Most amino acids were used more quickly in glutamate-based medium than in glutamine-based medium. Glutamate was a major substrate rather than glutamine, and glutamine became a produced amino acid. Experimental assays showed that there was no obvious difference in the intracellular glutamine concentrations in the two media, and CHO-GS cells could provide sufficient glutamine from glutamate in the glutamate-based medium. Meanwhile, the activities of almost all key enzymes of cellular glutaminolysis in the glutamate-based medium were higher than those in the glutamine-based medium. In the glutamate-based medium the cellular metabolism of amino acids became more active.

As the glutamine metabolism in fish is different from that in mammals, it is interesting to investigate the glutamine metabolism of fish cells. Chen et al. [26] found in Chinook salmon embryo (CHSE) cell culture that glutamine-free medium-199 could support the growth of CHSE cells as well as medium-199 with glutamine. The maximum cell density, 16.19×10^5 cells mL^{-1}, was obtained at 288 h in the culture with 0.54 mmol L^{-1} glutamine. This result showed that the glutamine was not as necessary in fish cell culture as in mammalian.

2.1.3 Interaction Between Glucose and Glutamine Metabolism

Glucose and glutamine are main nutrients of animal cells in culture. Glucose is utilized in two ways, namely the pentose phosphate pathway to provide nucleotides for biosynthesis and the glycolysis pathway to provide metabolic intermediates and energy. Though a small portion of glucose enters the tricarboxylic acid (TCA) cycle, most of the glucose goes to the lactate production. Like glucose, glutamine can also be used in two ways, either as protein and peptide constituent or to be converted to the precursors of nucleic acids, purines and pyrimidines. However, more glutamine may take part in the deamination and transamination reactions. Cell viability in most mammalian cell lines declines immediately and rapidly once glucose is depleted. However, if glutamine is depleted, some cell lines find it difficult to survive but others can gradually adapt to the condition lacking glutamine.

Table 1 Growth and metabolism of CHO-GS cells at 96 h in glutamate- and glutamine-based cultures [25]

Medium	X (10^6 cells mL^{-1})	Viability (%)	QGlc (10^{-9} mmol $cell^{-1}$ d^{-1})	$Y_{Lac/Glc}$ (mol mol^{-1})
Glutamate-based	1.41 ± 0.09	96.0 ± 1.27	3.2	1.70
Glutamine-based	1.12 ± 0.14	94.8 ± 1.19	2.8	1.25

Chen et al. [15] found that in the culture of marine fish Chinook salmon embryo (CHSE) cells, they could be passaged into the medium lacking both glucose and glutamine, which was completely different from mammalian cell culture. The metabolism parameters of glucose and glutamine showed in their work that lack of glutamine did not expedite glucose consumption but made it shift to lower lactate production to increase energy metabolism efficiency. At the same time, hexokinase activity remained constant, and lactate dehydrogenase (LDH) activity decreased. In the glucose-free culture, the specific consumption rates of glutamine and other amino acids increased greatly, suggesting that lack of glucose forced the cells to utilize amino acids as energy sources. Simultaneous increase of glutaminase activity and of specific ammonia production rate suggested an increased flux into the glutaminolysis pathway, and increases of both glutamate dehydrogenase activity and yield coefficient of ammonia to glutamine showed an increased flux into deamination pathway. When glucose and glutamine were both lacking, the specific consumption rates of most of amino acids increased markedly. ATP plays an important part in cell growth, while the metabolism of glucose and glutamine are the main sources of energy. Chen et al. [15] also researched the intracellular ATP under conditions of low glucose, glutamine or both of them. They found the ATP content was almost unchanged with lack of glutamine or glucose, and only decreased by 11% even when lacking both glucose and glutamine.

As mentioned in Sect. 2.1.1, Lu et al. [14] investigated the metabolism of CHO cells at low glucose concentration in the pulse culture and found the relationship among glucose, ATP and amino acids metabolism. In the pulse culture of CHO cells, when the glucose concentration was less than 1.2 mmol L^{-1}, the consumption rate of glutamine decreased and the intracellular ATP concentration increased with the increase of glucose concentration. However, when the glucose concentration was above 1.2 mmol L^{-1}, the change of glucose concentration had no distinct effect on the glutamine metabolism and the ATP level. Glucose concentrations had a greater effect on intracellular ATP concentrations than did glutamine concentrations, which was evidence that glucose was the main energy provider for CHO cells. Different from glucose, when glutamine concentration was less than 1.2 mmol L^{-1}, the increase of glutamine concentration could promote the utilization of glucose and the intracellular ATP concentration increased. When glutamine concentration was above 1.2 mmol L^{-1}, further increase of its concentration would not promote the utilization of glucose and the intracellular ATP concentration remained unchanged. Lu et al. demonstrated that glutamine as well as some other amino acids could be utilized to make up the shortage of glucose although amino acids cannot completely replace glucose as the only energy source.

Sun et al. [27] researched the effects of glucose and glutamine on the growth and maintenance of recombinant Chinese hamster ovary (CHO) cells in batch cultures. In batch cultures at low glucose concentrations, viable cell densities declined in pace with glucose exhaustion at 108 h although glutamine was available at that time. However, in the cultures at high glucose concentrations, a decline of viable cell densities did not occur while 11 mmol L^{-1} glucose remained in the media at 132 h. These indicated that cell viability rapidly declined with the depletion of glucose even though glutamine was available. In the work of Sun et al., it was found that glucose was an obligatory nutrient for cell growth and maintenance, and was not limiting to cell growth when its concentration was above 6 mmol L^{-1}. On the other hand, glutamine was another essential nutrient, and was not limiting when its concentration was above 0.26 mmol L^{-1}. Glutamine was not critical to cell maintenance since cells grew a little and cell viability was as high as 90% for 40 h after the depletion of glutamine provided glucose was available. rCHO cells could not grow or maintain themselves with the exhaustion of glucose. Glutamine could not substitute for glucose as the sole energy or carbon source.

2.1.4 Amino Acids Metabolism

In animal cell culture, amino acids metabolism is very critical for the cell growth and protein production. It contains the anabolism to form proteins, polypeptides and other nitrogen related active compounds, and catabolism to supply essential energy for supporting cell growth and survival. Because of its close coupling with the energy metabolism, more concern has been focused on this field to find a way to achieve high cell density and to improve the protein production. Thus, the control of amino acid metabolism is considered to be one of the most important aspects in animal cell culture.

Twenty kinds of amino acids constitute various proteins needed for the cell growth and proliferation. However, higher eukaryotic cells have lost the ability to synthesize some of these basic amino acids which are called essential amino acids, while those synthesized by cell are called non-essential. However, in fact, all these 20 amino acids play their part in a cell culture.

In animal cell culture, the medium is often supplemented with all the above-mentioned amino acids, and therefore the cell itself need not synthesize these so-called non-essential amino acids to maintain its normal proliferation, but would rather conveniently make use of them from the well-designed medium. So when talking about amino acids metabolism, we usually refer to how the catabolism occurs and the process by which an amino acid is degraded into other compounds.

In general, there are two stages in amino acids catabolism in in vitro cell culture. The first stage includes the removal or transfer of ammonia, the formation of α-keto acids, and the synthesis of other amino acids. In vivo, the by-product ammonia can further be transferred and removed outside, while in in vitro animal cell culture ammonia accumulation is serious and so harmful that it may reduce the growth rate and maximal cell density in batch culture, and cause changes in metabolism and inhibition of protein production. The second stage is mainly about the degradation of α-keto acids into CO_2 and H_2O and the generation of energy for cell growth. In this stage, α-keto acids can also be transformed into polysaccharides or lipids for energy storage. In fact, most amino acids would be metabolized this way, resulting in the formation of by-products such as ammonia and alanine. The accumulation of ammonia has been clearly proven to be toxic for the cell proliferation while the alanine accumulation may be less toxic, but affect the cell behavior in a different way.

The effect of these by-products on mammalian cell culture will be discussed in detail in the following part of this review.

Most in vivo amino acids catabolism starts with the removal of ammonia. Generally speaking, there are about four mechanisms in this kind of reaction, including oxidative deamination, transamination, associated deamination, and non-oxidative deamination.

Oxidative deamination refers to the removal of amine by oxidation. This kind of reaction is often catalyzed by certain oxidases. For instance, most cells possess glutamate dehydrogenase, which catalyzes the removal of amine from glutamate to form α-ketoglutarate.

Transamination means the removal of amine by transferring from one amino acid to another molecule, usually an α-keto acid. During the process, the former is degraded into an α-keto acid and a new amino acid is formed at the same time. Most of the 20 kinds of indispensable amino acids release their ammonia by transamination except glycine, lysine, threonine and proline. In animal transamination reactions, the key enzyme is transaminase. There also exist many amine recipients, of which the α-ketoglutarate, pyruvate and oxaloacete are the most important. After transamination, they can be transformed into glutamate, alanine and aspartate respectively. Biosynthesis of these products is closely linked to the central intermediary metabolism.

Associated deamination describes an ammonia removal process combined with more than two different kinds of mechanisms. One example is the combination of glutamate dehydrogenase and transaminase. At first, amine is transferred to α-ketoglutarate to form glutamate and then, under the effect of glutamate dehydrogenase, the ammonia of glutamate is removed and leads to the formation of a new α-ketoglutarate which can transfer another batch of ammonia. The whole associated transamination is reversible. Another example is related with the metabolisms of amino acids and nucleic acids, and ammonia is transferred between these two systems. Thus, the transfer of ammonia has built a sophisticated relationship among amino acids metabolism, glycolysis, lipid metabolism and nucleic acid metabolism in all.

Non-oxidative deamination occurs relatively infrequently in amino acids metabolism. This refers to the ammonia removal process without oxidation. For example, serine can be catalyzed to remove ammonia by a kind of dehydratase SDH while cysteine can do the same thing by desulfuration. In addition, unlike other amino acids, aspartate can remove its amine directly by aspartase.

After the first stage of transamination or deamination, α-keto acids metabolism begins. α-Keto acids are a kind of intermediate metabolism product, which can be transformed back to form amino acids from associated ammonification, or become CO_2 and H_2O through further degradation processes closely linked to TCA. Thus, it is easily deduced that amino acid is not only a nitrogen source but also a source of carbon.

Among all the amino acids, glutamine may be the key aspect in animal cell culture as mentioned above. In glutamine metabolism, many studies have proven that glutamine is first transformed to glutamate by glutaminase, but it is doubtful whether the amine is removed from glutamate by transamination or deamination in the second step, which is dependent on the cell specific characteristics and culture environments. Many researchers are interested in the nitrogen transfer process here. It was showed in hybridoma cells that ammonia was removed from glutamate mainly by transamination, while the deamination process was nearly prohibited. Also, the same situation was found in HeLa and CHO cells by other researchers. Huang et al. [28] used $^{1}H/^{15}N$ NMR spectroscopy to investigate the ammonia removal process of Vero cells in glutamate metabolism. In the study, ^{15}N-labelled glutamate or asparagine was added into the culture medium, and their incorporation into nitrogenous metabolites was monitored by heteronuclear multiple bond coherence (HMBC) NMR spectroscopy. In cells incubated with L-[^{15}N]-glutamate, the ^{15}N label was subsequently found in a number of metabolites including alanine, aspartate, proline, but no detectable $^{15}NH_4^+$ signal occurred. This indicated that the glutamate was utilized mainly through transamination rather than deamination. In cells incubated with L-[2-^{15}N]-asparagine the ^{15}N label was subsequently found in aspartate, the amine group of glutamate/glutamine, and two unidentified compounds. Incubation of cells with L-[4-^{15}N]-asparagine showed that the amide nitrogen of asparagine was predominantly transferred to glutamine amide, and there was also no detectable production of $^{15}NH_4^+$, indicating that most of asparagine amide was transaminated by asparagine synthetase rather than deaminated by asparaginase. In addition, the activities of some nitrogen metabolism related enzymes were also investigated. In the enzymes, the activities of phosphate-activated glutaminase, glutamate dehydrogenase, and alanine aminotransferase decreased dramatically, and the activity of aspartate aminotransferase decreased slightly.

In general, during the batch culture of animal cells, certain amino acids are often found to decrease, including glutamine, arginine, histidine, leucine, isoleucine, lysine, methionine, phenylalanine, threonine, valine, aspartate and asparagine. Among them, arginine, leucine, isoleucine and valine are degraded quite fast. On the other hand, some amino acids usually accumulate in the medium, such as alanine, proline, glutamate and glycine. Different cell culture strategies may induce different amino acids metabolism characteristics. For example, alanine, the most common by-product during the cell culture, often accumulates in the medium, but its accumulation rate will decrease when glucose concentration in the medium is low [14]. Indubitably, the amino acids metabolism may change in different cell culture conditions.

2.1.5 Application in Fed-Batch Culture

In fed-batch culture the amount of one nutrient to be added depends on its consumption rate which, in turn, is a function of cell density, growth and metabolism. Through establishing the kinetics of the nutrients metabolism of animal cells, the viable cell density and product concentration in fed-batch culture can be greatly improved. However, the metabolisms of glucose, glutamine and other amino acids in animal cells are interrelated. Nutrient mixtures can be so complex that it is impossible to understand fully their inter-relations and too time consuming to optimize each nutrient. So it is important not only to establish the kinetics of individual nutrient consumption but also to determine their relationship.

Zhang et al. [4] presented a model of cell metabolism and monoclonal antibody production in hybridoma cell culture and according to the model a fed-batch culture in serum-free medium was conducted in a bioreactor with an optimized feeding strategy. In the culture process, nutrient concentrations, especially glucose and glutamine, were maintained at relatively constant and low levels. The formation of toxic by-products, such as ammonia and lactate, was greatly reduced in comparison with that in batch culture. The concentration of monoclonal antibody reached 350 mg L^{-1}, which was seven times higher than in batch culture.

In the work, on the basis of the morphology of the cultivated cells, specifically the proportion of round, smooth, and transparent cells, an initial medium was chosen, and the concentrations of glucose and glutamine in the culture process were 500 mg L^{-1} and 58.4 mg L^{-1}, respectively. First, the fed-batch culture process was divided into many sections. The feeding volume of each nutrient added into the bioreactor in every section was calculated by a kinetic model based on the nutrient concentration in feeding medium, the specific consumption rate of the nutrient, the average culture volume in the section, and the viable cell density. Among the variables, the key problem is to estimate the specific consumption rate of the nutrient. Cells can use substrates to synthesize new cell materials and extra-cellular products and to supply energy. Taking all influential factors into account, the model of substrate utilization is an equation with the specific growth rate of cells and the specific production rate of monoclonal antibody as variables. In turn, the specific production rate is the function of the specific growth rate of cells. Thus, the feeding volume of concentrated feeding medium in one section will depend on the increment of total cell density and the average viable cell density in the section. At any time, cell density could be measured by sampling and the feeding volume of the medium in the section was estimated by the model. A fed-batch culture was conducted in the section. At the end of the section, a sample was taken to compare the measured value with the estimated value so as to guide the feeding in the next section.

As the result, the fed-batch culture of WuT3 hybridoma cells in a 5-L CelliGen bioreactor was conducted in serum-free medium with the help of the control model. The maxima of viable and total cell densities were 6.1×10^6 cells mL^{-1} and 9.4×10^6 cells mL^{-1}, respectively, and the maximum concentration of monoclonal antibody was 350 mg L^{-1}. In the culture the nutrient concentrations in the supernatant were relatively steady, which demonstrated that the feeding medium was well formulated and that the feeding strategy was successful. Although the yields of both lactate to glucose and ammonia to glutamine in the culture were lower than those in both the batch culture and the fed-batch culture without model control (Table 2), the concentrations of lactate and ammonia were too high to keep the high cell viability for a longer period of time. The kinetic model of cell metabolism and monoclonal antibody production, presented by Zhang et al. [4], correctly described the relationship between the cellular metabolism and the cell density and growth rate.

Table 2 Comparison of parameters of batch culture and fed-batch cultures of WuT3 hybridoma cells with and without model control [4]

Parameter	Batch culture	Fed-batch culture	
		Without model control	With model control
Culture span, h	96	156	348
Initial glucose concentration, $g\ L^{-1}$	20	2.8	2.8
Initial glutamine concentration, $mmol\ L^{-1}$	4.0	1.0	0.4
Maximal viable cell density, 10^6 cells mL^{-1}	1.1	2.1	6.1
Total cell density, 10^6 cells mL^{-1}	1.3	2.8	9.4
Final antibody concentration, $mg\ L^{-1}$	48	119	350
Glucose consumption, $mmol\ L^{-1}$	11.7	12.2	61.1
Lactate production, $mmol\ L^{-1}$	21.0	14.0	63
Glutamine consumption, $mmol\ L^{-1}$	3.1	3.8	10.0
Ammonia production, $mmol\ L^{-1}$	4.0	2.5	5.0
Lactate/glucose, $mmol\ mmol^{-1}$	1.80	1.15	1.03
Ammonia/glutamine, $mmol\ mmol^{-1}$	1.29	0.66	0.50

2.2 *Effect of Metabolic By-Products on Cell Culture*

Cellular metabolism of glucose and amino acids produces various kinds of by-products in mammalian cell culture. In China, much research has been focused on this field, especially on the effects of ammonia, lactate and alanine on hybridoma and CHO cell culture in recent years.

It is widely accepted that ammonia and lactate are two key inhibitive factors in the processes of cell growth, metabolism and product synthesis. As a glycolysis product, the accumulation of lactate seriously affects the cell growth mainly by changing pH and osmotic pressure of the culture environment. High concentration of lactate in culture medium will increase the osmotic pressure, which is considered to lead to the increase of cell maintenance energy. Then the pathway of cell metabolism will be changed, and thus the cell growth will be inhibited. Therefore, it is very important and necessary to control the lactate production, particularly in the high-density cell culture. The effects of ammonia on glycometabolism are different in different cell lines. The accumulation of ammonia in hybridoma and CHO cell culture can severely inhibit the cell growth. Besides, it will affect pH regulation, protein glycosylation, and amino acids metabolism as well. With the development of gene manipulation technology, a CHO-GS cell line has been successfully constructed [29]. Because of its enhanced synthetic ability of glutamine from glutamate by foreign glutamine synthatase, CHO-GS can utilize glutamate instead of glutamine to decrease ammonia formation, so that it has a tremendous potential in the animal cell culture industry. Apart from the two key factors above, alanine also plays a role in the cell metabolism which accumulated more than any other amino acids in culture medium. Although alanine is not critical in animal cell culture, the addition of alanine to a medium can inhibit the metabolism of amino acids, reduce the production of ammonia and certain amino acids, as well as increase the utilization of glucose and the production of lactate as well.

2.2.1 Lactate Effect on Cell Growth

Under anaerobic conditions, pyruvate is converted to lactate by lactate dehydrogenase (LDH) and the lactate is transported out of the cell into circulation. Most of glucose is converted to lactate via the glycolysis in the process of cell growth in vitro and nearly 15% glutamine is metabolized to lactate.

Sun and Zhang [30] investigated the effects of lactate and osmolarity on cell growth, metabolism and EPO expression of recombinant CHO cells in pulse cultures with different lactate concentrations and osmolarity. It was demonstrated that the lactate molecule itself had little effect on final cell density, metabolite accumulation, and EPO productivity. With the increase of sodium lactate concentration from 0 to 69 mmol L^{-1}, corresponding to the increase of osmolarity from 300 to 420 mOsm kg^{-1}, the specific growth rate (μ) of recombinant CHO cells was reduced by 33%. As control, the increase of NaCl concentration from 0 to 80 mmol L^{-1} corresponded to the increase of osmolarity from 300 to 420 mOsm kg^{-1}; the specific growth rate (μ) of recombinant CHO cells was also reduced by 33% (Table 3). It was clear that increased osmolarity inhibited cell growth while lactate concentration increased.

The specific consumption rates of both glucose and glutamine increased by 23% at 69 mmol L^{-1} lactate. The yields of cell growth to glucose consumption and glutamine consumption were reduced by 45% and 47%, respectively, and the specific production rates of lactate almost kept constant (Table 4). It was evident that more nutrients were consumed via aerobic metabolism pathway for more energy production to meet the increasing demand of maintaining energy of recombinant CHO cells at high osmolarity. The yield of ammonia production to glutamine consumption increased with the osmolarity while the yield of alanine decreased. It was shown that the flux of glutamine consumed via glutamate dehydrogenase pathway increased in response to the increase of osmolarity. On the other hand, EPO production rate decreased with the increase of osmolarity because of low specific growth rate.

In the culture of anchorage-dependent cells, the cell growth rate usually relies not only on the concentrations of nutrients and byproducts, but also on the availability of surface for cell growth. Under low cell density condition, cell growth rate mainly relies on the concentration of nutrients and by-products. In batch culture, the short exponential growth phase mainly results from the contact inhibition, but not the limitation of nutrient nor the inhibition of by-products.

Sun and Zhang [31] established a novel method to estimate recombinant CHO cell density in a packed-bed bioreactor by lactate production rate. The specific growth rate of the cells gradually decreased from 0.036 h^{-1} at 60 h to zero at 96 h

Table 3 Effects of lactate and osmolarity on specific growth rate of recombinant CHO cells in pulse cultures with different lactate concentrations and osmolarity [30]

Lactate (mmol L^{-1})	NaCl (mmol L^{-1})	Osmolarity (mOsm kg^{-1})	μ (d^{-1})
0	0	300	1.41
23	0	330	1.37
46	0	370	0.99
69	0	420	0.95
92	0	470	0.83
0	0	300	1.48
0	20	330	1.37
0	40	360	1.17
0	60	390	1.01
0	80	420	0.99

Table 4 Effects of lactate and osmolarity on glucose metabolism and EPO production in pulse cultures with different lactate concentrations and osmolarity (data from [30])

Lactate (mmol L^{-1})	NaCl (mmol L^{-1})	Osmolarity (mOsm kg^{-1})	$Y_{X/Glc}$ (10^8 cell $mmol^{-1}$)	$Y_{Lac/Glc}$ (mol mol^{-1})	Q_{Glc} (10^{-8} mmol $cell^{-1}$ d^{-1})	Q_{lac} (10^{-8} mmol $cell^{-1}$ d^{-1})	Q_{EPO} (10^{-3} IU $cell^{-1}$ d^{-1})
0	0	300	1.1	2.8	1.3	3.7	1.9
23	0	330	1.0	2.5	1.4	3.5	1.7
46	0	370	0.6	2.3	1.6	3.6	1.8
69	0	420	0.6	2.3	1.6	3.7	1.7
92	0	470	0.5	2.1	1.7	3.6	1.5
0	0	300	1.2	3.0	1.2	3.6	1.9
0	20	330	1.0	2.6	1.4	3.7	1.8
0	40	360	0.8	2.5	1.5	3.7	1.6
0	60	390	0.6	2.3	1.6	3.6	1.7
0	80	420	0.6	2.5	1.5	3.7	1.6

while lactate production increased corresponding to the cell growth. The lactate production rate equation can be obtained:

$$\frac{dC_{Lac}}{dt} = \frac{(Q_{Lac})_g - (Q_{Lac})_m}{\mu_g} \times \frac{dX}{dt} + (Q_{Lac})_m X$$

where C_{Lac} is lactate concentration, mmol L^{-1}; X is viable cell density, cells mL^{-1}; $(Q_{Lac})_g$ and $(Q_{Lac})_m$ are specific production rates of lactate for growing cells and non-growing cells, respectively, mmol $cell^{-1}$ h^{-1}; μ_g is specific growth rate of growing cells. The identity of lactate productions between model calculations and experiment values in batch culture of rCHO cells showed that the lactate production rate method would be useful in tracing the cell growth in packed-bed bioreactors.

In the study by Chen et al. [32], main growth and metabolism features of Chinook salmon embryo (CHSE) cells were not affected at 1.1~ 6.14 mmol L^{-1} lactate, corresponding to a medium osmotic pressure of 260~ 272 mOsm kg^{-1}. High lactate concentrations such as 13.1~ 41.6 mmol L^{-1}, corresponding to a medium osmotic pressure of 283~ 381 mOsm kg^{-1}, inhibited CHSE cell growth and changed cellular metabolism. Inhibition of CHSE cell growth by lactate was mainly caused by the decline of pH and the increase of osmotic pressure in medium. CHSE cells could bear the lactate level as high as 7 mmol L^{-1} and grow regularly under a stable and neutral pH condition.

Zhang et al. [33] reported a different result in the study of lactate influences on growth and metabolism of WuT3 hybridoma cells in serum-free batch culture. Lactate below 5.3 mmol L^{-1} hardly influenced cell growth, and maximum cell density of 7.5×10^5 cells mL^{-1} was achieved, comparable to the control. The maximum cell density decreased to 6.0×10^5 cells mL^{-1}, while lactate increased to 10.5 mmol L^{-1}. Lactate concentration increased to 24.8 mmol L^{-1}, exponential phase of growth extended to 50 h (about twice the control). It was showed that lactate between 10 and 25 mmol L^{-1} inhibited cell growth and metabolism a little, and lactate above 34 mmol L^{-1} inhibited cell culture strongly. Viability of WuT3 cells corresponded well to the growth. The viability declined slowly during culture at lactate concentration below 24.8 mmol L^{-1}. However, cell death was quick at a lactate concentration above 30 mmol L^{-1}.

2.2.2 Production and Toxicity of Ammonia

Ammonia, one of the main by-products in animal cell culture, can seriously inhibit cell growth when it accumulates, albeit gradually, in the medium. Compared with other accumulated low- to medium-size molecular weight substances in the medium, such as alanine and lactate, ammonia is thought to be more damaging. Extensive research has shown that ammonia is very toxic to cell proliferation and protein expression through the reduction of growth rate and maximal cell density in batch culture, the change in metabolic rate, the perturbation of protein processing and so on. Besides, as the ammonia concentration increases, the pH of cultural environment changes and the normal physiology of the cell is negatively affected as well. For all the above-mentioned reasons, the control of ammonia accumulation is always regarded as a big challenge in the cell culture industry, and needs more of our attention.

In cell culture, ammonia is released mainly through amino acids catabolism which involves the utilization of 20 amino acids. Among them, glutamine is the most important because it is not only a main nitrogen source but also a main carbon source. Its metabolism can generate more ammonia than that of other amino acids. A successful animal cell culture process has to involve a reasonable control on the glutamine utilization.

In fact, the problem of ammonia accumulation has been studied for many years, and many strategies have been conceived to overcome it. Among them, some are based on environmental modification, like the substitution of glutamine by glutamate [20], the control of glutamine concentration at low levels [22], and some are based on the modification of the cell line, like the application of the rCHO-GS system [3]. However, the problem is so tough that no single method can resolve it completely.

Here, attention is just focused on the study of the negative effects of ammonia accumulation on cell culture. Actually, much work on this point has been published during recent years [18, 34, 35]. For most cell lines, just 1–5 mmol L^{-1} ammonia in the medium can seriously hamper normal cell growth although different cell lines may show a little difference in ammonia-bearing characteristics. The phenomenon has triggered the interest of many scientists to investigate how and to what extent the accumulation of ammonia would affect the growth of different cell lines.

The effects of ammonia on the WuT3 hybridoma cells were investigated in a serum-free culture [36]. Only a small amount of ammonia could seriously inhibit the growth of WuT3 cells, just like other cell lines. The inhibition effect of ammonia was concentration-dependent. The lowest ammonia concentration that could influence cell growth was 0.8 mmol L^{-1}. At that concentration, the maximum viable cell density was about 6.4×10^5 cells mL^{-1} compared with 8.5×10^5 cells mL^{-1} in control culture. Cell growth was inhibited completely at 6.5 mmol L^{-1} ammonia, and cells directly entered the death phase at 9 mmol L^{-1} ammonia. Cell viability was in good agreement with cell growth when manipulating WuT3 hybridoma cell culture in media with different initial ammonia concentrations. Before the cell growth reached its maximum density, cell viability was always very high. However, as ammonia concentration increased, cell viability decreased sharply.

Different cell lines have different ammonia-bearing characteristics. The effects of ammonia on recombinant CHO cell culture were also investigated [37]. Just like hybridoma cell culture, the addition of ammonia to the medium could seriously inhibit the growth of CHO cells. When the initial ammonia concentration was between 0.21 and 1.74 mmol L^{-1}, CHO cell viability was almost the same with a little fluctuation. However, when the initial ammonia concentration was increased to 3.09 mmol L^{-1}, 4.28 mmol L^{-1}, and 5.66 mmol L^{-1}, cell viability decreased sharply, and the cultures lasted 108 h, 84 h, and 60 h, respectively. In the analysis of CHO cell metabolism, it was found that, under low concentration of ammonia, glucose utilization was almost the same as the control. However, when the concentration was as high as 5.66 mmol L^{-1}, glucose utilization slowed down. The phenomenon implies the change of cellular glucose metabolism in different ammonia concentrations. In general, the accumulation of ammonia will influence the functions of some enzymes in glycolysis pathway and interfere with the normal glucose metabolism.

The effects of ammonia on rCHO-GS cells have been especially concerned because the cell obtains the ability to synthesize directly glutamine from glutamate and ammonia by gaining the foreign gene of glutamine synthetase. As an important feature, the cell line will utilize the ammonia of the medium and alleviate the by-product accumulation. Zhang et al. [38] investigated the effects of ammonia accumulation on rCHO-GS cell metabolism in serum-free culture. Unlike the above-mentioned common CHO cell lines, the rCHO-GS cell could endure much higher ammonia concentration. The maximum cell density of 15.6×10^6 cells mL^{-1} was obtained in the culture with 1.42 mmol L^{-1} ammonia. The growth of rCHO-GS cell was inhibited at an increased ammonia concentration. However, a cell density of 8.9×10^5 cells mL^{-1} was obtained even at 12.65 mmol L^{-1} ammonia.

The intracellular metabolic pathways were affected due to the decrease of the toxic effect of ammonia on the rCHO-GS cell [38]. With the increase of initial ammonia concentration from 0.36 to 12.65 mmol L^{-1}, the yield coefficients of cell to glucose and lactate to glucose decreased to 66% and 84%, respectively, and the activities of hexokinase, pyruvate kinase, and lactate dehydrogenase increased by 43%, 140% and 25% respectively. This would increase the glucose utilization. An increased activity of glutamate-pyruvate aminotransferase showed that the conversation from glutamate to α-ketoglutarate was shifted to glutamate-pyruvate transamination pathway. The deamination pathway was inhibited due to the decreased activity of glutamate dehydrogenase.

Since the accumulation of ammonia is very toxic for cell culture, its negative effect has to be prevented efficiently, mainly to control its formation. The methods include the change of medium constituents, the control of glutamine concentration and so on. If the approaches mentioned above are not effective, then a number of ammonia removal systems can be tried. Anyway, the methods applied to solve the problem of ammonia accumulation should be on the basis of thorough understanding of the cell line.

2.2.3 Alanine Effect

Alanine is the only amino acid that is consistently found to accumulate in the medium of mammalian cell culture. However, the effect of alanine has rarely been studied systematically. Alanine can be formed by the transamination reaction between pyruvate and other amino acids, although lactate is the main by-product of the glucose metabolism [39]. On other hand, alanine can be transformed to pyruvate by alanine transaminase in some conditions.

The influences of alanine on cell growth and metabolism were investigated in the batch cultures of WuT3 hybridoma cells in serum-free medium with different alanine concentrations [40]. In the range of alanine concentration of 0–8.9 mmol L^{-1}, alanine had little influence on cell growth and viability. However, as a by-product of glucose and amino acids metabolism, it brought about some changes in cell metabolism. Glucose consumption increased with alanine concentration, and 2.5 g L^{-1} glucose was consumed during the batch culture with 8.9 mmol L^{-1} alanine supplementation, in comparison with 1.76 g L^{-1} glucose consumption in the control without alanine supplementation. In contrast, glutamine consumption decreased in alanine supplementation. As the by-product of glucose metabolism, lactate production in different alanine concentrations corresponded to glucose consumption. Lactate accumulation increased to 21 mmol L^{-1} in 8.9 mmol L^{-1} alanine supplementation from 12 mmol L^{-1} without alanine supplementation. At the same time, the formations of ammonia and alanine decreased by alanine supplementation. In the yield of by-products, as alanine supplementation increased, the yields of lactate production to glucose consumption and ammonia production to glutamine consumption increased, and the yields of alanine production to glucose consumption and alanine production to glutamine consumption decreased. Alanine brought little effect on antibody production and, combined with the little influence on cell growth, it could be assumed that the specific production rate of antibody was unaffected by alanine.

3 Protein Expression Improvement in Animal Cell Culture

Because of its affinity for correct post-translation modification for recombinant protein, the mammalian cell, compared with bacteria and yeast, is an excellent host cell for the production of recombinant proteins, such as erythropoietin (EPO), hepatitis B surface antigen (HBsAg), von Willebrand factor (vWF) and so on. However, the productivity of mammalian cell expression system is unsatisfactory. The process of protein expression by mammalian cell is much more complicated, including transcription, post-transcriptional process, translation, post-translational process, secretion and stabilization of protein. Each step may be the bottleneck limiting the productivity of the expression system. So the way to improve the productivity plays a key role in this field.

At present, many strategies have been tested to improve the expression efficiency, such as the supplementation of foreign additives, the optimization of culture conditions, and the selection of preferable promoters. For an established cell line, the strategy of foreign additives supplementation is much more convenient and effective. Actually, there are many additives used successfully in the mammalian cell expression system, such as butyrate, adenosine 5′-monophosphate (AMP), dimethyl sulfoxide (DMSO), and chemical chaperones. On the other hand, the optimization of cell culture conditions is also very important because various cell lines differ in their cultural characteristics. According to some recent research, temperature and osmolarity can seriously affect the productivity of mammalian cell expression system.

3.1 Protein Expression Improvement by Foreign Additives

Among the strategies used to improve the productivity of animal cell expression systems, the supplementation of chemical agents is an effective, convenient and low-cost method because of their availability, convenience in use and inexpensive cost. The researches focus on some chemical compounds of low molecule weight. In China, many chemical agents have been investigated, and some of them have been proven to be useful to improve the protein production in animal cell culture.

Butyrate has been used to increase the production of EPO, large hepatitis delta antigen and monoclonal antibody in mammalian cell cultures [40–44]. Liu et al. [45] reported that sodium propionate and sodium pentanoate had a similar effect to butyrate. DMSO improved specific production rates of monoclonal antibody in hybridoma cells and a fusion protein in CHO cells [46, 47]. The productivity of

HBsAg in CHO cells was enhanced with DMSO [48]. Other additives, such as cysteine, adenosine 5′-monophosphate and rapamycin, are able to act as "enhancing agents" of protein production in animal cell culture [49].

The mechanisms of chemical agents to enhance protein production vary from different chemical agents. In recent years the mechanisms of chemical agents increasing protein production were also researched in China. Trichostatin A and sodium butyrate enhanced the human IL-5 expression by altering histone acetylation status at its promoter region [50]. Wang et al. [51] reported that the effect of DMSO on increasing HBsAg production was mainly due to the increase of HBsAg gene transcription rate.

The concentration and addition time of chemical agents are also very important for protein production, especially for cell growth. The most appropriate concentration and addition time of agents for certain cell lines may need to be determined by small scale tests. In addition, the effects of chemical agents on the protein production have cell specificity.

3.1.1 Butyrate

Butyrate, a four-carbon fatty acid, has multiple effects on cultured mammalian cells, including inhibition of cell growth, induction of differentiation and cellular apoptosis. It also causes histone hyperacetylation by decreasing histone deacetylase activity, which facilitates access of the general transcription factors [52]. Thus it can be used for improving gene expression. Furthermore, butyrate can influence the quality of foreign glycoprotein, which is very important for therapeutic application.

Sodium butyrate (NaBu) has been widely used in the cultures of hybridoma cells and recombinant Chinese hamster ovary (rCHO) cells to enhance the productions of monoclonal antibodies and recombinant proteins. The effects of NaBu on anti-A type blood antibody expression in hybridoma cells were investigated by Yao et al. [53]. NaBu of 0.25 mmol L^{-1} or 0.5 mmol L^{-1} added to RPMI 1640 medium could significantly increase the titer and affinity of antibody. Yin et al. [54] reported the effects and mechanism of NaBu in the expression of human clotting factor VIII (hFVIII) in recombinant mouse NIH/3T3 cells. The cells were incubated in DMEM medium containing NaBu for 24 h. After stimulation of NaBu, the levels of hFVIII increased by 70% in comparison with the control without NaBu addition. They also showed that NaBu enhanced the transcription of cDNA which encoded the heavy chain of hFVIII.

The effects of NaBu on cell growth, glucose metabolism and EPO expression were investigated [42]. The growth of recombinant CHO cells decreased with the addition of NaBu in the serum-free medium CHO-S-SFMII and was entirely inhibited at 2 mmol L^{-1} NaBu. The specific glucose consumption rate was decreased by 30% in the presence of NaBu. In addition, EPO production was not facilitated by the addition of NaBu, but it may be helpful to improve EPO productivity. So it was suggested that NaBu was unnecessary in cell growth phase, and it could be added at proper time to prolong EPO production period. There are different effects of NaBu on the quality of recombinant glycoproteins expressed in CHO cells. The effects depend on cell lines and proteins. Wu et al. [55] investigated the influences of NaBu on the glycosylation of EPO expressed in CHO cells. They found that NaBu significantly decreased the content of sialic acid but slightly increased the occupancy of N-linked oligosaccharides. Furthermore, the type of oligosaccharides was not changed significantly, although their total amount decreased. The plausible cause for sialic acid reduction is that high specific protein productivity induced by NaBu makes incomplete protein glycosylation.

In conclusion, a high concentration of NaBu used in recombinant glycoprotein production will play not only the beneficial role in the productivity but also a detrimental role in cell growth and glycoprotein quality. When NaBu is chose to enhance protein production, its detrimental effect on protein glycosylation has to be considered.

3.1.2 AMP

In recent years, a few studies have shown the effect of extracellular nucleotides on enhancing protein production [56]. Luo et al. [49] found that adenosine 5-monophosphate (AMP) can effectively enhance HBsAg production in the culture of rCHO cells, but it markedly inhibited cell growth. In order to increase HBsAg production and avoid the AMP-induced inhibition of cell growth, a culture strategy with periodical medium changes was established.

Luo et al. first studied the influence of different AMP concentrations on HBsAg production by rCHO cells. Generally, 4 mmol L^{-1} AMP is optimal for enhancing HBsAg production in the culture of rCHO cells. At concentrations above 4 mmol L^{-1}, AMP was toxic to cell growth and protein production. However, the marked inhibition of cell growth was observed in the presence of 4 mmol L^{-1} AMP in a batch culture and no obvious enhancement in HBsAg production was achieved, but specific HBsAg production rate was still greatly enhanced. To understand the AMP-induced increase in productivity, the intracellular ATP pool content and its fluctuations were determined. In the first 24 h after AMP addition, the intracellular ATP content significantly increased to 14.5×10^{-6} nmol $cell^{-1}$, 2.5-fold higher than that of the control, indicating that AMP could significantly and instantaneously increase intracellular ATP level. AMP could stimulate catabolic metabolism, leading to higher ATP production. So the effect of AMP on energy metabolism could be one of the reasons for the increased HBsAg production.

To increase HBsAg production and avoid the AMP-induced inhibition of cell growth, cells were changed to a fresh medium containing 4 mmol L^{-1} AMP at 96 h post-inoculation and periodically carried out every 48 h afterwards. Cell growth inhibition by AMP was observed again. However, in the first 48 h of cultivation, no increase in specific productivity and HBsAg production was observed. The second medium change to a medium with AMP showed a sharp increase in both HBsAg production and specific productivity by 90% and 150%, respectively. After the third medium change, HBsAg concentration increased to 2.2-fold of control and specific productivity to 3.3-fold. These enhancements were maintained until the end of the culture. The results demonstrate that the application of AMP in the late exponential stage of cell growth could markedly enhance HBsAg production.

3.1.3 DMSO

Dimethyl sulfoxide (DMSO), an amphipathic molecule soluble in both aqueous and organic media and a well-known cryoprotectant and a lipophilic solvent, has wide applications. However, there have been few reports about DMSO on protein production and cell behaviors in recombinant animal cell cultures. Recently it was found that DMSO could regulate the protein production in gene, protein, and cell levels.

At gene level, DMSO may prevent cell differentiation, inhibit cell growth, and reduce cell apoptosis. The addition of DMSO as an antioxidant can inhibit the induction of DNA damage and increase the productivity of protein. Seki et al. [57] showed that 1% DMSO decreased the amount of human immunodeficiency virus type 1 (HIV-1) required to establish similar infection by 90%, and the enhanced infection effect of DMSO was mainly at the step of transcription of viral RNA.

Zangar et al. [58] investigated the effect of DMSO in protein level. When cytochromes P450, such as CYP3A, CYP2E1 and CYP2B, were expressed, DMSO increased the formation of CYP3A and CYP2E1 proteins without the increase of their mRNA levels. It was deduced that DMSO altered the post-transcriptional regulation of these cytochromes P450. Neither CYP2B nor NADPH cytochrome P450 reductase expression was altered in response to DMSO treatment, and the effects of DMSO on CYP3A and CYP2E1 were selective. Also, CYP3A and CYP2E1 protein levels were increased ~2.5-fold at 12 h and returned to basal levels at 32 h, indicating that the effect of DMSO was time-dependent.

At cell level, DMSO enhances the membrane permeability. Therefore, the productivity of extracellular proteins increases. DMSO was recently found to exhibit three distinct modes of action, each over a different concentration range [59]. At low concentrations, membrane thinning and increased fluidity of the membrane hydrophobic core were induced by DMSO. At higher concentrations, transient water pores into the membrane were induced by DMSO. At still higher concentrations, individual lipid molecules were desorbed from the membrane followed by disintegration of the bilayer structure. DMSO also has complex effects on cell physiological state. The arrest of murine erythroleukemia at G_1 phase of growth and the induction to terminally differentiate by DMSO resulted in a 10- to 50-fold increase in the amount of adult α- and β-globins productivities [60].

Certainly, DMSO may have different effects on cell lines or proteins produced by one cell line. However, DMSO increases protein production in many pieces of research. Ling et al. [61] found that when adding 0.2% DMSO to shake flask cultures at maximal viable cell density, specific monoclonal antibody production was increased twofold without great effects on cell density and viability. Liu et al. [62] cultured several rCHO cell strains to produce different recombinant proteins with 1% DMSO addition. They found that the proliferation rates of all the cell strains decreased and the productivities of β-galactosidase and a fusion protein, composed of a B-cell surface antigen CD20 and an immunoglobulin G-Fc γ4 fragment as the amino-terminal fusion partner, were increased by 1.4- and 1.6-fold, respectively. The simultaneous addition of DMSO and pentanoic acid increased fusion protein production by up to 2.8-fold. Moreover, there are two contrary roles of DMSO in the rCHO cell culture. DMSO enhances recombinant protein productivity, and however, it induces G_0/G_1 phase growth arrest and reduces cell growth rate. So a two-stage process was presented in other work by Liu and Chen [63]. In the first stage, cells were incubated in a DMSO-free medium for a certain period to obtain a high cell density. In the second stage, DMSO was added to achieve high productivity. With the addition of 1% DMSO, volumetric productivity of macrophage colony-stimulating factor (M-CSF) was increased by 57% compared with the control.

The effects and mechanism of DMSO on HBsAg production have been investigated in recombinant CHO cells. Li et al. [64] studied CHO cell growth and HBsAg production at different DMSO concentrations (0.5–2.0%). They found 1.5% DMSO was the optimum concentration for the HBsAg production. HBsAg production and specific HBsAg production rate were improved by 56% and 270%, respectively, in the 4-day repeated medium change culture with 1.5% DMSO. Although 2.0% DMSO improved the specific HBsAg production rate more effectively than 1.5% DMSO, the HBsAg production was lower. The enhancement of HBsAg synthesis by DMSO was always accompanied with the inhibition of cell growth.

DMSO also alters cell metabolism. In the batch culture of recombinant CHO cells with 1.5% DMSO treatment, glucose consumption decreased by more than 50% compared with the control. The average specific glucose consumption rate at 144 h decreased over 40% compared with the control. Lactate production and its average specific rate in the culture with DMSO were lower than those in the control culture. Moreover, metabolism of amino acids was changed with the addition of DMSO. The consumption of many amino acids decreased, including glutamate, histidine, threonine, arginine, tyrosine, tryptophane, phenylalanine, and lysine. However, glutamate was accumulated in the culture with DMSO treatment. Also cystine consumption was enhanced by DMSO.

From the flow cytometer analysis of recombinant CHO cells, DMSO arrested cells at G_1 phase of cell cycle. The cell percentage at G_1 phase increased to 68.3% in the culture with 1.5% DMSO from 57.6% in control culture, in correspondence with the improvement of HBsAg production. Therefore, the improvement of HBsAg production by DMSO in CHO cells was connected with the arrest of cells at G_1 phase. The result of the flow cytometer analysis also showed that DMSO did not cause cell apoptosis in the concentration range of 0–2.0%. A large quantity of HBsAg was found inside cells.

In order to disclose the molecular mechanism behind these phenomena, a proteomic approach together with matrixassisted laser desorption/ionization-time of flight mass spectrometry (MALDI-TOF-MS) was applied to identify the proteins related to the cellular response to the DMSO presence [65]. Four enzymes related to glycolysis, including aldolase, triosephosphate isomerase, glyceraldehydes 3-phosphate dehydrogenase, and phosphoglycerate kinase, were identified to be down-regulated in the presence of DMSO. The redistribution of substrate metabolism flux by DMSO in CHO cells was believed to be contributed to the enhancement of HBsAg expression by DMSO. On the other hand, HSP70 and ERP57, two important chaperone proteins which play an important role in protein post-translational modification and secretion, were not up-regulated by DMSO, and the unmatched protein processing with the increase of protein expression could result in the accumulation of HBsAg in cells. The post-translational modification and secretion of HBsAg potentially became the limiting-step in further improvement of protein expression with DMSO.

In order to investigate the mechanism of DMSO effect on HBsAg production improvement in gene level, Wang et al. [51] used real-time PCR to detect the mRNA level of HBsAg. They found HBsAg mRNA content increased by about 1.5-fold at 1.5% DMSO. Also analysis of mRNA from cells treated with 1.5% DMSO and control at different culture time showed that the increase of HBsAg production was related to the enhanced mRNA levels. The increase of mRNA content could emanate from the increase of HBsAg gene copy number, the improvement of HBsAg mRNA stability, or the acceleration of HBsAg gene transcription. Using absolute quantification real-time PCR, they found that HBsAg gene copy number was not significantly changed in the cells stimulated with DMSO, and the HBsAg gene copy numbers of cells in addition of different DMSO concentrations were about 550 copies per cell. The half-life of HBsAg mRNA was measured by calculating the decay of RNA accumulation after transcription was blocked with actinomycin D. By real-time PCR, they found HBsAg mRNA stability of cells with DMSO treatment was also not obviously different from the control, and the mRNA half-life of 5.58 h in the cells at 1.5% DMSO was comparable to that of 5.36 h in the control culture. DMSO did not have an effect on HBsAg mRNA at the post-transcriptional level. Using nuclear run-on assay based on real-time PCR, they discovered DMSO resulted in 80% increase in HBsAg gene transcription activity. Therefore, it could be deduced that the acceleration of HBsAg gene transcription is the main cause of HBsAg production improvement.

Ma et al. [66] studied the HBsAg accumulation within DMSO-stimulated rCHO cells. Intracellularly accumulated HBsAg was localized in the dilated areas in which HBsAg was assembled into virus-like particles. The markedly enhanced formation of dilated areas in the DMSO-stimulated rCHO cells resulted from the significantly increased HBsAg expression, rather than the inducement of DMSO itself. Furthermore, it was discovered that the failure of HBsAg virus-like particle assembly within cells blocked the effective secretion of HBsAg from rCHO in response to DMSO, thus to limit the further improvement of HBsAg production.

3.1.4 Other Additives

Methotrexate

Dihydrofolate reductase (DHFR) is a commonly used target for selection pressure in recombinant mammalian cell, and protein expression has been greatly improved by cotransfection and coamplification of *dhfr* and exogenous gene sequences. Exogenous genes could be amplified effectively when relative concentration of methotrexate (MTX) increased gradually. During perfusion culture in a disc reactor, the specific EPO production rate was improved. However, MTX might be ineffective in improving the expression of some macromolecule recombinant proteins. Accordingly, MTX amplification shows the specificity of cell line and expressed protein.

Reductant

The expression of a secretory protein cannot be improved by increasing mRNA level when the secretory pathway is saturated. Correct folding in endoplasmic reticulum is important for protein secretion and it is necessary to optimize protein post-translational modification in the secretory pathway. Yoon and Ahn [67] found that secretion ability could be enhanced by optimizing redox environment in endoplasmic reticulum and raising expression level of PDI and Bip. Alberini et al. [68] reported that β-mercaptoethanol could improve secretion of immunoglobulin aggregation intermediate by B cell and plasma cell. Further studies indicated that reductant addition might destroy normal redox potential gradient of the secretory pathway, so that surplus proteins were secreted. There were also experiments proved that changing redox environment interrupted the formation of disulfide bonds.

Yoon and Ahn [67] studied reductant effect on recombinant EPO expression, and found that the specific EPO production rate (q_{EPO}) was decided by different concentrations and different kinds of reductant. Cysteamine was the most important reductant of all the experimental ones. No increase of intracellular EPO expression level by cysteamine also caught highly attention with the explanation that reductant improved EPO secretion by affecting the protein secretory pathway. In addition, each reductant has its own optimizing culture oxidation-reduction potential corresponding with the maximal q_{EPO}. According to this, there is not direct relationship between culture oxidation-reduction potential and q_{EPO}.

Chemical Chaperone

Chemical chaperones are a kind of low molecular weight compounds to stabilize the structure of proteins in vitro, including the organic solvent dimethyl sulfoxide and cellular osmolytes glycerol, glucose, and trimethylamine *N*-oxide. Chemical chaperones have been used to prevent the formation of the aggregates and the subsequent packaging into inclusion bodies in the process of in vitro folding. Sachiko et al. [69] reported that the use of chemical chaperones, including glycerol and trimethylamine *N*-oxide, could correct the defective folding of ΔF508 cystic fibrosis transmembrane conductance regulator protein (CFTR) in cystic fibrosis. Glycerol treatment also stabilized immature (core-glycosylated) ΔF508 and CFTR molecules that were normally degraded rapidly.

Phosphatidic Acid

Phosphatidic acid (PA) could improve growth of recombinant CHO, and also increased hIFN-g secretion [70]. When nonionic wetting agent Tween-80 and PA were added with 0.5% concentration into serum-free medium, extracellular hIFN-γ concentration increased by 240%, whereas the specific growth rate and the specific production rate did not increase. In this study, relationship between hIFN-γ production and cell growth was not determined. So mechanism of PA improving protein expression needs further researches.

3.2 *Effect of Culture Condition on Protein Expression*

The effect of culture conditions on cell growth and protein expression is important. Temperature, osmolarity, pH condition, dissolved oxygen (DO), carbon and nitrogen source, etc. are all considered to be optimized during a culture. Several strategies have been addressed to improve protein production, including temperature control of process optimization [71–73], medium exchange (partial or total) for osmolarity and pH maintenance [74], specific nutrient supplementations and different feeding strategies [75]. Some of these approaches have been efficient for improving specific protein production to a certain extent.

Temperature and osmolarity are two important factors to be optimized in the process. On one hand, they have an effect on increasing the cell percentage in G_0/G_1 phase and decreasing in G_2/M phase obviously. On the other hand, they can slow down the cell growth by lowering the temperature [71]. Therefore, the protein production phase can be extended. Cell growth was seriously inhibited under high osmotic pressure as well as the expression of protein [75].

3.2.1 Temperature

Various cell lines grow in different temperatures while the mammalian cell is commonly cultured at 37 °C. In most cases, both higher and lower temperatures other than 37 °C will retard cell growth. However, generally speaking, the cell can bear lower temperature rather than higher one. In the cell culture field, temperature has been widely investigated to find an optimized one for the cell growth and protein expression. For its complicated functions on cell cycle, apoptosis, metabolism and protein synthesis, temperature is the significant factor worth close investigation to ameliorate the cell culture process.

Recombinant baby hamster kidney (rBHK) cells have been widely used for production of recombinant proteins such as Factor VIII and von Willebrand factor (vWF). The effect of temperature on cell growth was investigated in the batch culture of rBHK cells [71]. At the lower temperature (33 °C), specific growth rate of rBHK cells declined, and the maximal density obviously decreased as well. Similarly, the cell grew slowly when the culture temperature was as high as 39 °C. However the cell culture at a lower temperature (33 °C) could lead to long maintenance of high cell viability. In further investigations on cell growth, it was found that growth retardation by temperature change was attributed mainly to the prolonging of lag phase, and the specific growth rate of rBHK cells was almost independent on the variation of temperature in exponential phase (Table 5). In the cell cycle, temperature changes had little effect on the cell percentage of S phase, and the cell percentage of G_0/G_1 phase increased from 43.8% at 37 °C to 51.4% at 33 °C and 52.5% at 39 °C while that of G_2/M phase decreased from 15.4% at 37 °C to 9.5% at 33 °C and 10.0% at 39 °C. Besides, the apoptotic cells of the whole cell population increased with temperature. When rBHK cells were cultured for 24 h, then cultured at different temperatures for further 40 h after medium change, apoptotic rBHK cells at 37 °C was 12.0%, higher than 10.8% at 33 °C and lower than 16.0% at 39 °C.

Temperature can influence protein expression in animal cell culture too. Several reports showed that lower culture temperature increased the cellular productivity of recombinant protein in some recombinant cell lines. In the same work by Yi et al. [71], in order to increase the vWF productivity of rBHK cells, the expression and the specific production rate of vWF were compared when culturing the cell at different temperatures. Clearly, the specific production rate of vWF at 33 °C would increase by 45% compared with 37 °C. It may be partly related to the cell arrest in the G_0/G_1 phase.

3.2.2 Osmolarity

Osmotic pressure is one of the major environment factors in animal cell cultures. It can influence the kinetics of enzyme reactions in cells, cell metabolism, and finally cell growth and protein production. So far, some reports from China have been published about the effects of osmotic pressure on animal cell cultures.

Various cell lines are different in the tolerance for osmotic pressure. Generally speaking, 260–320 mOsm kg^{-1} is suitable for the majority of animal cells. Sun and Zhang [30] reported the results in the culture of recombinant CHO cell. With the increase of lactate concentration from 0 to 69 mmol kg^{-1}, corresponding to the increase of osmolarity from 300 to 420 mOsm kg^{-1}, the specific growth rate of CHO cells was reduced by 33%, and the yields of cell growth to glucose consumption and glutamine consumption were reduced by 45% and 47%, respectively. Yi et al. [74] investigated the effect of osmotic pressure on cell growth in the batch culture of rBHK cells. When the osmotic pressure increased from 330 to 350 mOsm kg^{-1},

Table 5 Specific growth rates of rBHK cells cultured at different temperatures. μ_{24}, μ_{24-48}, average specific growth rates in the batch cultures during 0–24 and 24–48 h, respectively; μ, average specific growth rates in temperature pulse cultures or culture at 37 °C [71]

Temperature (°C)	μ_{24} (d^{-1})	$\mu_{(24-48)}$ (d^{-1})	μ (d^{-1})
33	0.69	1.5	1.3
37	1.2	1.5	1.5
39	0.77	1.4	1.4

no difference in cell growth was found, showing that the rBHK cells were tolerable for the change of osmotic pressure up to 350 mOsm kg^{-1}. However, when it increased directly to 370 mOsm kg^{-1}, there was an obvious decrease in maximum cell density. At 470 mOsm kg^{-1}, maximum cell density further declined. These results indicated that high osmotic pressure could inhibit rBHK cell growth, which was also proven by the average specific growth rates of cells at 24 h (Table 6). A similar phenomenon was also viewed by Chen et al. [32] in the culture of Chinook salmon embryo (CHSE) cells.

However, it was reported that hybridoma cells could adapt the change of osmolarity to a certain extent [76]. Similarly, when practicing the strategy by gradually increasing osmotic pressure from 330 to 370–470 mOsm kg^{-1}, rBHK cells could adapt to higher osmotic pressure [74]. Noticeably, no significant improvement in cell growth rate was observed when the stepwise increase of osmotic pressure from 370 to 470 mOsm kg^{-1} underwent a shorter process than 3 h in comparison with that when increasing to 470 mOsm kg^{-1} directly. However, when the stepwise increase of osmotic pressure underwent 6 h or longer, average specific growth rates increased by at least 40% than that when increasing to 470 mOsm kg^{-1} directly.

Osmotic pressure could influence protein expression in animal cell culture. The enhancement of monoclonal antibody production has been widely demonstrated when increasing osmotic pressure in the culture of hybridoma cells. The expression level and average specific production rate of vWF in rBHK cells under different osmotic pressures were investigated (Table 6) [74]. With the control under 330 mOsm kg^{-1}, the vWF expression level increased by 16% and 12% under 370 and 420 mOsm kg^{-1}, respectively, and the average specific production rate of vWF increased 47% under 370 mOsm kg^{-1} and 44% under 420 mOsm kg^{-1}. The result showed that the protein production of single rBHK cell was enhanced by a big margin under high osmotic pressure. The mechanism of protein production enhancement was found to be different from hybridoma cells. In hybridoma cells, low cell growth rate under high osmotic pressure resulted in an increased percent of cell population in G_0/G_1 phase of cell cycle, and average production rate of monoclonal antibody per cell would be higher because the cell in G_0/G_1 phase could synthesize antibody more rapidly. Yi et al. [74] found in rBHK cell culture that under enhanced osmotic pressure there were few changes in the cell percent in

Table 6 Specific growth rates at 24 h in the pulse cultures (μ_P) and batch cultures (μ_b) and vWF productions at 96 h in rBHK cell culture under different osmotic pressures (data from [74])

Osmotic pressure (mOsm kg^{-1})	μ_P (d^{-1})	μ_b (d^{-1})	vWF (IU mL^{-1})	Specific production rate of vWF (10^{-8} IU $cell^{-1}$ h^{-1})
330	1.54	1.15	4.3	4.5
350	1.50	1.09	-	-
370	1.41	0.7	5.0	6.6
420	1.28	0.61	4.8	6.5
470	0.76	0.51	2.2	3.6

G_0/G_1 phase and an obvious increase in the cell percent in S phase. In the recombinant BHK cell strain, foreign vWF gene is regulated by SV40 promoter which is S-phase specific to control vWF gene expression. The high cell percent of S phase can increase the specific production rate of vWF. For the culture under 470 mOsm kg^{-1}, the decrease of average specific production rate of vWF relative to that under 420 mOsm kg^{-1} might be derived from the strong inhibition of hyperosmotic pressure on the integrated activity of rBHK cells, and vWF expression level declined largely because of the superposed effect of sharp decrease of cell density.

3.2.3 Other Environmental Factors

Besides temperature and osmolarity, dissolved oxygen (DO), static pressure, pH value, stirring, etc. can also effect the expression of recombinant protein.

Generally, the optimal DO level is 10–60% air saturation. Its effect on mammalian cells in culture may vary in different cell lines and different conditions. Some cell lines can be adapted to withstand high oxygen concentration while for some cell lines high oxygen concentration environment may damage the cell membranes, reduce cell viability and induce cell death. Expression of recombinant protein may be changed with the different DO levels. Hu et al. [77] showed that prourokinase (pro-UK) had good yield with DO level varied from 20 to 45% when CHO cells were cultured on porous microcarriers in a 30-L stirred tank bioreactor. From day 42 to 72, when DO kept 20%, pro-UK production in bioreactor was 1.7–2.5 g day^{-1}, and when DO level decreased to 7–9%, the production was 0.75–1.7 g day^{-1}. But after 72 days, when some new microcarriers were added and DO level was maintained between 15 and 30%, pro-UK production could not recover to the best yield. At the same time, the lactate concentration in low DO level increased greatly. They concluded that low DO levels could influence the expression of protein in CHO cell and lactate may be one of the main reasons for pro-UK production decrease in low DO level.

Static pressure may also effect the protein expression [78–80]. Yu et al. [80] found that high static pressure could improve hGM-CSF production in CHO cell. They cultured CHO cells in normal pressure (0.1 MPa) for 24 h, then transformed into pressure containers with static pressures of 0.15, 0.4, 0.75, 0.9 MPa, respectively. After 96 h, the specific growth rate did not change nearly as much as the increase of the static pressure. However, the production of expression protein was improved from 1.23×10^{-13} mg $cell^{-1}$ h^{-1} at 0.1 MPa to 1.62×10^{-13} mg $cell^{-1}$ h^{-1} at 0.9 MPa, and increased by 32%. The mRNA level of hGM-CSF was also detected. The expression level of hGM-CSF mRNA was 0.33 ± 0.05 in 0.1 MPa and 0.40 ± 0.07 in 0.9 MPa. At the same time, β-actin, as the reference, did not have significant deviation. So, the increased production of hGM-CSF could emanate from the improvement of hGM-CSF mRNA level in high static pressure.

Regulation of pH value is essential for the survival of mammalian cells. The pH is important not only for maintaining the appropriate ion balance, but also for maintaining the optimal function of cellular enzymes and for optimal binding of hormones and growth factors to cell surface receptors. Changes in pH can alter cell growth conditions, and then may effect the production of recombinant protein (protein quantity and quality). The optimal pH of cell growth may be different from that of protein expression. So, the effect of culture condition should be investigated with different cell lines and different recombinant proteins.

All the animal cells are sensitive to the shear effect of stirring. As one of the regulation factors, stirring not only effects the DO level and cell growth, but also destroys the cells in intense stirring. When animal cell culture was scaled up, stirring should be designed cautiously.

4 Other Research Areas in China

4.1 Serum-Free Medium

Serum-free medium is designed to grow a specific cell type or perform a specific application in the absence of animal serum, but may contain serum constituents or substitutes. Serum-free medium represents an important tool, which allows cell culture to be done with a defined set of conditions. In China, the use of animal serum in biologicals preparation for human application is under more and more strict regulations. It is particularly important in industrial production where the use of serum presents a safety problem, especially a source of unwanted contaminations in the production of biopharmaceuticals.

The use of serum-free medium is common for many cell types [81–91] and many constituents can be obtained through a number of vendors. Nevertheless, there are a few drawbacks that should be considered when using serum-free medium. First, an accommodation time is necessary for cell adaptation to serum-free medium. The cells have to be weaned from serum slowly [81, 82]. Moreover, some cell lines may require the addition of growth factor or inorganic ion [83–85] to overcome deficiencies or promote growth in a certain medium. It is advisable to begin with a serum-free medium which has some kinds of growth factors, such as pituitary extract, which can be incrementally removed if necessary.

An extracellular matrix on the growth culture is essential for attachment-dependent cell lines. Serum provides some components playing the role as this matrix. Therefore, when using serum-free medium, the substratum (plastic dishes) should be pre-coated with fibronectin [86], laminin [87, 88] and so on.

To overcome these obstacles, efforts have been made to develop new kinds of serum-free medium for different demands [89–91]. Methods commonly used are cultivation process analysis, molecular biotechnology and statistical experiment.

In the work by Zhang et al. [90], MTT (3-(4,5-dimethylthiazol-2-yl)-2,5-diphenyltetrazolium bromide) assay was used to screen for potential nutritional factors in bringing about a growth medium for primary cell culture of *Suberites domuncula*. Ferric iron and pyruvate were found to significantly improve cell viability in a dose-dependent manner, while silicon and glutamine showed limited improvements at certain concentrations. Therefore, several improved media could be formulated to maintain high cell viability in a short-term culture of primary sponge cells.

In 2004, Yi et al. [89] formulated a new serum-free medium named SFMA using HADAMARD statistical design for the cultivation of rBHK cells and the expression of rec-vWF. In SFMA, the growth of rBHK cells and the expression of rec-vWF were higher than those in serum-supplemented culture.

Using factorial design combined with the steepest ascent method, Liu et al. [91] systematically built up a serum-free medium for recombinant CHO cell growth and fusion protein expression. The optimal composition of serum replacement for specific fusion protein production was 1% SITE (selenium, insulin, transferrin, ethanolamine), 0.3 g L^{-1} yeast extract, and 0.09% linoleic acid-BSA. These serum substitutes could also promote CHO cell growth and fusion protein production in nine kinds of commercial media.

4.2 *Fed-Batch and Perfusion Cultures*

The main techniques for mass cultivation of animal cells are fed-batch and perfusion cultures in stirred tank bioreactor. The models based to analysis of cell metabolism can facilitate the performance of fed-batch and perfusion cultures. Zhang et al. [4] presented a model of cell metabolism and monoclonal antibody production in hybridoma cell culture to conduct a fed-batch culture in serum-free medium in a bioreactor with an optimized feeding strategy. The concentration of monoclonal antibody reached 350 mg L^{-1}, seven times greater than that in a batch culture. Feng et al. [92] executed a fed-batch process in chemically defined protein-free medium with a glucose uptake coupled feeding of balanced amino acids together with groups of nutrients and a feeding of $CaCl_2$ and $MgCl_2$ concentrate. The maximal monoclonal antibody concentration achieved was approximately eight times that in serum supplemented batch culture. As for perfusion, factors such as temperature and cell aggregation are of vital importance to cell culture performance and need further investigation. Chen et al. [93] examined the potential of temperature shift as a tool in the optimization of the perfusion culture of CL-11G cells for the production of pro-urokinase. The specific pro-urokinase productivity of CL-11G cells increased by 74% at 34 °C compared with controls at 37 °C in batch cultures. Liu et al. [94] observed the perfusion culture of HEK 293 cells grown as suspended aggregates in a 7.5-L stirred tank bioreactor, and a maximum viable cell density of 1.2×10^7 cells mL^{-1} was achieved in 17 days culture. No significant difference in cell performance was found between HEK 293 cell populations grown as suspended aggregates and those grown as anchored monolayer. Their results demonstrate the feasibility and proof-of-concept for using aggregates as an immobilization system in the perfusion culture in large-scale stirred bioreactors.

4.3 Cell Damage by Shear Stress

Hydrodynamic shear stress has been known as an important factor influencing cell phenotype and function in animal cell or tissue cultures. For the cells adhering to microcarriers, it is the diameter of the microcarrier that is of major importance. When turbulent eddies are smaller than or the same size as the microcarrier, cells can be damaged. The smallest size of these eddies can be estimated using the Kolmogoroff eddy-length scale. Shear-induced death of animal cells is supposed to relate to a range of parameters including shear stress, shear time, power dissipation and the growth phase of the cells. On the other hand, for suspended animal cells, the important differences in liquid flow (pressure and momentum) are those that occur on the scale of cells. The death of cells caused by shear forces is related to the way of oxygen supply which could damage cell, influence cell morphology, cell aggregation, cell viability, in addition, escape the volatilizing gas from culture medium. The use of Pluronic F68, a component to protect cells against sparging damage, and the minimization of sparging-gas flow in reactor with a low superficial velocity could reduce the damage caused by bubble. Liu et al. [95] described the use of hydrodynamic shear stress derived from the proper agitation in controlling the formation and size distribution of HEK293 cell aggregates and providing sufficient mass transfer to support the normal growth and metabolism of HEK293 cells in suspended aggregates. Qiao et al. [96] also related shear stress to cell apoptosis. When reaching 15 dyn cm^{-2}, shear stress could inhibit LPS-induced apoptosis, which was supposed to inhibit the flow of signal transfer of LPS-induced apoptosis.

4.4 Bioreactors for Animal Cell Culture

Bioreactors play a key role in achieving the large-scale culture of animal cells. A few studies on the design of new type bioreactor and the performance improvement of bioreactor were conducted in China. The stirred-tank bioreactor is the most commonly used one. This bioreactor was combined with microcarriers or porous micro-spheres techniques for the production of u-PA [97]. The air-lift bioreactor was also applied in animal cell culture. In an airlift bioreactor, a novel perfusion system with a setting tank and a flat settler was designed. With the application of the system to hybridoma cell cultures, the maximum viable cell density, monoclonal antibody concentration and average productivity were improved [98]. A novel method for the scale-up culture of CHO cells in a packed-bed bioreactor was developed wherein microcarriers attached with CHO cells were inoculated directly into the packed-bed bioreactor [99]. The method provides a new technique for decreasing the labor cost and ensuring the safety of operation. Recently, a new cell-detaching reactor was developed for inoculation of anchorage-dependent cells between stepwise-expanded bioreactors [100]. The equipment consisted of a trypsinization zone and a separation zone. After the cells have been trypsinized and detached from microcarriers, a 200-mesh screen only allowed the cells to pass through to the next bioreactor. The operating feasibility of the cell-detaching reactor was tested with anchorage-dependent rCHO and Vero cells. rCHO and Vero cells were first cultured in a small microcarrier bioreactor, and then inoculated via the cell-detaching reactor into either a packed-bed bioreactor (rCHO) or a larger microcarrier bioreactor (Vero). For rCHO cells, the cell density reached 1.3×10^7 cells mL^{-1} in a perfusion culture, and Vero cells reached 1.3×10^6 cells mL^{-1} in a batch culture.

4.5 Stem Cell Culture

The studies on stem cells have been hot in recent years because of their high self-renewal ability and extensive characteristics of proliferation in vitro. Both the embryonic stem cells and the adult stem cells can be used as seed cells for reconstruction of tissues and organs. The Chinese government has paid much attention to the stem cell researches in National Basic Research Program of China (973 Program) and Hi-Tech Research and Development Program of China (863 Program). In 2002, the first Chinese ESC line was established by He et al. [101].

It is necessary to expand stem cells in vitro for the increased demand in treatment and research because of the limited amount of stem cells itself. The medium requirements of stem cells are different from those of most mammalian cell lines and have to be balanced to keep the cells from differentiation. This purpose can be achieved by adding suitable cytokine. Li et al. [102] gained high purified neural stem cells by culturing the brain cells in the present of bFGF. In addition, cell differentiation inhibition factor is important and necessary. There are three popular methods to keep stem cells from differentiation, namely co-incubation with mesenchymal cells, using conditioned medium (CM) which is added extrinsic cell differentiation inhibition factor and using CM prepared by cells capable of secreting cell differentiation inhibition factor. Wang et al. [103] showed that human umbilical cord blood mesenchymal stem cells co-incubated with the supernatant of human marrow mesenchymal stem cells could promote the culture and proliferation of umbilical cord blood mesenchymal stem cells in vitro.

Expanding stem cells in vitro while keeping their undifferentiation state is not a simple process; strict culture conditions are often necessary. Wang et al. [104] found that it was very important for proper culture conditions such as inoculum density and culture medium to cultivate successfully human umbilical cord blood mesenchymal stem cells in vitro. Ouyang et al. [105] investigated the effect of low oxygen tension (LOT) on the in vitro life span of mouse epidermal keratinocytes, and the results showed that a longer replicate life span was achieved under the reduced oxygen conditions. As some stem cells is difficult to survive when exposed to a shear stress environment, microcapsule technology has been used. Li et al. [106] first explored the feasibility of expanding neural stem cells in three-dimensional calcium alginate beads (Ca-Alg-Bs), and the results demonstrated the feasibility of NSC expansion within Ca-Alg-Bs and provided further possibilities for NSC expansion in a bioreactor with the scale of clinical relevance.

Recently, the expansion of stem cells in vitro has being researched in molecular level. To investigate underlying mechanisms for functional changes of expanded $CD34^+$ hematopoietic stem and progenitor cells (HSPCs), Li et al. [107] systematically examined gene expression of cultured and fresh $CD34^+$ HSPCs using SMART-PCR and cDNA array. In cultured $CD34^+$ HSPCs 20 genes were up-regulated while 25 genes were down-regulated. The cues implicated in differentially expressed genes gave some useful insights into understanding biological behaviors of cultured $CD34^+$ cells. In addition, to understand the genetic changes provoked by culture microenvironments, the gene expression profiles of $CD34^+$ HSPCs grown in static and stirred culture systems were also compared using SMART-PCR and cDNA arrays [108]. The results revealed that 103 and 99 genes were significantly expressed in $CD34^+$ cells from static and stirred systems, respectively, 91 with similar levels of expression, while 12 showed different transcription levels, which would give new insights into optimizing culture strategies to produce hematopoietic cells.

5 Conclusions

With the development of biotechnology, more and more recombinant proteins have been expressed in animal cells. The optimization and scale-up of animal cell culture processes are very important issues for the commercial production of the recombinant proteins. In recent years, several aspects of the research advances on animal cell culture process development have been accomplished in China. The understanding of the metabolic regulation, including glucose and glutamine metabolism and effects of metabolic by-products (lactate and ammonia) on cell culture, lays a foundation for medium optimization and process design. Strategies for protein expression improvement in animal cell culture were proved effective and implemented in the cultures of different cell lines, mainly concerning additives (AMP, DMSO, reductant, chemical chaperone, phophatidic acid) and culture conditions (osmolarity, temperature, DO, static pressure, pH value, stirring). Moreover, some researches were conducted in other areas, such as serum-free medium, fed-bath and perfusion cultures, cell damage by shear stress, bioreactor for animal cell culture, and stem cell culture. From all these achievements, a promising development of the animal cell culture technology in China can be expected. By animal cell culture, the large-mass production of vaccine, antibody and other recombinant proteins is a rapidly ascending industry in China. To meet it, the medium design and process optimization based on cell metabolism regulation and protein expression improvement will play an important role.

Acknowledgement I am grateful to the support from the Shanghai Leading Academic Discipline Project, Project Number: B505.

References

1. Kretzmer G (2002) Appl Microbiol Biotechnol 59:135
2. Merten OW (2006) Cytotechnology 50:1
3. Bell SL, Bebbington C, Scott MF, Wardell JN, Spier RE, Bushell ME, Sanders PG (1995) Enzyme Microb Technol 17:98
4. Zhang L, Shen H, Zhang YX (2004) J Chem Technol Biotechnol 79:171
5. Ljunggren J, Haggstrom L (1994) Biotechnol Bioeng 44:808
6. Xie L, Wang DIC (1997) Trends Biotechnol 15:109
7. Mendonca RZ, Arrózio SJ, Antoniazzi MM, Ferreira Jr JMC, Pereira CA (2002) J Biotechnol 97:13
8. Wlaschin KF, Hu WS (2007) J Biotechnol 131:168
9. Zhou F, Bi JX, Zeng AP, Yuan JQ (2006) Process Biochem 41:2207
10. Lu SB, Sun XM, Shi CO, Zhang YX (2003) J Chromat A 1012:161
11. Reitzer LJ, Wice BM, Kennell D (1980) J Biol Chem 255:5616
12. Barnabé N, Butler M (2000) Cytotechnology 34:47
13. Fitzpatrick L, Jenkins HA, Butler M (1993) Appl Biochem Biotechnol 43:93
14. Lu SB, Sun XM, Zhang YX (2005) Process Biochem 40:1917
15. Chen JX, Sun XM, Zhang YX (2005) Biotechnol Lett 27:395
16. Capiaumont J, Legrand C, Carbonell D, Dousset B, Belleville F, and Nabet P (1995) J Biotechnol 39:49
17. Ljunggren J, Haggstrom L (1990) Biotechnol Lett 12:705
18. Genzel Y, Ritter JB, Konig S, Alt R, Reichl U (2005) Biotechnol Prog 21:58
19. Hassell T, Gleave S, Butler M (1990) J Cell Sci 96:501
20. Christie A, Butler M (1999) Biotechnol Bioeng 64:298
21. Christie A, Butler M (1994) J Biotechnol 37:277
22. Wang MD, Yang M, Huzel N, Butler M (2002) Biotechnol Bioeng 77:194
23. Huang HY, Yi XP, Zhang YX (2006) Process Biochem 41:2386
24. Zhou J, Zhang Y, Wang W, Ma JY, Zhang HA, Guo X, Ma XJ (2006) Chin J Biotechnol 22:162
25. Zhang F, Sun XM, Yi XP, Zhang YX (2006) Cytotechnology 51:21
26. Chen JX, Sun XM, Zhang YX (2004) Chin J Appl Environ Biol 10:442 (in Chinese)
27. Sun XM, Zhang YX (2004) Process Biochem 39:719
28. Huang HY, Yu YH, Yi XP, Zhang YX (2007) Appl Microbiol Biotechnol 77:427
29. Hodgson J (1993) Bio/Technol 11:887
30. Sun XM, Zhang YX (2002) J Chem Ind Eng (China) 53:1034 (in Chinese)
31. Sun XM, Zhang YX (2003) Biotechnol Lett 25:853
32. Chen JX, Sun XM, Zhang YX (2005) J E China Univ Sci Technol 31:290 (in Chinese)
33. Zhang L, Jin YH, Zhang YX (2002) J E China Univ Sci Technol 28:157 (in Chinese)
34. Cruz HJ, Freitas CM, Alres PM, Moreira JL, Carrondo MJT (2000) Enzyme Microb Technol 27:43
35. Yang M, Butler M (2000) Biotechnol Bioeng 68:370
36. Zhang L, Jin YH, Zhang YX (2002) Chin J Biologicals 15:83 (in Chinese)
37. Sun XM, Zhang YX (2001) Chin J Biotechnol 17:304 (in Chinese)
38. Zhang F, Yi XP, Sun XM, Zhang YX (2006) Chin J Biotechnol 22:94
39. Sun XM, Zhang YX (2001) Chin J Process Eng 1:387
40. Zhang L, Yifeng J, Zhang YX (2002) J East China Univ Sci Technol 28:357 (in Chinese)
41. Liu XP, Wang Y, Zhu K, Cao YX, Lu DR (1997) Chin J Biotechnol 13:269 (in Chinese)
42. Zhou L, Sun XM, Zhang YX (2001) J E China Univ Sci Technol 27:310 (in Chinese)
43. Chiang YW, Wu JC, Wang KC, Lai CW, Chung YC, Hu YC (2006) World J Gastroenterol 12:1551
44. Ma J, Dong YQ, Zou M, Cheng LS, Liu J (2005) China Biotechnol 25(10):12 (in Chinese)
45. Liu CH, Chu IM, Huang SM (2001) J Biosci Bioeng 91:71

46. Chen ZH, Ke YZ, Chen YL (1993) Cytotechnology 11:169
47. Liu CH, Chu IM, Hwang SM (2001) Biotechnol Lett 23:1641
48. Li JH, Sun XM, Zhang YX (2005) Process Biochem 41:317
49. Luo WW, Sun XM, Zhang YX (2005) J Biosci Bioeng 100:475
50. Han SY, Lu J, Zhang Y, Cheng C, Li L, Han LP, Huang BQ (2007) Immunol Lett 108:143
51. Wang WY, Yi XP, Zhang YX (2007) Biotechnol Bioeng 97:526
52. Lee DY, Hayes JJ, Pruss D, Wolffe AP (1993) Cell 72:73
53. Yao YY, Zhang LL, Wang XQ (2000) Immunol J 16:204 (In Chinese)
54. Yin J, Wang HL, Wang XF (2002) China J Hematol 23:463 (in Chinese)
55. Wu MQ, Sun XM, Zhang YX (2001) China J Biologicals 14:217 (in Chinese)
56. Carvalhal A, Santos S, Calado J, Haury M, Carrondo M (2003) Biotechnol Prog 19:69
57. Seki J, Ikeda R, Hoshino H (1996) Biochem Biophys Res Commun 227:724
58. Zangar RC, Hernandez M, Novak RF (1997) Biochem Biophys Res Commun 231:203
59. Gurtovenko AA, Anwar J (2007) J Phys Chem B 111:10453
60. Partington GA Patient RK (1999) Nucl Acids Res 27:1168
61. Ling WL, Deng L, Lepore J, Cutler C, Cannon-Carlson S, Wang Y, Voloch M (2003) Biotechnol Prog 19:158
62. Liu CH, Chu IM, Hwang SM (2001) Biotechnol Lett 23:1641
63. Liu CH, Chen LH (2007) J Biosci Bioeng 103:45
64. Li JH, Sun XM, Zhang YX (2006) Process Biochem 41:317
65. Li JH, Huang ZY, Sun XM, Yang PY, Zhang YX (2006) Enzyme Microb Technol 38:372
66. Ma ZY, Yi XP, Zhang YX (2008) Process Biochem 43:690
67. Yoon SK, Ahn YH (1998) Biotechnol Lett 20:101
68. Alberini CM, Bet P, Milstein C (1990) Nature 347:485
69. Sato S, Ward CL, Krouse ME, Wine JJ, Kopito RR (1996) J Biol Chem 271:635
70. Sakai K, Matsunaga T, Hayashi C, Yamaji H, Fukuda H (2002) Biochem Eng J 10:85
71. Yi XP, Sun XM, Zhang YX (2007) J E China Univ Sci Technol 33: 28 (in Chinese)
72. Shi M, Xie ZG, Yu M, Shen BF, Guo N (2005) Biotechnol Lett 27:23
73. Chen ZL, Wu BC, Liu H, Huang PT (2004) J Biosci Bioeng 97:239
74. Yi XP, Sun XM, Zhang YX (2004) Process Biochem 39:1817
75. Zhang SX, Li DX, Zhu ML, Zhou Y, Tan WS, Zhang SL (2004) J Chem Ind Eng 55:247
76. Lin JQ, Takagi M, Qu YB, Gao PJ, Yoshida T (1999) Cytotechnology 29:27
77. Hu XW, Xiao CZ, Gao LH, Li ZH (2001) Lett Biotechnol 12:198 (in Chinese)
78. Gong H, Takagi M, Moriyama T, Ohno T, Yoshida T (2002) J Biosci Bioeng 94:271
79. Gong H, Takagi M, Yoshida T (2003) J Biosci Bioeng 96:79
80. Yu YG, Fu SL, Gong H, Gao MM (2005) Chin J Biologicals 18:526 (in Chinese)
81. Liu CH, Chang TY (2006) Process Biochem 41:2314
82. Shi J, Chang X, Feng J, Cheng Y, Cheng H, Guo H, Ye X, Cui H (2007) Hybridoma 26:289
83. Zhang C, Li XL, Lian XH, Wang Y, Zeng YJ, Yang K, Yu J, Gao QG, Yang T (2007) Acta Histochem 109:461
84. Zhang Y, Zhai SQ, Shou JY, Song W, Sun JH, Guo W, Zheng GL, Hu YY, Gao WQ (2007) J Neurosci Methods 164:271
85. Yan F, Hui YN, Li YJ, Guo CM, Hao M (2007) Ophthalmologica 221:244
86. Xia RS, Hao F (2006) Chin J Dermatology 39:335 (in Chinese)
87. Zhang X, Wang S, Yang S, Li T, Ji S, Chen H, Li B, Jin L, Xie Y, Hu Z, Chi J (2006) Reprod BioMedicine Online 13:412
88. Lu YY, Zhou RL, Zhang S, Jiang XN (2000) Chin Med J 113:466 (in Chinese)
89. Yi XP, Sun XM, Zhang YX (2004) J E China Univ Sci Technol 30:254 (in Chinese)
90. Zhang XY, Pennec GL, Stefen R, Müller WE, Zhang W (2004) Biotechnol Prog 20:151
91. Liu CH, Chu IM, Hwang SM (2001) Enzyme Microb Technol 28:314
92. Feng Q, Mi L, Li L, Liu R, Xie L, Tang H, Chen ZN (2006) J Biotechnol 122:422
93. Chen ZL, Wu BC, Liu H, Huang PT (2004) J Biosci Bioeng 97:239
94. Liu XM, Liu H, Wu BC, Li SC, Wu BC, Ye LL, Wang QW, Chen ZL (2006) Appl Microbiol Biotechnol 72:1144

95. Liu H, Liu XM, Wu BC, Ye LL, Ni XP, Huang PT, Chen ZL (2006) Chin J Biotechnol 22:101
96. Qiao YH, Zeng YJ, Yan GT, Zhang XJ, Xu XH, Xu H (2005) Chin J Biomed Eng 24:122 (in Chinese)
97. Hu XW, Xiao CZ, Huang ZC (2000) Cytotechnology 33:13
98. Wen ZY, Teng XW, Chen F (2000) J Biotechnol 79:1
99. Cong C, Chang Y, Deng J, Xiao C, Su Z (2001) Biotechnol Lett 23:881
100. Sun XM, Zhang YX (2007) Biotechnol Lett 29:697
101. He ZX, Huang SL, Li Y, Zhang QX (2002) Natl Med J China 82:1314 (in Chinese)
102. Li Y, Yu Y, Shen HY, Yu D (2002) Sichuan J Anatomy 10:140 (in Chinese)
103. Wang Q, Ha XQ, Bai H, Wu T, Lu JH, Yang XL, Si ZG, Xue ZW (2006) Med J Ndfnc 27:447 (in Chinese)
104. Wang LP, Jiang H (2007) J Clin Rehabilitative Tissue Eng Res 11:502 (in Chinese)
105. Ouyang AL, Zhou Y, Hua P, Tan WS (2002) Enzyme Microb Technol 30:817
106. Li XQ, Liu TQ, Song KD, Yao LS, Ge D, Bao CY, Ma XH, Cui ZF (2006) Biotechnol Prog 22:1683
107. Li QL, Cai HB, Liu QW, Tan WS (2006) Biotechnol Lett 28:389
108. Li QL, Liu QW, Cai HB, Tan WS (2006) Cell Mol Biol Lett 11:475

Adv Biochem Engin/Biotechnol (2009) 113: 217-254
DOI: 10.1007/10_2008_32

Published online: 17 April 2009

Approaches to High-Performance Preparative Chromatography of Proteins

Yan Sun, Fu-Feng Liu, and Qing-Hong Shi

Abstract Preparative liquid chromatography is widely used for the purification of chemical and biological substances. Different from high-performance liquid chromatography for the analysis of many different components at minimized sample loading, high-performance preparative chromatography is of much larger scale and should be of high resolution and high capacity at high operation speed and low to moderate pressure drop. There are various approaches to this end. For biochemical engineers, the traditional way is to model and optimize a purification process to make it exert its maximum capability. For high-performance separations, however, we need to improve chromatographic technology itself. We herein discuss four approaches in this review, mainly based on the recent studies in our group. The first is the development of high-performance matrices, because packing material is the central component of chromatography. Progress in the fabrication of superporous materials in both beaded and monolithic forms are reviewed. The second topic is the discovery and design of affinity ligands for proteins. In most chromatographic methods, proteins are separated based on their interactions with the ligands attached to the surface of porous media. A target-specific ligand can offer selective purification of desired proteins. Third, electrochromatography is discussed. An electric field applied to a chromatographic column can induce additional separation mechanisms besides chromatography, and result in electrokinetic transport of protein molecules and/or the fluid inside pores, thus leading to high-performance separations. Finally, expanded-bed adsorption is described for process integration to reduce separation steps and process time.

Keywords Affinity ligand, Bioseparation, Chromatography, Electrochromatography, Expanded bed adsorption, Protein, Stationary phase

---

Y. Sun (✉), F.-F. Liu, and Q.-H. Shi
Department of Biochemical Engineering, School of Chemical Engineering and Technology, Tianjin University, Tianjin, 300072 China
e-mail: ysun@tju.edu.cn

Contents

1 Introduction

Over the past three decades, bioscience and biotechnology have developed rapidly as marked by the advances in genetic engineering and cell fusion techniques. These have made it possible to produce various biological macromolecules such as proteins in sufficient quantities for therapeutic and diagnostic uses. Therefore, bioseparation sciences and technology have also developed remarkably to meet the requirement of large scale production of biologically active substances [1, 2]. On the other hand, progress in the separation and purification technology of biomolecules has been a prerequisite for many of the advances made in bioscience and biotechnology. As we move into the twenty-first century of the postgenome era, the correlation between advances in biotechnology and breakthroughs in separation science has strengthened. One distinct example is that the studies of systems biology and biotechnology such as proteomics and metabolomics are becoming more dependent on the progresses of bioseparation technology [3, 4].

It is notable that chromatographic technology has played the most important role in the separation and purification of biomolecules [5]. More importantly, chromatography is a separation methodology of great diversity; it is based on various interactions of a target solute with a ligand coupled to solid surface (except size exclusion chromatography), and also links to various other separation methods such as adsorption, membrane separation, liquid–liquid extraction and centrifugation.

Hence, chromatographic separations can be operated at many different modes based on different interactions and separation mechanisms [5–7]. As a result, the technology is so powerful that various substances, including small molecules, biopolymers and particulate materials such as viruses and whole cells, can be purified by the methodology. For this reason, chromatography has been the most widely used separation technique in the studies of systems biology and biotechnology. In the postgenome era it is expected that more and more functional proteins will be discovered and identified by proteomics studies, which requires more powerful chromatographic separation and purification in their identification and towards their applications. Hence, it is essential to improve continuously chromatographic technology for both academic studies and industrial production of biological substances. In this review we will focus on the development of preparative chromatography towards high-performance separation of biomacromolecules. Proteins are the primary products of discussion, whereas other biological macromolecules and particulates such as plasmid DNA and viruses will also be mentioned.

To begin with, we need to clarify the term 'high-performance preparative chromatography (HPPC)' that we focus on. We all know high-performance liquid chromatography (HPLC). HPLC is definitely a powerful tool for the analysis of liquid samples containing many different components, so it is indispensable for the technology to show high-resolution in the separation of *many components* by a single chromatographic run. Namely, HPLC for analytical applications should have high column efficiency (large plate number or small value of height equivalent to a theoretical plate, HETP) [6, 7] and peak capacity [8] at minimized sample loading. For preparative separation, however, the major objective of chromatography is to reach maximal loading and fastest operation while maintaining defined resolution or selectivity for a *single product*. Hence, the so-called HPPC herein means high feedstock loading capacity and high resolution or selectivity of a target product while operating at a flow rate as high as possible. This is the goal of preparative chromatography, but how to get it?

2 Approaches to HPPC of Proteins

For biochemical engineers, the conventional method for improving chromatographic separation of proteins is to model and optimize a purification process to make it exert its maximum capability. To this end, the fundamentals of chromatographic processes such as equilibria and mass transport kinetics should be well elucidated. Since the last decade, there have been significant advances in the fundamental studies of protein chromatography, including modeling of protein adsorption equilibria [9–17] as well as characterization of protein adsorption equilibrium [18–22] and mass transfer kinetics [23–29]. Based on fundamental theories, process modeling can be realized for the analysis and optimization of protein chromatography [30–35]. In recent years, confocal laser scanning microscopy (CLSM) has been widely employed for the direct and noninvasive measurement and characterization of intraparticle concentration distributions of proteins on a single bead scale due to its outstanding depth-discriminating capability in detecting the property of specific structures or molecules inside a specimen [36–40]. With the help of CLSM, the microscopic phenomena of protein adsorption inside porous particles in a chromatographic process, including the competitive adsorption of multicomponent proteins [37], can be quantitatively depicted. With the intraparticle concentration profiles obtained by direct visualization, mass transport models can be evaluated for their suitability for description of protein adsorption dynamics [41, 42]. Thus, the technique can offer a better understanding of protein chromatography, helping matrix design and establishment of more sophisticated equilibrium and kinetic models for process analysis and optimization.

Besides the fundamental studies aiming at optimization of protein chromatography towards high-performance purification of desired products, the most direct way to HPPC of proteins is the development of novel materials and methods with enhanced separation performance. The first approach we can think of is definitely the development of novel matrices because packing material is the central component of chromatography. To meet the demand of high-performance, the matrices should have high specific surface area that are readily accessible for target molecules, so high loading capacity is available while operating at high flow rate. In almost all chromatographic methods, proteins are separated based on their interactions with the ligands attached to the surface of solid media. A target-specific ligand can result in the selective separation of desired proteins. Hence, the second approach to HPPC of proteins is considered as the design or discovery of affinity ligands for proteins. With a highly specific affinity ligand, the target solute can be purified even though the solid medium is not of high performance. Third, besides the medium and ligand composing a chromatographic stationary phase, we can also develop methods for process intensification towards the objectives of high-performance separation. It is well recognized that diffusive mass transfer in porous materials is the rate-limiting step of chromatographic separations. To facilitate mass transport in porous media, external electric field can be applied to a chromatographic column, resulting in electrokinetic transport of protein molecules and/or the fluid inside pores. This is a separation method called electrochromatography that we will discuss as the third approach. Finally, process integration is effective to reduce separation steps and process time of the purification train for a target solute. Since the 1990s, expanded bed adsorption has been well studied as a promising approach to this end, and we will introduce it as the fourth approach. In this review chapter we will discuss the above four approaches to HPPC of proteins.

3 Innovation of Solid Matrices

3.1 Diffusive Mass Transfer in Porous Chromatographic Matrices

It is known that the diffusivity of a small solute in liquid phase (e.g., water) is about four orders of magnitude lower than that in gas phase (e.g., air) under normal temperature and atmosphere because of the much higher viscosity and density of an aqueous phase compared to a gas phase. This implies that the mass transfer rate in aqueous solutions is much lower than in gas. Therefore, biochemical processes involving liquid phase often suffer from mass transfer limitations. For the biomacromolecules such as proteins, plasmid DNA and viruses, they have much smaller diffusivities than small molecules (Table 1) [2, 43, 44], so the problem becomes even more serious. Furthermore, in chromatography with a porous stationary phase, the intraparticle diffusion of macromolecules is significantly hindered [25–29, 45], so the intraparticle diffusivity is much lower than in bulk liquid phase and decreases more with increasing molecular size [26, 46]. Hence it is widely recognized that intraparticle diffusive mass transport is the rate-limiting step in chromatographic processes of biomacromolecules. To offer high adsorption capacity, however, solid matrices for preparative chromatography must be made of porous structure to provide sufficient surface area for the accommodation (binding) of biomolecules. This leads to low column efficiency and in turn reduced dynamic binding capacity due to the intraparticle mass transfer limitation, as described by the van Deemter equation [7, 47] for packed-bed chromatography:

$$\mathrm{HETP} = \frac{2\varepsilon D_z}{U} + \frac{mFUd_p^2}{30\varepsilon D_e (1+mF)^2}, \quad (1)$$

where D_z represents axial dispersion coefficient, m the partition coefficient of solute, F the volumetric ratio of solid to liquid phase in the column, U the superficial flow velocity of mobile phase, d_p the particle size, e the bed voidage and D_e the effective diffusivity of solute in the solid particles. Consequently, it is necessary to reduce particle size to increase the column efficiency, which is the basic idea of HPLC.

As illustrated in Fig. 1, however, the decrease in particle size would result in the decrease of bed permeability or the increase of back pressure as described by the Darcy's law in a laminar flow region [48]:

$$\frac{\Delta p}{L} = \frac{150\eta U(1-\varepsilon)^2}{d_p^2 \varepsilon^3}, \quad (2)$$

Table 1 Molecular size and diffusion coefficients (or order of magnitude) of representative biomolecules (20 °C) [2, 43, 44]

Biomolecule	Molecular mass	Characteristic length (nm)	Diffusivity ($m^2\ s^{-1}$)
Small molecule	<1,000	–	0.5×10^{-9}–2×10^{-9}
Lysozyme[a]	14,300	1.5	11.8×10^{-10}
Serum albumin, bovine[a]	66,000	3.0	5.93×10^{-11}
γ-Globulin, human serum[a]	153,100	7.0	4.0×10^{-11}
Fibrinogen[a]	390,000	14.2	1.86×10^{-11}
Virus	–	~50[b]	~10^{-12}[c]
Plasmid DNA	2×10^6–13.2×10^6 (3–20 kbp)	150–250[b]	10^{-12}–10^{-13}[c]

[a]Data from [43], the characteristic length represents radius of gyration
[b]Data from [2]
[c]Data from [44]

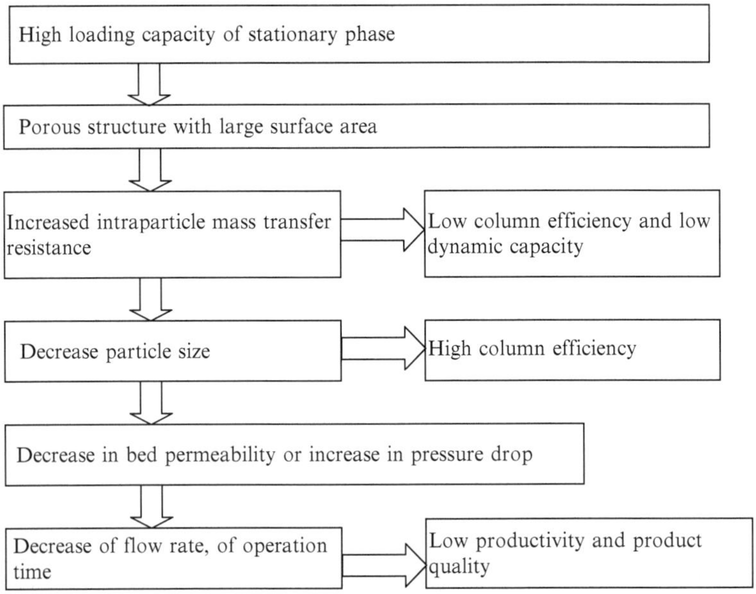

Fig. 1 The effort to high loading capacity results in low productivity: a conflict encountered in conventional chromatography

where Δ is the column pressure-drop, L the column length, and η is the viscosity of mobile phase. The flow rate must therefore be kept at a low level due to pressure-drop limitation. This in turn results in the decrease of overall productivity. Moreover, because of the lengthening of separation time at low flow rate, more denaturation or conformational changes of the bioproduct would occur during the separation process, leading to the lowering of product quality. So, the effort made to enhance loading capacity has finally resulted in the decrease of chromatographic productivity and separation performance. This is a conflict caused by diffusive mass transfer resistance within particles, which is often encountered in conventional chromatography with porous media. Thus, there is often a trade-off between resolution/capacity and flow rate in conventional chromatography.

Therefore, chromatographic matrices need evolution to overcome this problem. Since the conflict comes from the intraparticle mass transfer limitation, the primary way to be considered should be elimination or at least alleviation of the diffusive mass transfer resistance. In the past 20 years, various efforts were made to reduce mass transfer limitations and to realize HPPC of biomacromolecules.

3.2 *Superporous Microspheres*

A direct way to the goal of elimination or alleviation of intraparticle diffusive mass transfer resistance is to open convective flow channels at sizes of submicron to microns in porous particles (Fig. 2). The mobile phase can flow through the channels in chromatographic operations, so intraparticle mass transport is greatly enhanced due to a shortened diffusive path. This kind of chromatography is called flow-through chromatography or perfusion chromatography and the materials with intraparticle convection is called flow-through or perfusion medium [49, 50]. As compared to the diffusive pores of conventional media (usually, 10–200 nm [51–54]), the flow-through pores are over 600 nm in size, one to two orders of magnitude larger the diffusive pores. So, the wide pores that the mobile phase can flow through can also be called superpores or gigapores, and the corresponding particles are called superporous (or gigaporous) microspheres. Moreover, one characteristic of the material is the bimodal pore size distribution, so this kind of microsphere is also called biporous bead or bidisperse porous bead. The micropores (i.e., diffusive pores) of 10–200 nm offer large specific surface area for protein binding, so high adsorption capacity can be maintained in the adsorbents based on the biporous geometry.

There have been various biporous media available [50], and well-known is that produced by poly(styrene-*co*-divinylbenzene) copolymer with the commercial name of POROS (Applied Biosystems, CA, USA). In the particles, superpores of 600–800 nm are connected to each other by diffusive pores with a path length <1 μm, so high resolution and dynamic binding capacity can be obtained at flow velocity higher than 300 cm h^{-1} [55, 56]. The high column efficiency is definitely caused by the intraparticle convection that eliminates diffusion limitation to large extent. According to the simple model derived by Afeyan et al. [55], the intraparticle convection can be dominant for mass transfer, leading to the independence of column efficiency on flow velocity in a broad range (e.g., 500–5,000 cm h^{-1}).

In recent years, continuous efforts have been made to develop different superporous media. One excellent example is the superporous agarose gel prepared by a double emulsification [(oil-in-water) in oil, (o/w)/o] protocol [57]. Using this method, superpores created by the interior oil phase could be of microns to tens of

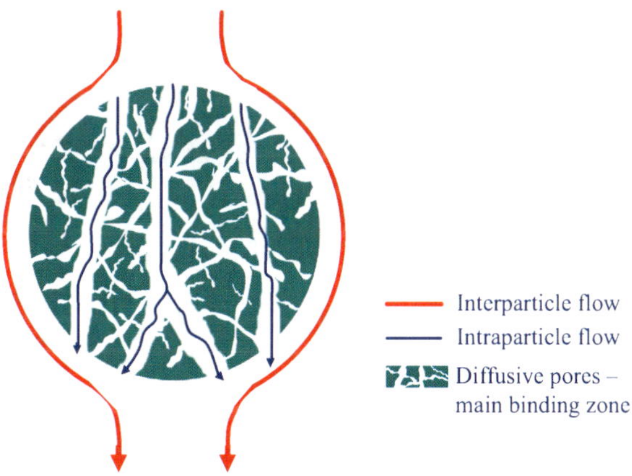

Fig. 2 Illustration of biporous bead for chromatography: convective flow and diffusive mass transfer

microns. Since its invention, the material has been extensively characterized for various applications such as immobilization of human red blood cells [58] and plasmid isolation [59]. Inspired by the research of superporous agarose, Sun et al. introduced the double emulsification technique to fabricate rigid polymeric biporous medium using glycidyl methacrylate (GMA) and ethylene glycol dimethacrylate (EDMA) as monomers and benzoin ethyl ether (BEE) as initiator [60]. Because the monomers are oil-soluble, the biporous polymer bead was fabricated by (w/o)/w emulsification and subsequent radical suspension photo-polymerization by ultraviolet irradiation (Fig. 3). The biporous bead of 43 mm in mean size had a bimodal pore distribution, i.e., micropores (20–100 nm) and superpores (300–4,000 nm). The convective flow of mobile phase through the superpores lowered the backpressure by about 70%, at a flow velocity up to 3,000 cm h^{-1}, and the HETP value was only about 40% that of the column packed with microporous beads at this high flow velocity. After modification with diethylamine to prepare anion exchangers, frontal analysis with bovine serum albumin (BSA) was carried out. It proved that the dynamic binding capacity of the biporous anion exchanger column changed less and was 1.6–2.4 times higher than that of the microporous anion exchanger column at flow velocity range as high as 1,200–3,000 cm h^{-1}. Moreover, the biporous anion exchanger column offered better separation of protein mixtures (myoglobin and BSA) at mobile phase velocities up to 3,000 cm h^{-1}, indicating that it was promising for high-speed protein chromatography.

In addition to the double emulsification procedure, Sun and coworkers have developed a novel porogenic mode to create superpores of high connectivity in rigid polymeric microspheres [61]. The method involves application of superfine salt granules (0.5–5 μm) such as sodium sulfate crystals and calcium carbonate as

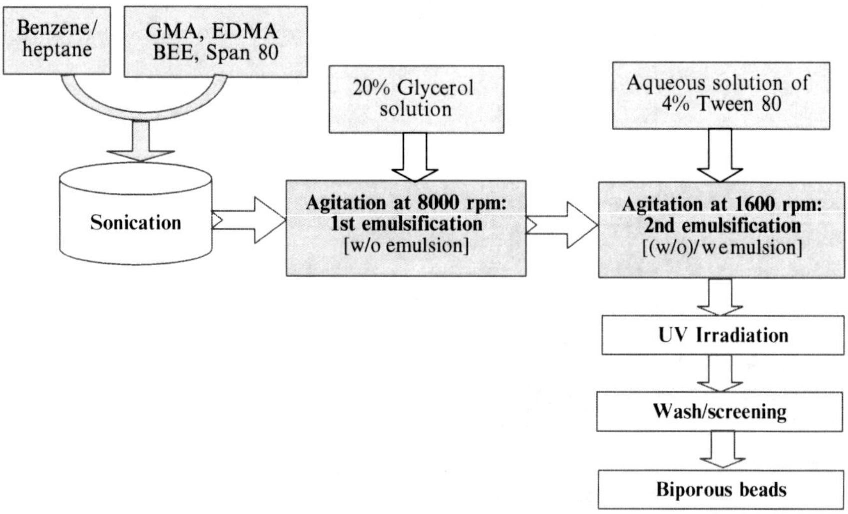

Fig. 3 Fabrication procedure of biporous polymer beads of GMA and EDMA by (w/o)/w double emulsification and photo-polymerization

solid template to produce superporous beads by suspension–polymerization. After polymerization reaction, removal of solid granules by dissolving them in water (for sodium sulfate) or in dilute hydrochloric acid (for calcium carbonate) created wide pores interconnected inside a particle. Consequently, by cooperation of superfine salt granules as solid porogen and a mixture of solvent as liquid porogen, biporous microspheres for high-speed chromatography of biomolecules were fabricated [62–65]. Figure 4 shows the pictures of biporous and microporous beads of poly (GMA–EDMA) observed by scanning electron microscopy (SEM) [64]. In the preparation of microporous particles, a mixture of cyclohexanol and dodecanol (90/10 in v %) was used as liquid porogen, with a volume ratio of 60/40 with respect to the monomers (GMA and EDMA). In the fabrication of biporous beads, however, calcium carbonate granules of 0.8 mm in mean size were introduced to the liquid phase at a volume ratio of 40/50/10 (monomers/solvent/solid granules). The matrix was modified with diethylamine to afford ionizable weak base 1-*N,N*-diethylamino-2-hydeoxypropy functionalities that are required for ion exchange chromatography. Results from the SEM pictures and mercury intrusion porosimetry measurements revealed that the biporous matrix contained two families of pores, that is, micropores (10–90 nm) and superpores (180–4,000 nm). Because of the

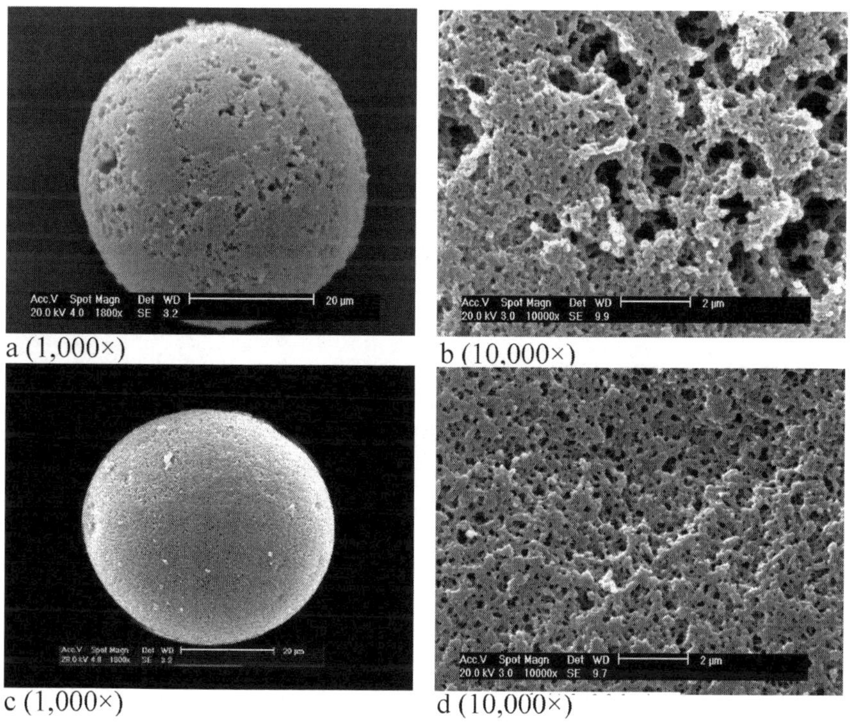

Fig. 4 a–d SEM photographs of biporous bead (**a,b**) and microporous bead (**c,d**) at different magnifications. Figure from [64]

presence of the superpores that offered convective flow channels for the mobile phase, the backpressure of the biporous bead column was only about one fifth that of the microporous bead column. The convective flow of the mobile phase promoted intraparticle mass transfer, so, as shown in Fig. 5, the HETP value of the biporous beads remained nearly unchanged in the flow velocity range of 300–3,000 cm h^{-1}, while that of the microporous beads increased linearly, in good agreement with the van Deemter equation (1) that depicts HETP as linearly proportional to flow rate in conventional chromatography [7, 47]. Moreover, the dynamic adsorption capacity of a 2-mL biporous bead column for BSA was found to be as high as 35 mg mL^{-1} bed at 300 cm h^{-1}, over 60% of its static capacity. It was much higher than that of the column packed with microporous anion exchanger (7 mg mL^{-1} bed). Moreover, as displayed in Fig. 6, the dynamic capacity decreased only slightly with increasing flow velocity up to 3,000 cm h^{-1}. The results indicated that the biporous structure of the media offered enhanced intraparticle mass transfer that led to the high-performance of protein chromatography at high flow rate. Recent studies proved that this kind of biporous material was suitable for high-speed purification of proteins [65] and plasmid DNA [66, 67].

Using calcium carbonate granules as a porogenic agent, we have also developed a superporous agarose (SA) gel by water-in-oil emulsification with 5 vol.% calcium carbonate granules suspended in 6% agarose solution [68]. After cross-linking, the solid granules were removed by dissolving in hydrochloric acid. Then the gel was modified with diethylaminoethyl (DEAE) groups to produce an anion exchanger,

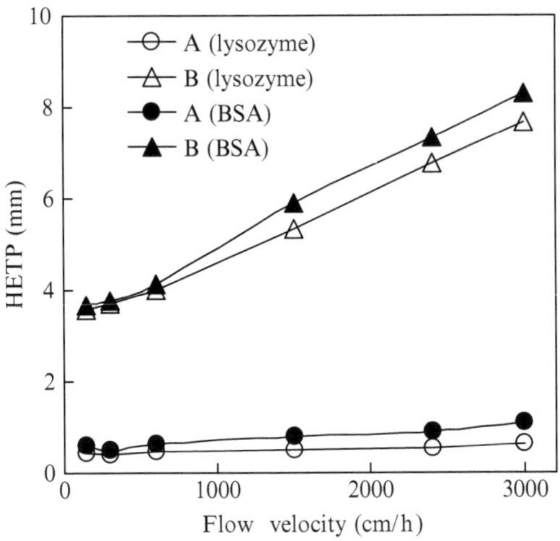

Fig. 5 Column efficiency detected with lysozyme and BSA as a function of mobile phase flow velocity. Column HR5/5 (5 mm I.D., 5 cm in length) was used. Injection size was 25 mL and the concentrations of lysozyme and BSA were 5 and 2 mg mL^{-1}, respectively. Mobile phases were 10 mmol L^{-1} Tris-HCl buffer (pH 7.6) for lysozyme and 10 mmol L^{-1} Tris-HCl buffer (pH 7.6) plus 0.1 mol L^{-1} NaCl for BSA. Figure from [64]

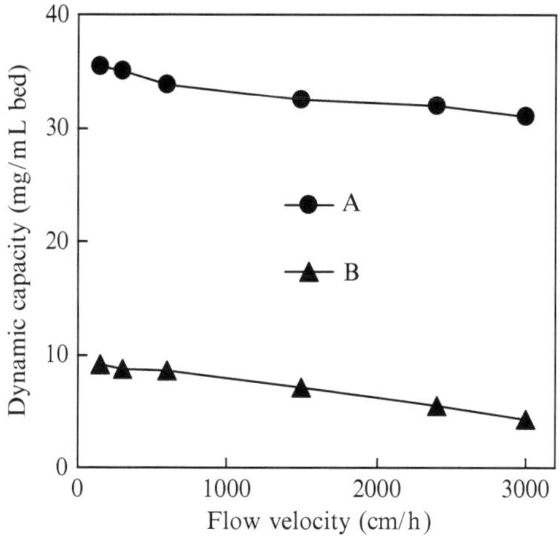

Fig. 6 **a,b** Comparison of the dynamic capacities of BSA on the columns packed with biporous beads (**a**) and microporous beads (**b**) at 10% breakthrough as a function of flow velocity. Column HR5/10 (5 mm I.D., 10 cm in length) was used. Figure from [64]

SA-DEAE. A homogeneous agarose (HA) gel was also prepared and modified with DEAE for comparison. Optical microscopic observation revealed that the beads entrapping calcium carbonate were opaque due to the presence of calcium carbonate granules. After removal of the solid granules, the gel became transparent, but not yet as homogeneous and bright as the HA gel: gray craters were visible in the SA gel. It is considered to be the superpores left by the calcium carbonate granules. To compare further the difference between SA-DEAE and HA-DEAE, fluorescein isothiocyanate labeled immunoglobulin G (FITC-IgG) was adsorbed to the ion exchangers by batch adsorption for 24 h and the beads were observed by CLSM. Figure 7 shows the two-dimensional images of the two ion exchangers in an in-focus plane around the central cross-sections perpendicular to optical axis. From the picture, we can see many bright spots inside the SA-DEAE, which shows the clusters of adsorbed FITC-IgG. The bright spots were distributed in the whole cross section (or in other words, in the whole particle) (Fig. 7a). Moreover, it can be seen that the sizes of such spots are uneven, which is considered to be due to the size distribution of the calcium carbonate granules used for creating the superpores. In contrast to the SA-DEAE, a relatively uniform halo can be seen for the HA-DEAE (Fig. 7b). The fluorescence intensity of the labeled protein in the images shown in Fig. 7 was analyzed and the results along particle diameter are exhibited in Fig. 8. It is evident that there exist drastic fluctuations of the fluorescence intensity in the SA-DEAE, while a relative smooth distribution along the radial direction is present in the HA-DEAE. The results indicated the high adsorption capacity and fast mass transfer kinetics in the SA-DEAE, which was further demonstrated by adsorption equilibrium and dynamic experiments. Due to the presence of the wide pores, more channels were available for protein transport and, furthermore, more diffusive pores in the

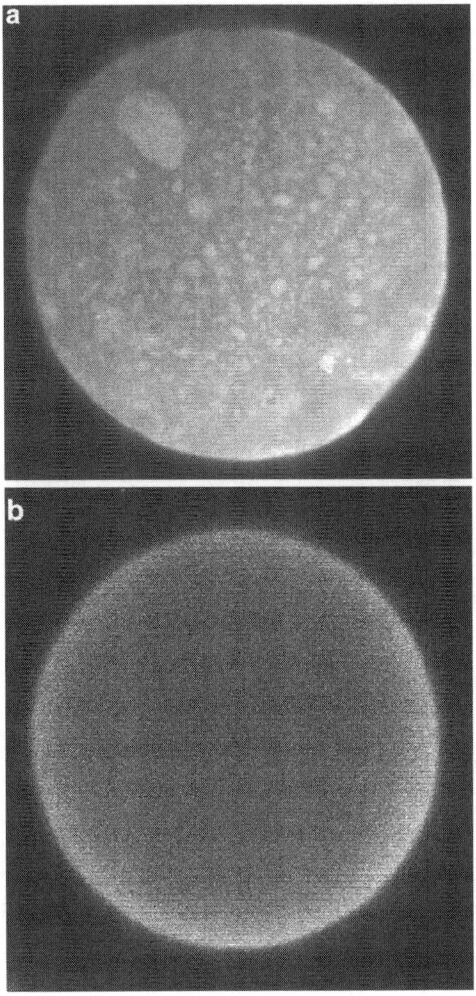

Fig. 7 **a,b** CLSM images of SA-DEAE (**a**) and HA-DEAE (**b**) with adsorbed FITC- IgG. Figure from [68]

agarose network were accessible for the protein approach from different directions. This led to 40% higher protein capacity and twice as high effective pore diffusivity in the SA-DEAE than in HA-DEAE. Moreover, an increase of the efficiency of the SA-DEAE column up to a flow velocity of 300 cm h^{-1} and independence of the column efficiency at flow velocities from 300 to 1,070 cm h^{-1} were found, indicating that intraparticle mass transfer was intensified by convective flow at elevated flow rates. Therefore, the chromatographic resolution of IgG and BSA was little affected up to a flow velocity of 1,070 cm h^{-1}. The results indicated that the superporous agarose medium was favorable for high-speed protein chromatography. Recently, the method was extended to produce superporous cellulose beads for application in protein chromatography [69, 70].

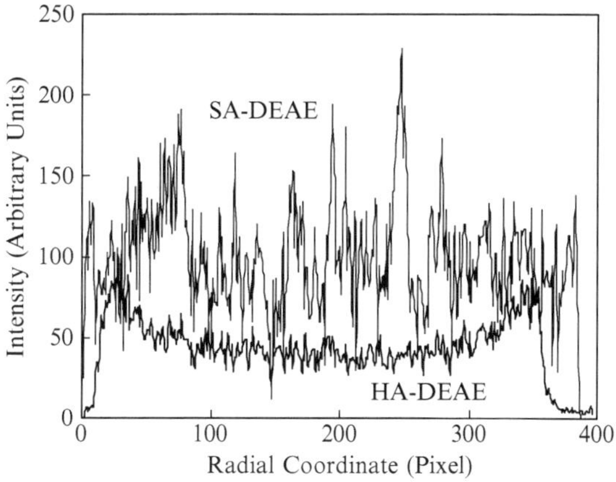

Fig. 8 Fluorescence intensity (protein concentration) distributions across the SA-DEAE and HA-DEAE with adsorbed FITC- IgG. Figure from [68]

3.3 *Membranes and Monoliths*

To maximize chromatographic throughput, mass transfer limitations need to be eliminated for the fast uptake of target substance, and the flow rate should be as high as possible at a given pressure drop across the bed. This leads to the efforts to design short bed of large diameter and other column configurations such as radial-flow chromatography. To this end, it is obvious that an ideal chromatography column is a piece of filter because a porous membrane with thickness in the 100-m range is the shortest bed available in reality [71]. Microfiltration membranes primarily contain flow-through pores, so the main feature of membrane-based chromatography is the absence of pore diffusion, which is the main transport resistance in conventional packed-bed chromatography using porous particles. In membrane chromatography, the target solute binds to the ligands attached to the inner surface of the through-pores when it flows through the pores with feedstock, so only the surface film diffusion resistance is left [72]. For this reason, membrane chromatography can be operated at high flow rate and low pressure drop at maximum efficiency of ligand utilization, so membrane chromatography offers high-speed purification of biomacromolecules such as proteins and plasmid DNA [44], and now it is particularly popular for antibody purification [73, 74].

However, a drawback of membrane chromatography is its low chromatographic efficiency (small plate number) resulting from low 'bed height'. So it can be claimed that membrane chromatography is primarily suitable for affinity-based adsorption because the high-selective adsorption relies less on plate number. In addition, membrane chromatography often suffers from low binding capacity for proteins due to the lack of 'diffusive pores' as in conventional porous particles, although it shows high capacity for larger biomolecules such as DNA [44, 75]. These have limited its wide application in protein chromatography.

A solution to the low efficiency of membrane chromatography is using a disc membrane stack packed in a column, but it would certainly compromise the advantage of low flow resistance of membrane chromatography. Hence, use of a monolithic column or continuous bed instead of a membrane stack would be more straightforward. A monolith can be regarded as a piece of very 'thick' membrane. Besides a larger plate number than membrane chromatography, a monolithic column can offer higher binding capacity than membrane because it can be made to contain both through pores and diffusive pores [76]. As compared to a packed bed of porous particles, moreover, monolithic columns have the following advantages. First, the volume fraction of through pore in a monolith can be much less than a packed bed (usually 0.3–0.5), so a monolith provides a higher fraction of stationary phase for solute binding, and in turn higher binding capability. Second, in a packed bed of biporous beads, the convection through the superpores accounts for less than 5%. In a continuous bed, however, there is relatively even flow rate distribution because the flow-through pores can be controlled in a narrow range. So, with a controlled pore structure, a monolith can offer lower mass transfer limitation in protein chromatography. Because of these advantages, monolithic materials have been widely studied for diverse applications, especially for the analytical and preparative chromatography of biomolecules such as proteins and DNA [76–78].

The first step to a successful monolith preparation is to reduce flow resistance and in turn to obtain a high flow rate [79]. In the synthesis of rigid polymeric continuous rod, this can be realized by adjusting some key variables such as reaction temperature, composition of the pore-forming solvent mixture and the content of cross-linker [76]. At the same time, a monolith should have plenty of micropores connecting the through pores for high binding capability for use in preparative chromatography. Zhang and Sun have synthesized an anion exchange monolith with GMA as a monomer, triallyl isocyanurate (TAIC) and divinylbenzene (DVB) as cross-linkers, and a mixture of toluene and heptane as porogen [80]. The capacity of the anion exchanger reached 76 mg g^{-1} for BSA and the resolution of proteins was not compromised up to 900 cm h^{-1}. However, the hydrodynamic property was not as good as expected; the backpressure was over 9 MPa at 720 cm h^{-1}. Recently, Sun and coworkers introduced the coporogens of sodium sulfate granules and organic solvent mixtures described above to monolith preparation [81]. The experimental conditions and the main results are listed in Table 2. In the table, B_0 is the hydraulic permeability of the monolith defined by Darcy's law [48] [see also (2)]

$$\frac{\Delta p}{L} = \frac{\eta U}{B_0}. \tag{3}$$

By adjusting the volume ratio of dodecanol (DoOH) and cyclohexanol (CyOH), the back pressure of the column (MLs) can be readily decreased, as shown in Fig. 9, leading to an increase of the column permeability (Table 2). By introducing sodium sulfate granules, the bed permeability was further increased to 7.8–8.8 × 10^{-14} m^2. It is interesting to see that in the three monoliths produced with solid porogen (MLSs), the bed permeabilities were similar, regardless of the difference in the compositions of pore-forming solvents. The result indicated that the solid porogen played the main role in creating through pores in the monoliths. It is notable that the change in the mechanical strength of the monolithic column prepared by using the solid porogen was not observed. Moreover, the dynamic binding capacities of the anion exchanger MLSs were not significantly compromised by the introduction of through pores using solid crystals, and it was found that with the increase of flow rate, the DBC of MLS-3 decreased slightly at the beginning, and then changed little

Table 2 Effect of porogen composition on the permeability B_0 and DBC of the Poly(GMA-EDMA) monoliths prepared at different condition (data from [81])[a]

Monolith	DoOH/CyOH[b] (v/v)	Na_2SO_4[c] granules (g)	B_0 (10^{-14} m^2)	DBC[d] (mg mL^{-1})	Graft
ML-1	1/9	0	1.8	33.4	No
ML-2	6/19	0	4.2	29.3	No
ML-3	½	0	5.0	27.5	No
MLS-1	1/9	0.25	7.8	26.4	No
MLS-2	6/19	0.25	8.1	26.0	No
MLS-3	½	0.25	8.8	23.9	No
MLS-3T	½	0.25	7.2	74.7	Yes

[a] Reaction condition: GMA, 26 vol.%; EDMA, 14 vol.%; organic porogenic solvent, 60 vol.%; benzoyl peroxide (BPO), 1 wt% with respect to GMA monomer; temperature, 55 °C; polymerization time, 24 h; column, stainless-steel tube of 50 × 4.6 mm I.D

[b] Volume ratio of dodecanol (DoOH) and cyclohexanol (CyOH)) in the polymerization mixture

[c] Na_2SO_4 crystal was 0.7–1.5 µm

[d] The monoliths were derivatized to anion exchangers by modification with diethylamine. Dynamic binding capacity (DBC) was measured by frontal analysis at 50% breakthrough at 360 cm h^{-1}

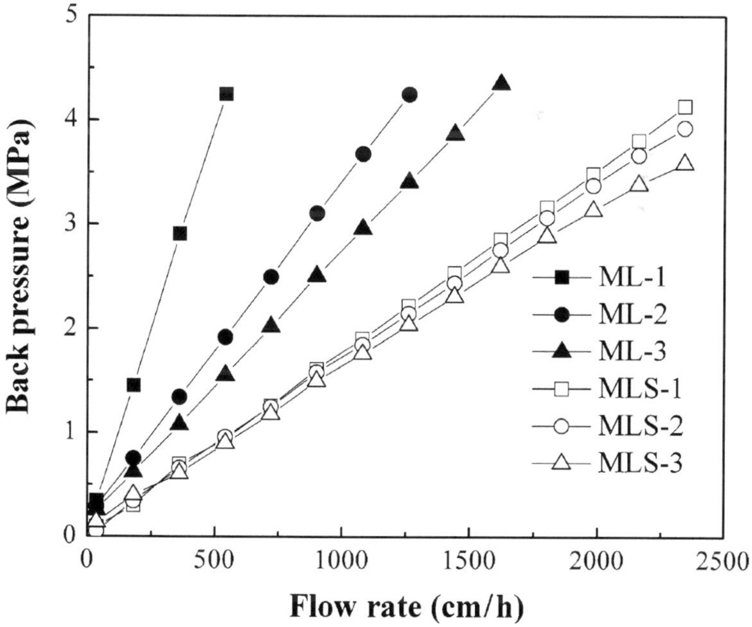

Fig. 9 Effect of porogen composition on the back pressure as a function of flow velocity for monoliths MLs and MLSs (see Table 2). Conditions: column 50 × 4.6 mm I.D.; mobile phase, distilled water. Figure from [81]

in the flow velocity of 500–3,600 cm h^{-1} [81]. Then poly(glycidyl methacrylate-diethylamine) tentacles were grafted onto the pore surface of MLS-3 monolith to prepare MLS-3T. The DBC of BSA for the MLS-3T was about three times higher than the MLS-3 (Table 2). Moreover, the column efficiency (HETP value) of the MLS-3 or MLS-3T was independent of flow velocity up to over 3,000 cm h^{-1}. Thus, monolithic columns of high column permeability, high binding capacity and high column efficiency can be produced by the cooperative porogenic method with solid granules and solvent mixture.

3.4 Perspectives

Mass transfer resistance is a barrier on the way to high-speed chromatography of biomacromolecules such as proteins, DNA, polysaccharides and viruses. The barrier cannot be completely removed, but can be lowered to some extent by the efforts made to fabricate new types of solid materials mentioned above. In addition, for preparative applications at maximized throughput, the materials need to be made to offer a capacity for the biomacromolecules as high as possible. For this objective, the materials must be of porous structure with high specific surface area that is accessible for the biomolecules. Namely, the pore size should be properly controlled to offer both large surface area and accessibility for the target molecule. So, we would like to emphasize that chromatographic media need to be customized according to the properties of the target molecule and its existing environment. Otherwise, various media should be provided for users' selection.

However, current media available for preparative chromatography have never fulfilled the requirement of HPPC at high loading capacity, high resolution and low pressure drop while operating at high flow rate. Herein, we suggest an ideal chromatographic medium as illustrated in Fig. 10. It is a monolith with straightforward and evenly distributed uniform through pores (flow channels). The through pores look like a honey comb supported by a skeleton full of diffusive pores of proper size that provide a large surface area accessible for target solute molecules. The straightforward through pores offer low flow resistance as predicted by the Poiseuille equation for laminar flow in a tube:

$$\frac{\Delta p}{L} = \frac{32\eta U}{\varepsilon_D D_c^2}, \tag{4}$$

where D_c is the diameter of through pores (1–5 μm) and ε_D the volume fraction of the flow channels. As compared to (2) for packed beds, the pressure drop in a straightforward tube is one to two orders of magnitude lower. Moreover, the short distance between the flow channels (1–5 μm) permits fast approach of the target molecule to the binding sites. Thus, this chromatography can really fulfill the requirement of HPPC. Creation of this ideal medium for diverse applications may require a great effort from chemists and materials scientists.

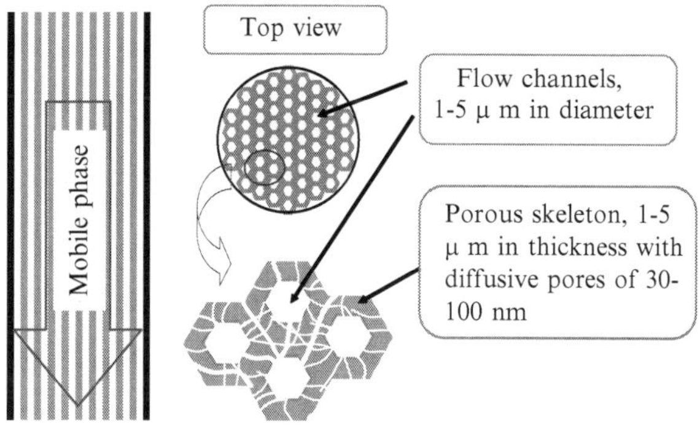

Fig. 10 Ideal configuration of chromatography: monolithic column with a porous skeleton and straightforward and ordered convective flow channels

4 Discovery and Design of Affinity Ligands

4.1 Advances in Affinity Ligand Studies

Affinity chromatography is considered the most selective method for protein purification from complex mixtures. It relies on the molecular recognition, that is, the ability of affinity ligand to form specific and reversible complex with the target protein [82]. Due to its high selectivity and purification power, affinity chromatography has become one of the best ways to separate biomacromolecules [83]. However, the difficulty in the use of affinity chromatography is the lack of specific ligands for most proteins. One solution to this problem is to screen and discover affinity ligands from combinatorial libraries [84, 85]. Currently, there are two common methods to construct large combinatorial libraries, synthetic and phage surface display techniques. Using combinatorial synthesis, a large number of structurally distinct molecules are synthesized at a time, which can provide a great many novel compounds for random screening. Several laboratories have recently reported the application of combinatorial methods to select affinity ligands from the libraries based on substituted triazine [86] and peptides [87]. Phage surface display has also rapidly developed and evolved as a tool for discovering high affinity ligands for affinity chromatography [88]. However, these approaches often suffer from several drawbacks, such as pseudo-positives, high screening cost and long screening time. Moreover, a limitation of phage display is that peptides often function only when the peptide is an integral part of the phage coat protein but not when immobilized to a chromatographic matrix.

Rational design is another approach to discover affinity ligands. The method uses the information of the structure of natural ligands or its target protein to upgrade or create a new ligand. Now, some software packages have been developed to calculate, visualize, formulate and hypothesize about the energy and orientation of candidate ligands in the pocket of a target protein. Since Lowe et al. [89] applied rational design to upgrade dyes to enhance its affinity, a great number of ligands for affinity chromatography have been successfully selected by this approach. With the rapid advance in computational tools and the availability of more three-dimensional structures of proteins obtained by X-ray crystallography or NMR technology, rational design of affinity ligands has become faster, more feasible and powerful. This method has also been extended to peptides and their derivatives [90].

There are two distinct rational design methods: structural template approach and functional approach [91]. The first is based on the knowledge of the protein structure and the interactions between the target protein and its natural ligands or counterparts. The natural ligand–protein complex is investigated, and the conformation of the bound molecule is used as a template to design a new ligand. For example, Platis et al. [92] applied the lock-and-key motif to design affinity ligands for human antiHIV monoclonal antibody 2F5. The second is employed when insufficient structural data are available for the target protein and a reliable protein model cannot be build by homology modeling methods. The approach is based on preexisting knowledge for the interaction of the protein with functional groups, moieties and molecular shapes on natural ligands, e.g., substrates and inhibitors of enzymes. The ligand can be designed either by exploiting a recognizable molecular shape and properties, such as hydrophobicity and electrostatic potential [93, 94], or by introducing a specific functional group. For example, the benzamidine group is often used as a ligand for trypsin-like proteases [95].

Recently, the combination of rational design and combinatorial library technique emerged as a new and promising approach for ligand selection [96]. The method involves the following steps: (1) selection of an appropriate site on the target biomolecule, (2) design of a complementary ligand compatible with the candidate pocket using modeling techniques, (3) synthesis of a limited ligand library of structures resembling the rational designed lead ligand, and (4) screening of the library against the target protein. The method has already led to the development of a series of triazine scaffolded molecules with affinity to different proteins [97]. In addition, Melissis et al. [98, 99] have applied the combination of molecular modeling and combinatorial approach for designing nucleotide-mimetic ligand for *Pfu* DNA polymerase and Taq DNA polymerase.

4.2 Rational Design of Peptide Ligands by Molecular Simulation

Small peptides are completely biocompatible and are considered as better ligand candidates for affinity chromatography, as compared to antibodies, because they are composed of only a few amino acids, which are not likely to cause any immune response in case of leakage into the product. Moreover, peptide ligands are more stable and less expensive than antibody ligands. Therefore, peptide ligands have been a research focus in recent years [87].

Recently, Sun and coworkers have focused on the discovery of peptide ligands based on the rational design approach. An example is the rational design of affinity peptide ligand by flexible docking simulation [100]. There are two key parts in any docking programs, namely searching of the conformation and the scoring function. The evaluation and ranking of predicted ligand conformations is a crucial aspect during docking. Even when the binding conformations are correctly predicted, the calculations may not ultimately succeed if they cannot differentiate correct poses from incorrect ones, and thus appropriate ligands will not be identified. So, the selection of reliable scoring functions is of fundamental importance. There are many scoring functions in current application: Dscore, Gscore, Chemscore and PMFscore [101]. In order to decide which scoring function should be used to evaluate the affinity between the candidate ligand and the target protein, the heptapeptides screened from phage display library with lysozyme [102] and insulin [103] were used to give an evaluation. The results showed that the values of Dscore [104], a function of the charge and van der Waals interactions between the protein and the ligand, could denote the affinity between a ligand and its target protein. Molecular surface analysis with the MOLCAD program further validated that the Dscore could be used as an index of the affinity between a ligand and its target protein. Therefore, the Dscore was used as a measure to evaluate the affinity of a peptide for the target protein in the docking simulation to find a small peptide ligand of α-amylase [100].

The structure of α-amylase complex with a substrate (arabinose) was obtained from the Protein Data Bank (PDB) (http://www.rcsb.org/pdb/home/home.do) [105]. The candidate pocket of α-amylase was identified on the basis of the substrate. It comprises aromatic amino acid residues, including two tryptophan residues and one phenylalanine residue. Aromatic π–π stacking interactions were considered to contribute to the affinity between the ligand and the target protein. So, tryptophan and phenylalanine residues were selected. As we know, hydrogen bonds (H-bonds) play a key role in the interactions between ligand and the target protein. Thus, three amino acids (histidine, glutamic acid, and asparagine) which can readily donate or accept protons were chosen. Finally, the five amino acids were used to build manually a library of 120 peptides.

An attempt was made to dock all the peptides into the candidate pocket of α-amylase using FlexX [106]. The result showed that 101 of the peptides were successfully docked into the candidate pocket. The FHENW peptide showing the highest absolute Dscore value (−253.96) was selected for further analysis. However, analysis with the MOLCAD program showed that the C-terminal of FHENW was embedded in the candidate pocket, which would greatly weaken its affinity for α-amylase when the peptide was coupled to a solid matrix. So, serine was linked to the C-terminal in order to elongate the ligand, improve flexibility and enhance specific interactions. According to the analysis by the MOLCAD program, the affinity interactions between this peptide to α-amylase were mainly electrostatic interactions (including 10 H-bonds) and van der Waals forces. Chromatographic experiments with the immobilized peptide gave further evidence for this observation. Adsorption equilibrium studies with the affinity adsorbent showed that the respective peptide–enzyme complex had a moderately strong binding affinity (dissociation constant = 4.0×10^{-6} mol L^{-1}), falling within the range of desired affinity interactions applicable to affinity chromatographic separation of proteins (10^{-4}–10^{-8} mol L^{-1}). Finally, affinity chromatography with the peptide ligand was successfully applied to purify α-amylase from the fermentation broth of *Bacillus subtilis*.

Although high throughput screening using docking programs is useful as a fundamental rational design tool, there are three drawbacks of the docking methods, i.e., the oversimplification of the docking process (e.g., rigid proteins, single conformation of ligands), low predictive accuracy of the scoring functions, and explicit solvent effects. Therefore, a possible strategy that could include both the solvent effects and the flexibility of both partners is the use of molecular dynamics (MD) simulations with an accurate force field. Unfortunately, such calculations are applicable only to a few ligand–protein systems because of the large computational cost. So, their use is usually restricted to a few selected ligand candidates.

Liu et al. [107] recently proposed the combination of docking and MD simulations for the rational design of affinity ligand, and the method was demonstrated in the design of a peptide ligand for affinity chromatography of tissue-type plasminogen activator (t-PA). First of all, the crystal structure of t-PA was obtained from the PDB and its active site was divided into three parts, the interior site, the middle site, and the exterior site [108]. An attempt was made to dock each of the 20 amino acids into the candidate pocket of t-PA, and Dscore was used to evaluate the affinities. As a result, it was found that arginine, asparagine, glutamine (Q), histidine, phenylalanine, tryptophan and tyrosine had favorable interactions with the interior pocket site, glycine and aspartic acid (D) with the middle site, and glutamate (E) with the exterior site.

A peptide library with 14 ($=7 \times 2 \times 1$) tetrapeptides with serine (S) linked to the C-terminal of each peptide was built from the above ten amino acids. Here serine was linked to the C-terminal to lengthen the peptides. Molecular docking led to the selection of six tetrapeptides which have Dscore values lower than −200 and C-terminals outside the candidate pocket. Because a selected peptide ligand would be coupled to EAH Sepharose 4B through a spacer arm [$-NH(CH_2)_6NH_2$] for affinity chromatography, the molecular group of the spacer arm was linked to the C-terminal of the six tetrapeptides for further docking simulation. As a result, the QDES linked with the spacer arm showed the highest absolute Dscore value (−227.42). Molecular surface analysis suggested that charge interactions would occur in the affinity binding. In addition, inspecting its best conformation and orientation with respect to the candidate pocket, it was confirmed that the ligand forms nine H-bonds with t-PA.

The QDES peptide and t-PA complex obtained by docking was then solvated and analyzed by MD simulations. First, the movement of the protein and ligand was investigated in order to examine whether the structures of the peptide-t-PA complex remained stable. The results showed that the conformations of the candidate pocket of t-PA and the peptide changed smaller than the overall conformation of the protein during the MD simulations, indicating the stability of the candidate pocket and the peptide. Second, the interactions between the tetrapeptide and the candidate pocket were analyzed. The MD simulation confirmed that all of the nine H-bond interactions observed in the docking, and that six of them were very stable and existed during most of the whole MD simulations. Third, the bridging interactions of water molecules were studied. It was found that three water molecules were involved in the H-bonds interactions between the peptide and the candidate pocket of t-PA. Therefore, it can be concluded that water molecules played an important role in the protein–ligand interactions. Docking, molecular surface analysis and MD simulations have all revealed that QDES would be a candidate of high-affinity ligands of t-PA. Finally, high binding affinity and specificity of the peptide ligand were confirmed by the purification of t-PA from crude porcine heart extract using the immobilized-ligand column for affinity chromatography. The present work indicates that the combination of docking and MD simulations are promising for the rational design of affinity peptide ligands on the basis of the target protein structure. Docking can be used first to screen ligands from a large library. Once a few potential ligands have been identified, their complexes can be submitted to MD simulations for further analysis and validation to discover high-affinity ligands.

4.3 Perspectives

The importance of protein flexibility and solvent effects for accurate modeling of ligand–protein interactions is now well known. As most docking algorithms do not take into account the receptor flexibility or the effect of explicit water molecules, MD simulations can be used to verify, complement and improve a docking protocol. In the initial stage, MD simulations can be used to refine or explore the conformation of the target protein. The exact conformation of the target protein is expected to provide more realistic representation of the behavior of the protein in solution. In addition, MD simulations can be used to calculate the binding free energy, which is expected to provide much better ranking than the simple algorithms used in docking. In future, the increase in computer power and improved algorithms will allow the use of more realistic representations of protein systems with improved boundary conditions, and application of MD simulation to the docking of flexible ligands into the mobile candidate pocket of the target protein in aqueous environments. Therefore, versatility of the rational design approach can evolve into a routine method for the discovery of affinity ligands of various proteins.

5 Preparative Electrochromatography

5.1 Electrokinetic Phenomena in Electrochromatography

Section 3 of this chapter has described innovation of matrices to achieve enhanced mass transfer. Another way to lowering the barrier of mass transfer resistance is to use an external field that introduces additional mass transport mechanisms. Electrochromatography is an example of such an effect. In a typical electrochromatography, two electrodes are located at the two ends of a chromatographic column, so electric field is applied at the longitudinal direction of the column. Therein, charged solutes flow through packed columns or open tubes via three possible modes, i.e., convection, electrophoresis and electroosmosis. Except for convection of bulk liquid phase driven by pressure, the latter two electrokinetic migrations are driven by external electric field (eEF) [109]. Both electrophoresis and electroosmosis in an electric field can promote mass transfer, thus resulting in the increase of chromatographic performance.

In capillary and packed column electrochromatography used for analyses and separations, eEF is usually applied in the longitudinal direction of the column, thereby providing an additive driving force along the liquid phase streamline. Driven by combined hydraulic and electrokinetic flow, solutes can pass through bulk liquid phase onto the surface of porous media with hierarchically structured pores, and then transport through the pore space. In the process of electrochromatography, electric field strength, ionic strength of liquid phase, and the inherent structural heterogeneity of porous matrix affect electrokinetic velocity and its distribution [110]. Thus, the contribution of the electrophoresis and electroosmosis to mass transport in a packed column is rather complex. Based on the Nernst–Planck equation, the total mass transfer flux of a solute can be expressed as the summation of electrically induced mass flux (J_E) and diffusion–convection flux (J_{D+C}) as follows [111, 112]

$$J_T = J_E + J_{D+C} = -(u + \mu E)c - D\frac{\partial c}{\partial x}, \quad (5)$$

where u, μ, c, D and E are flow velocity, electrophoretic mobility, concentration and diffusivity of solute, and electric field strength, respectively. Under the influence of eEF, mass transfer flux in bulk liquid phase is larger than intraparticle mass transfer flux [112–114] due to the effect of diffusional resistance, size exclusion and net charge of solute. Such a difference of flux in bulk liquid phase and porous matrix can lead to more solute deposit on the surface of porous matrix, leading to electrically induced concentration polarization (CP). CP was first confirmed in protein electrochromatography by Cole and Cabezas Jr [113]. Systematic studies revealed that electrically induced CP originated from the charge-selective transport within the porous domains in a matrix [110, 111, 114]. Besides ion-permselectivity of particles, the selectivity of pore space to solutes also depends on the size-exclusion effect for charged proteins of high molecular weights [112, 115, 116]. Thus, CP may provide a new way to modulate the retention behavior of charged solutes in electrochromatography with porous media.

Another phenomenon needed to be taken into account in electrochromatography is the occurrence of intraparticle pore flow. It is known that intraparticle mass transfer could be significantly augmented even by a rather small pore flow [117]. Section 3 of this chapter has described pressure-driven pore flow in superporous matrices. Compared with pressure-driven pore flow, the introduction of eEF into chromatography packed with porous beads is also a feasible approach to obtaining relatively high pore flow. Much evidence from experimental data and theoretical analysis has shown that an electrically induced pore flow in particle can significantly increase intraparticle mass transfer, eliminating mass-transfer resistance, and increasing column efficiency [112, 118–121].

Thus, eEF can introduce additional electrokinetic mass transfer and solute retention mechanisms in electrochromatography, which can be employed to enhance the separation performance of preparative chromatography of proteins.

5.2 *Electrochromatography with Transverse Electric Field*

Electrochromatography has been studied for the preparative purification of biomolecules such as proteins and plasmid DNA [115, 116, 122–124], and most of the electrochromatography is established by applying an eEF in the longitudinal direction of a packed bed. It is noted that some overwhelming challenges have to be faced in the scaling-up of preparative electrochromatography, particularly in the generation of Joule heat and electrolysis gases [122]. It is well documented that Joule heat generated in electrochromatography is proportional to the cross sectional area and the square of applied voltage. In the scaling-up of electrochromatography, an increase in column height requires a high voltage to maintain suitable electric field strength, and an increase in diameter results in the increase of the cross-sectional area of column. When the Joule heating exceeds the heat dissipation, overheating and temperature rise in the column are inevitable. On the other hand, the application of a high voltage means that more electrolysis gases would form at the electrodes, which would decrease the stability of an electrochromatographic operation. Therefore, efficient removal of Joule heat and electrolysis gases becomes more important for a successful preparative electrochromatography [113, 122]. To resolve these problems often encountered in conventional electrochromatography with applied axial electric field, Sun and coworkers [112, 125] proposed a mode of electrochromatography with an oscillatory transverse column low-voltage electric field. That is, the electric field is perpendicular to liquid phase streamline in the electrochromatography. The column was designed with three compartments [112]. The central compartment was packed with chromatographic particles while in the two side compartments electrodes were mounted with circulating coolant (electrode solution). They were separated from the gel compartment by two porous ceramic plates filled with polyacrylamide gel. In electrochromatography operation, an oscillatory direct current with equal duration of positive and negative polarities was applied in the direction perpendicular to the mobile phase flow as in normal liquid chromatography. Therefore, a charged solute would pass through the column in a zigzag trajectory in a separation process under transverse electric field (Fig. 11). There are some distinct advantages of this electrochromatography. First, the electrolytic gases formed at the electrodes can be readily kept out from the packed-bed and removed by circulating electrode solutions. Second, the Joule heat generated in the packed-bed is relatively low and can be readily removed by the cooled mobile phase flowing through the column in a pressure-driven mode. Third, the relatively short distance between the two electrodes can create high electric field strength with a low-voltage power applied [112, 125].

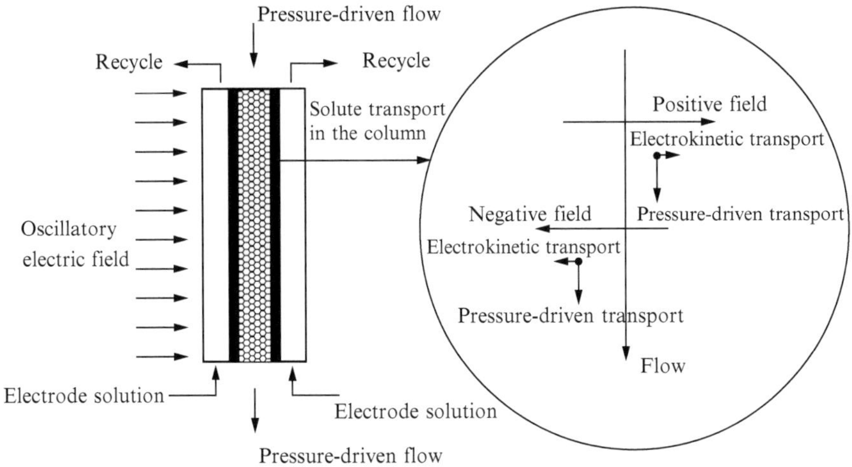

Fig. 11 Schematic diagram of the electrochromatography with an oscillatory low-voltage electric field perpendicular to the liquid phase streamline. The transport of a charged solute in chromatographic process is given on the right. Figure from [112]

With these advantages, the problems in preparative chromatography can be eliminated or alleviated to some extent.

The electrochromatography with an oscillatory low-voltage electric field perpendicular to liquid phase streamline was first applied to size-exclusion electrochromatography (SEEC) of proteins [112]. The retention behavior of proteins was extensively investigated under various conditions. As shown in Fig. 12, the partition coefficient of charged protein increased significantly with increasing the current strength as well as the difference between its isoelectric point and pH. For the gel-excluded protein like BSA, the CP on gel surface induced by the protein electromigration was the main reason for the increased retention. For a gel-permeable protein like myoglobin (Myo), both the CP and electrophoretic migration in the solid phase contributed to its increased retention. The method was compared with normal size-exclusion chromatography (SEC) in the separation of protein mixtures [112, 126]. Figure 13 [129] shows the separation of ovalbumin (Oval) and Myo at different sample loadings with and without an applied electric field. These two proteins could be partially separated by SEC because of their different molecular weights. However, it was obvious that the resolution by SEC decreased with increasing the sample size, and the separation became impossible when the sample loading size increased to 1.0 mL (14% of the bed volume) (see bottom panels in Fig. 13a–c). Upon application of the oscillatory transverse electric field, however, the resolution significantly increased and was little affected by increasing the sample size. In this case, both SEC and electromigration had positive effects on the resolution by the SEEC, so the resolution was improved greatly at increased sample loading. It should be noted that the SEEC resolution was very high at this high protein loading by considering that the column was only 12 cm long with a total packed volume of 7.2 mL Sephadex G75 gel

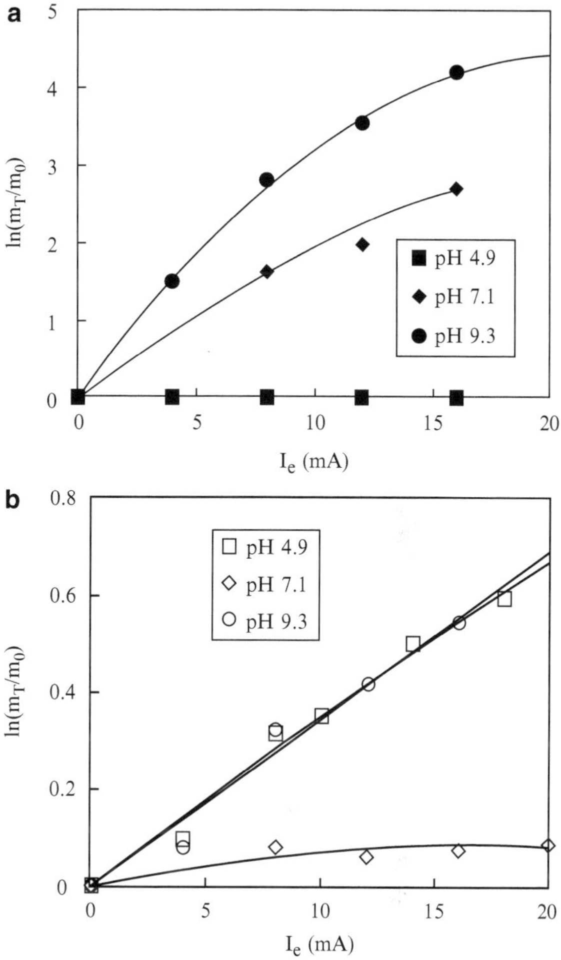

Fig. 12 a,b Dependence of In (m_T/m_0) for BSA (pI = 4.9) (**a**) and myoglobin (Myo, pI = 7.1) (**b**) on electric current at different pH values. The central compartment dimension was 12.0 × 0.5 × 1.2 cm, with 7-mL packed bed of Sephadex G75. The flow velocity was 60 cm h^{-1} and the current cycle was 120 s. In the figure, m_0 is the partition coefficient of protein in SEC; and m_T is the partition coefficient of protein in SEEC. Figure from [112]

Electrochromatography with an oscillatory transverse column electric field has also been demonstrated in adsorption chromatography such as ion exchange [125, 126] and dye-ligand affinity chromatography [130]. Tan et al. [125] examined the effect of electric field strength on the dynamic binding capacity (DBC) of ion exchange electrochromatography (IEEC) in packed beds of DEAE Sepharose FF. As shown in Fig. 14, the DBC at 10% breakthrough (Q_{10}) in IEEC increased linearly with increasing electric field strength. For example, with a packed-bed height of 15 mm and electric potential gradient of 38 V cm^{-1}, Q_{10} increased four times over that in normal ion-exchange chromatography. So, the transverse electric field has created significant electro-kinetic mass transports (electroosmosis and electrophoresis) that intensified exterior liquid-film and intraparticle mass transfers, leading to the increased protein binding capacity. Due to the increased capacity in the IEEC, separation of BSA and immunoglobulin under an overload condition (32 mL protein

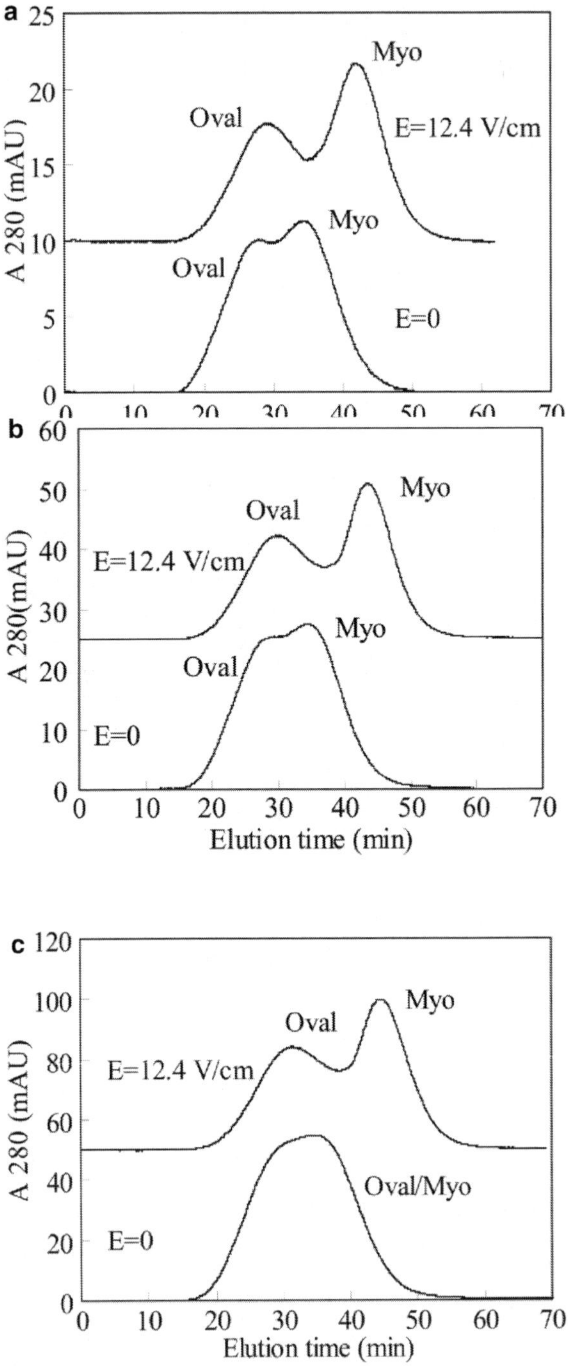

Fig. 13 **a–c** Separation of ovalbumin (Oval, pI = 5.08) and Myo by SEC and SEEC at different protein loadings. Mobile phase was 8.0 mmol L^{-1} acetate buffer (pH 4.9, conductivity = 0.47 mS cm^{-1}) at a linear flow velocity of 20 cm h^{-1}. Samples containing 0.5 mg mL^{-1} Myo and 0.5 mg mL^{-1} BSA were loaded at volumes of 0.2 mL (**a**), 0.5 mL (**b**), and 1.0 mL (**c**). Figure from [129]

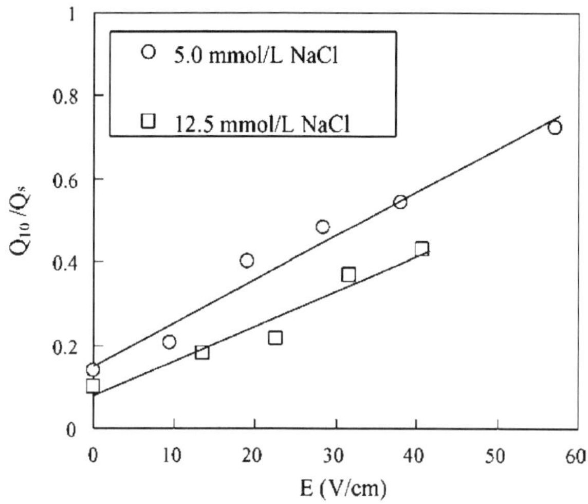

Fig. 14 Dependence of the ratio of dynamic binding capacity (Q_{10}) and static adsorption capacity (Q_S) on electric field strength at two different mobile-phase conductivities. The central compartment (12.0 × 0.5 × 1.2 cm) was packed with 0.85 mL DEAE Sepharose FF gel. Frontal analysis was carried out by loading 2 mg mL^{-1} BSA solution at 80 cm h^{-1}. The current cycle was kept constant at 20 s. Figure from [125]

solution containing 1.0 mg mL^{-1} BSA and 1.0 mg mL^{-1} immunoglobulin by 1.8-mL DEAE Sepharose FF column) was realized without any process optimization [125]. Moreover, in the purification of ovatransferrin and ovalbumin from crude hen egg-white solutions, the loading capacity of feedstock was increased 2.3 times with an oscillatory transverse column electric current of 30 mA at 1/20 Hz, as compared to IEC [126]. The results have revealed that an electric potential gradient of 20 V cm^{-1} was enough to enhance greatly the DBC in the IEEC, and when necessary, high electric field strength can be realized with a low applied voltage because the side distance of the column is usually an order of magnitude smaller than its height. The use of low voltage to carry out electrochromatography is a significant advantage over conventional electrochromatography with an axial electric field.

5.3 Perspectives

The most intriguing feature in protein electrochromatography is the occurrence of increased protein retention and intraparticle mass transport with an increase in electric field strength. These are attributed to the electrically induced CP and electrokinetic phenomena under the superimposed electric field. The experimental data and analysis presented in this chapter exhibits the importance of electrically induced phenomena on size-exclusion and adsorption chromatography. Up to now, however, much still remains unclear in the mechanisms of electrically induced CP and intraparticle electroosmotic flow in relation to the structure of porous media, even though both of them have been considered as the principle basis in the improvement of the separation efficiency in capillary and packed-bed electrochromatography. Regarding the utmost importance, a comprehensive insight and theoretical remark should be developed by further investigating the phenomena of the migration and retention of proteins in the electrochromatography. Moreover, the preferred architecture in the removal of Joule heating and electrolysis gases should be investigated in more detail for its capability of scale-up for HPPC of proteins and other biological macromolecules.

6 Expanded-Bed Adsorption

6.1 *Process Integration by Expanded-Bed Adsorption*

Biological products such as proteins produced by microbial fermentation or cell culture are usually of low content and highly contaminated by diverse soluble substances (e.g., other proteins, nucleic acids, lipids and polysaccharides) and even particulates (e.g., cells and cell debris). The cells or cell debris need removal by solid–liquid separation operations, and the soluble impurities have to be separated based on the differences of properties with the target protein. Consequently, the recovery and purification of proteins involve exclusively multistep processes including filtration, centrifugation, precipitation, adsorption and some chromatographic methods. The long separation train often results in low product recovery and quality, because more or less of the product is lost in each step, and a long-time exposure to a separation process can lead to conformational changes or even denaturation of proteins. Hence, it is necessary to increase the separation performance of each step by process optimization and/or separation material innovations as described above. Moreover, if some of the separation steps can be combined into a single one, the total step number can be significantly lowered, leading to the reduction of process time and the increase of product recovery as well as product quality. This approach is called process integration. Expanded bed adsorption (EBA) is such a technology that can combine solid–liquid separation and soluble product adsorption into a single step.

An expanded bed is a stable liquid–solid fluidized bed in which the stationary phase with controlled particle size and/or density distribution is fluidized in a liquid stream directed upward. Compared to conventional fluidized bed, there is a stable particle size and/or density classification in the axial direction, so expanded bed is a low-mixing fluidized bed with little solid phase mobility and reduced axial mixing of liquid phase. Hence, the chromatographic performance of an expanded bed can be comparable to a packed bed [131]. Since the increase in the upward flow velocity leads to the bed expansion and bed-voidage increase, thus allowing particulates in a liquid stream to pass through the bed, EBA is particularly suitable for application in the primary isolation of bioproducts from crude feedstock containing particulate materials such as whole cells and/or cell debris [131, 132]. It was claimed that the use of EBA reduced the cell concentration in the eluted fraction by a factor of 10^4–10^5 [133], which was at least one to two orders of magnitude higher than that achieved in an industrial centrifuge. Moreover, it has been demonstrated that EBA could be integrated with cell disruption and batch fermentation for direct product sequestration [134]. Hence, using the EBA technology a reduction in the number of process steps is achieved with particular advantages in terms of processing time and product yield, thus facilitating the establishment of a cost-effective bioseparation process.

Currently, EBA has been extensively studied in various aspects such as matrix development, column design, as well as process fundamentals and applications. Lots of review articles have been published on the different aspects of EBA [e.g., 131, 132], and a comprehensive one was recently reported by Hubbuch et al. [135]. Herein, we focus on the contributions of our group at the fabrication of solid matrices and solid phase classification studies.

6.2 Solid Matrices for EBA of Proteins

The terminal velocity of fluidized bed of slid particles can be calculated from the Stokes' equation:

$$U_\mathrm{t} = \frac{D^2(\rho_\mathrm{S} - \rho_\mathrm{L})g}{18\eta}, \tag{6}$$

where U_t is the terminal velocity, ρ_S the ρ_L solid and liquid densities, respectively, D the particle diameter, and g the gravitational constant. At a flow velocity higher than U_t, the solid phase would be elutriated from the bed, so the operation must be carried out at a flow velocity much lower than U_t, usually 10–20% of the U_t value to obtain a stable two- to threefold bed expansion. Towards high-performance EBA, we wish to increase the operational flow rate while keeping the process at high column efficiency. According to (1), small-sized adsorbent offers high column efficiency while (6) indicates that dense particles have high terminal velocity that permits operation at high flow. This implies that small-sized dense matrices can give better compromise between high flow rate and high column efficiency.

Hence, it is essential to develop small-sized dense matrices for stable EBA operation for particulates removal and target product adsorption [135, 136]. Moreover, to minimize the path length for product diffusion, an assembly of pellicular matrix is desired [134–139]. This particle design is characterized by a porous skin (or pellicle) of biocompatible polymer (e.g., agarose gel) cast upon a dense core (e.g., steel or glass beads). This design provides for the manufacture of a particle characterized by relatively large particle diameter and density to ensure a limited bed expansion response to rapidly flowing viscous feedstocks [139], while the thin porous layer offers rapid mass transfer to the binding sites. Sun and coworkers developed pellicular adsorbents comprising a 10–30 mm skin of agarose cast upon a zirconia-silica (ZS) core (density, 3.8 g cm^{-3}; particle size range, 63–125 mm, mean diameter, 99 mm) [138]. The pellicular beads were prepared by a water-in-oil emulsification with 4% agarose–ZS slurry. The particles were characterized with a mean particle diameter of 136 mm and a mean density of 2.39 g cm^{-3}. Derivatized with a dye ligand (Cibacron blue F3GA, CB), the pellicular particles were characterized by improved qualities of chromatographic behavior, particularly with respect to a threefold increase in the apparent effective diffusivity of lysozyme. The material was later successfully applied to the direct recovery of enzyme from yeast homogenates [134].

Sun and coworkers have also fabricated pellicular composite agarose–glass beads by coating 4% agarose gel onto glass beads by the water-in-oil emulsification method [140]. As listed in Table 3, two kinds of pellicular matrices, namely agarose-coated small glass beads (AG-S) and agarose-coated large glass beads (AG-L), were prepared (Fig. 15). The densities of AG-S and AG-L were as high as 1.77 and 1.98 g mL^{-1}, respectively, and the average depth of the agarose pellicle was estimated to be 20–30 μm. The matrices were modified with CB and lysozyme adsorption equilibrium and kinetic properties of the dye–ligand adsorbents were evaluated with the Langmuir isotherm and pore diffusion model, respectively. Stability of the matrices in basic solution and in recycled use was confirmed. It was found that the apparent pore diffusion coefficient of lysozyme in the pellicular beads was over twice as high as that in a homogeneous matrix due to their pellicular assembly. At nearly the same liquid velocity, the expanded bed of AG-S had an approximately twofold higher plate number than that of AG-L, so the expanded bed of the former was more efficient than that of the latter. Moreover, due to the rapid mass transfer property of the pellicular adsorbent CB-AG-S, its expanded bed showed a dynamic capacity over twice that of a commercial homogeneous medium designed for expanded bed application.

Besides expanded bed operation, a fluidized bed can be stabilized in a magnetic field by using magnetically susceptible matrices as the stationary phase. This kind of fluidized bed is called magnetically stabilized fluidized bed (MSFB) [140, 141]. With the stabilization effect of a magnetic field, the solid particles can be made in smaller size, while keeping the operation at higher flow rate. Hence, MSFB with small-sized dense magnetic adsorbents can offer better adsorption behavior for proteins than a normal expanded bed [141, 142].

Table 3 Physical properties of the pellicular composite agarose–glass matrices [140]

Matrix	Size range (μm)	d_p (μm)	d_i^a (μm)	f^b (–)	Mean density (g cm^{-3})
AG-S	70–190	120	61	0.45	1.77
AG-L	130–280	200	158	0.30	1.98
Small glass beads	35–85	61	0	0	2.40
Large glass beads	110–210	158	0	0	2.40

[a] Inner core mean diameter in the pellicular particles (i.e., mean diameter of glass beads)
[b] Agarose volume fraction in the pellicular particles

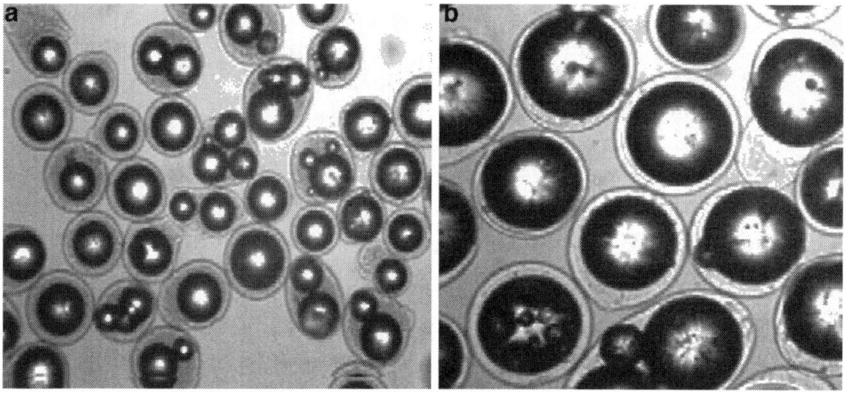

Fig. 15 a,b Pellicular agarose-glass beads of AG-S (**a**) and AG-L (**b**) observed by optical microscopy. Properties of the beads are listed in Table 3. Figure from [140]

6.3 Solid Phase Classification in Expanded Bed

EBA matrices are manufactured with controlled particle size and/or density distribution. The distribution of particle size and/or density within expanded bed system results in a distribution of terminal velocities, leading to a solid phase classification within the expanded bed. The particles with larger settling velocities are found at the bottom of the bed while those with smaller settling velocities are at the top end [131, 143]. Thus, lower liquid dispersion level is obtained in expanded bed because this classification can reduce the mobility of the adsorbents.

Sun and coworkers [127] investigated the size and density distributions of two EBA media, that is, Streamline quartz base matrix (Amersham Pharmacia Biotech, Uppsala, Sweden) and 6% agarose coated steel beads (6AS, UpFront Chromatography A/S, Copenhagen, Denmark), in an expanded bed system with a glass column (26 mm I.D.) modified by side ports. The Streamline particles were measured to show a broad size distribution (80–500 μm; volume-weighted mean diameter = 210 μm) but a relatively uniform density (1.135 g cm^{-3}), while the 6AS beads have both broad size (60–250 mm, volume-weighted mean diameter = 127 μm) and density distributions (about 1.5–4 g cm^{-3}, mean density = 2.843 g cm^{-3}). The effect of liquid-phase flow velocity, liquid viscosity and settled bed height on the particle size and density classifications was systematically investigated. It was found that the radial mean particle size and density of the two matrices were uniform, while axial classifications were obvious in the expanded beds. The volume-weighted mean particle size of Streamline decreased linearly with increasing the expanded bed height, as well correlated by the following equation:

$$\frac{D}{D_0} = 1.21 - 0.46\frac{h}{H}, \qquad (7)$$

where D is the volume-weighed mean diameter at different axial positions, D_0 the mean diameter of bulk particles, h the sampling position height from bed bottom, and H the expanded bed height. It is notable that the correlation was achieved with the experimental data measured at different liquid flow velocities, expanded bed heights, settled bed heights, and liquid phase viscosities (water, and 20 and 40 vol.% glycerol solutions). Moreover, the experimental data obtained by Bruce and Chase [143] could also be expressed by the correlation. Xue et al. [144] have established a polydisperse model to describe local particle size distributions and axial classification of solid particles like the Streamline of relatively uniform density.

In contrast to the Streamline particles, no axial particle size classification was observed for the dense 6AS beads. Instead, it was found that the mean particle density decreased exponentially with the increase of bed height, and correlated by (8) [127]:

$$\frac{\rho}{\rho_0} = 0.61\left(\frac{h}{H}\right)^{-0.42}, \tag{8}$$

where ρ is the local mean solid density at different axial positions, and ρ_0 the mean solid density of bulk particles (2.843 g cm^{-3}). The results indicated that density classification was dominant for the high-density particles with broad density and size distributions.

Zhou et al. [128] have studied the axial classification of the two dense pellicular agarose-glass matrices (AG-S and AG-L, see Table 3). Different from the Streamline and 6AS described above, it was found that the two AG particles showed both particle size and density classifications along the column axis. That is, both the size and density of each matrix decreased with increasing the axial bed height. However, the particle size classification of AG-L was less significant than AG-S, while the density classification of AG-L was more significant than AG-S. The phenomena were due to the density difference in AG-L and AG-S, as given in Table 3. AG-L was denser than AG-S, so the density classification of AG-L was more obvious than AG-S. The results suggest that axial density classification of solid phase becomes more significant as the solid phase density increases, and it becomes dominant when the particle density is extremely high, as in the case of 6AS, for which no axial particle size distribution was observed [127].

In terms of the Stokes' equation (6), the terminal velocity of a solid particle is proportional to $D^2(\rho_S - \rho_L)$. So Zhou et al. proposed a parameter to correlate axial mean particle size and density distributions as follows:

$$S = \frac{D^2(\rho - \rho_L)}{D_0^2(\rho_0 - \rho_L)}. \tag{9}$$

Then, the parameter was related to the dimensionless bed height in the following form:

$$S = a + b\frac{h}{H}, \tag{10}$$

where a and b are constants. By fitting the experimental data obtained with AG-S and AG-L [128] and those of Streamline and 6AS [127] to this equation, the coefficients were estimated to be $a = 1.5$ and $b = -1.2$. Figure 16 shows a comparison between the experimental data and the correlation with these coefficients [128]. A standard deviation (SD) of 0.11 between the experimental data and the correlation was obtained. This indicated that the correlation could express the axial classifications of the four matrices of different sizes and densities fairly well.

Based on the correlation for axial particle size distribution described by (7), Tong et al. [145] established a mathematical model for expanded bed protein adsorption by taking into account the axial size classification of Streamline DEAE (Amersham Pharmacia Biotech, Uppsala, Sweden) under various operating conditions, such as different feed protein concentrations, liquid velocities and liquid viscosities. Using independently determined model parameters, the model was found to fit the experimental breakthrough curves more precisely than that without considering the axial particle size distribution. Moreover, the correlation incorporating local particle size distribution was found useful for the analysis of breakthrough profiles [146, 147]; it was concluded that replacement of axial voidage distribution [146] and local particle size distribution [147] by their mean values resulted in little influence on the calculated breakthrough curves.

In EBA processes, there are often bed height changes during feed loading and washing processes due to the variations in solid and liquid properties. Using Streamline AC as the solid phase, Yang and Sun [148] studied the changes of local particle size distribution and bed voidage by changing mobile phase from water to 10 wt% glycerol solution or vice versa. It was found that the transient changes of

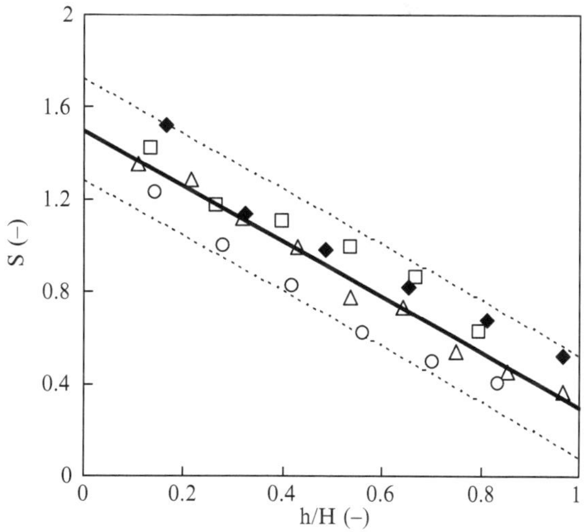

Fig. 16 Correlation of S as a function of h/H for the matrices of AG-S (*open squares*), AG-L (*open triangles*), 6AS (*open circle*) ($H_0 = 18$ cm, $H = 36$ cm) [127] and Streamline (*filled diamonds*) ($H_0 = 15$ cm, $H = 31$ cm) [127]. The *solid lines* are calculated from (10) with $a = 1.5$ and $b = -1.2$, and the *dotted lines* represent $S \pm 2$ SD (SD = 0.11). H_0 represents settled bed height. Figure from [128]

bed voidage at different axial positions were not unidirectional. That is, by changing the mobile phase to the high-viscosity glycerol solution, a constant increase of the bed voidage was observed in the bed bottom, while a distinct decrease of the bed voidage before its increase was involved at the middle and top positions. The phenomenon was caused by the compression effect of the upward movement of the lower part particles. The informaiton would be useful for the control of expanded bed during EBA processes.

6.4 Perspectives

EBA for single-step purification of proteins has made great progresses in recent years, and the technology is expected to find more applications in other areas such as recovery of nanoparticles (e.g., plasmid DNA and viruses) and protein refolding. Hence, as the principal pillar supporting the development of the EBA technology, diversity of matrices is required to meet various needs in different applications. It is essential to design small-sized dense microspheres of appropriate size distribution, hopefully in a pellicular structure to reduce mass transfer resistance. Moreover, EBA matrices should be designed to minimize the interactions with particulate contaminants such as cells and cell debris in biological feedstreams. The efforts would offer more robust adsorbents for selection in the purification of different biomolecules in repeated use, making the integrative separation technology more sophisticated for widespread applications.

7 Concluding Remarks

Chromatography of biomolecules is a major area of bioseparation research activities, and new achievements have been reported continuously. It is expected that chromatographic theory and technology will further develop to meet the demands in the research and development of biotechnology. In this process, more input needs to be offered to the innovation of solid matrices and the design and discovery of new types of ligands that have high affinity for proteins or can capture desired proteins under unfavorable feedstock conditions (e.g., extremely unneutral pH or moderate ionic strength). Moreover, a direct way is considered as the combination of the different approaches into the practice of chromatographic separation, realizing the objective of high-performance preparative chromatography of proteins: maximized feedstock loading and high separation selectivity of target products when operated at high flow rate. This can lead to reduced separation steps and process time, thus increasing the product quality and throughput, and in turn the process economics.

Acknowledgments We thank our colleagues and the researchers in our laboratory who contributed to the work described in this article. Financial supports from the Natural Science Foundation of China (No. 20636040) and the High-Tech Research and Development Program of China from the Ministry of Science and Technology of China (No. 2006AA02Z231) are also greatly appreciated.

References

1. Sadana A (1998) Bioseparation of proteins. Academic Press, San Diego, CA
2. Lightfoot EN, Moscariello JS (2004) Biotechnol Bioeng 87:259
3. Neverovaa I, Van Eykb JE (2005) J Chromatogr B 815:51
4. Chen C, Gonzalez FJ, Idle JR (2007) Drug Metabolism Rev 39:581
5. Janson JC, Rydén L (ed) (1998) Protein purification: principles, high-resolution methods and application, 2nd edn. Wiley, New York
6. García AA, Bonen MR, Ramírez-Vick J, Sadaka M, Vuppu A (1999) Bioseparation process science. Blackwell Science, MA, USA
7. Sun Y (2005) Bioseparation engineering (in Chinese). Chemical Industry Press, Beijing, China
8. Neue UD (2005) J Chromatogr A 1079:153
9. Brooks CA, Cramer SM (1992) AIChE J 38:1969
10. Li Y, Pinto NG (1994) J Chromatogr A 658:445
11. Raje P, Pinto NG (1997) J Chromatogr A 760:89
12. Zhang SP, Sun Y (2002) J Chromatogr A 957:89
13. Chen J, Sun Y (2003) J Chromatogr A 992:29
14. Zhang SP, Sun Y (2003) Ind Eng Chem Res 42:1235
15. Bosma JC, Wesselingh JA (2004) AIChE J 50:848
16. Su XL, Sun Y (2006) AIChE J 52:2921
17. Zhou XP, Su XL, Sun Y (2007) Biotechnol Prog 23:1118
18. Bosma JC, Wesselingh JA (1998) AIChE J 44:2399
19. Zhang SP, Sun Y (2001) Biotechnol Bioeng 75:710
20. Zhang SP, Sun Y (2004) Biotechnol Prog 20:207
21. Shi QH, Zhou Y, Sun Y (2005) Biotechnol Prog 21:516
22. Shen H, Frey DD (2005) J Chromatogr A 1079:92
23. Wesselingh JA, Bosma JC (2001) AIChE J 47:1571
24. Xue B, Sun Y (2001) J Chromatogr A 921:109
25. Zhang SP, Sun Y (2002) AIChE J 48:178
26. Chen WD, Dong XY, Sun Y (2002) J Chromatogr A 962:29
27. Carta G, Ubiera AR, Pabst TM (2005) Chem Eng Technol 28:1252
28. Chen WD, Dong XY, Bai S, Sun Y (2003) Biochem Eng J 14:45
29. Yang K, Sun Y (2007) Biochem Eng J 37:298
30. Jungbauer A, Kaltenbrunner O (1996) Biotechnol Bioeng 52:223
31. Xia F, Nagrath D, Cramer SM (2003) J Chromatogr A 989:47
32. Li W, Zhang SP, Sun Y (2004) Biochem Eng J 22:63
33. Li P, Xiu GH, Rodrigues AE (2004) Chem Eng Sci 59:3091
34. Jozwik M, Kaczmarski K, Freitag R (2005) J Chromatogr A 1073:111
35. Li W, Li Y, Sun Y (2005) Chem Eng Sci 60:4780
36. Ljunglöf A, Thömmes J (1998) J Chromatogr A 813:387
37. Hubbuch J, Linden T, Knieps E, Thömmes J, Kula MR (2002) Biotechnol Bioeng 80:359
38. Dziennik SR, Belcher EB, Barker GA, DeBergalis MJ, Fernandez SE, Lenhoff AM (2003) Proc Natl Acad Sci U S A 100:420

39. Hubbuch J, Linden T, Knieps E, Ljunglöf A, Thömmes J, Kula MR (2003) J Chromatogr A 1021:93
40. Hubbuch J, Linden T, Knieps E, Thömmes J, Kula MR (2003) J Chromatogr A 1021:105
41. Zhou XP, Li W, Shi QH, Sun Y (2006) J Chromatogr A 1103:110
42. Yang K, Shi QH, Sun Y (2006) J Chromatogr A 1136:19
43. Tyn MT, Gusek TW (1990) Biotechnol Bioeng 35:327
44. Teeters MA, Conrardy SE, Thomas BL, Root TW, Lightfoot EN (2003) J Chromatogr A 989:165
45. Zhang SP, Sun Y (2002) J Chromatogr A 964:35
46. Boyer PM, Hsu JT (1992) AIChE J 38:259
47. Guiochon G, Shirazi SG, Katti AM (1994) Fundamentals of preparative and nonlinear chromatography. Academic Press, Boston
48. Bird RB, Stewart WE, Lightfoot EN (1960) Transport phenomena. Wiley, New York, pp 180–199
49. Afeyan NB, Fulton SP, Gordon NF, Mazsaroff I, Várady L, Regnier FE (1990) Biotechnol 8:203
50. Collins WE (1997) Sep Purif Methods 26:215
51. DePhillips P, Lenhoff AM (2000) J Chromatogr A 883:39
52. Zhou X, Xue B, Sun Y (2001) Biotechnol Prog 17:1093
53. Zhou X, Xue B, Bai S, Sun Y (2002) Biochem Eng J 11:13
54. Yao Y, Lenhoff AM (2004) J Chromatogr A 1037:273
55. Afeyan NB, Gordon NF, Mazsaroff I, Várady L, Fulton SP (1990) J Chromatogr 519:1
56. Afeyan NB, Fulton SP, Regnier FE (1991) J Chromatogr 544:267
57. Gustavsson PE, Larsson PO (1996) J Chromatogr A 734:231
58. Gottschalk I, Gustavsson PE, Ersson B, Lundahl P (2003) J Chromatrogr A 784:203
59. Tiainen P, Larsson PO (2007) J Chromatogr A 1138:84
60. Sun GY, Shi QH, Sun Y (2004) J Chromatogr A 1061:159
61. Zhang ML, Sun Y (2001) J Chromatogr A 922:77
62. Shi Y, Dong XY, Sun Y (2002) Chromatographia 55:405
63. Shi Y, Sun Y (2003) Chromatographia 57:29
64. Wu L, Bai S, Sun Y (2003) Biotechnol Prog 19:1300
65. Chen JL, Bai S, Sun Y (2003) Chromatographia 58:701
66. Li Y, Dong XY, Sun Y (2005) Biochem Eng J 27:33
67. Li Y, Dong XY, Sun Y (2007) J Appl Polym Sci 104:2205
68. Shi QH, Zhou X, Sun Y (2005) Biotechnol Bioeng 92:643
69. Wang DM, Hao G, Shi QH, Sun Y (2007) J Chromatogr A 1146:32
70. Wang DM, Sun Y (2007) Biochem Eng J 37:332
71. Brandt S, Goffe RA, Kessler SB, O'Connor JL, Zale SE (1988) Biotechnology 6:779
72. Thömmes J, Kula MR (1995) Biotechnol Prog 11:367
73. Zhou JX, Tressel T (2006) Biotechnol Prog 22:341
74. Boi C (2007) J Chromatogr A 848:19
75. Haber C, Skupsky J, Lee A, Lander R (2004) Biotechnol Bioeng 88:26
76. Svec F, Fréchet JMJ (1999) Ind Eng Chem Res 38:34
77. Jungbauer A, Hahn R (2006) J Sep Sci 27:767
78. Svec F, Tennikova TB, Deyl Z (2003) Monolithic materials: preparation, properties and applications. J Chromatography Library, Elsevier, Amsterdam, Netherlands, pp 1–773
79. Hjertén S (1999) Ind Eng Chem Res 38:1205
80. Zhang ML, Sun Y (2001) J Chromatogr A 912:31
81. Du KF, Yang D, Sun Y (2007) J Chromatogr A 1163:212
82. Azarkan M, Huet J, Baeyens-Volant D, Looze Y, Vandenbussche G (2007) J Chromatogr B 849:81
83. Roque ACA, Silva CSO, Taipa MÂ (2007) J Chromatogr A 1160:44
84. Lowe CR (2001) Curr Opin Chem Biol 5:248
85. Falciani C, Lozzi L, Pini A, Bracci L (2005) *Chem Biol* 12:417

86. Roque ACA, Taipa MÂ, Lowe CR (2004) J Mol Recognit 17:262
87. Jacobsen B, Gårdsvoll H, Funch FJ, Østergaard S, Barkholt V, Ploug M (2007) Protein Exp Purif 52:286
88. Bellofiore P, Petronzelli F, Martino TD, Minenkova O, Bombardi V, Anastasi AM, Lindstedt R, Felici F, Santis RD, Verdoliva A (2006) J Chromatogra A 1107:182
89. Lowe CR, Burton SJ, Burton NP, Alderton WK, Pitts JM, Thomas JA (1992) Trends Biotechnol 10:442
90. Mondal K, Gupta MN (2006) Biomol Eng 23:59
91. Labrou NE (2003) J Chromatogr B 790:67
92. Platis D, Sotriffer CA, Clonis Y, Labrou NE (2006) J Chromatogr A 1128:138
93. Katsos NE, Labrou NE, Clonis YD (2004) J Chromatogr B 807:277
94. Lin DQ, Yao SJ (2007) *Biotechnol Prog* 23:904
95. Clonis YD (2006) J Chromatogr A 1101:1
96. Rupasinghe CN, Spaller MR (2006) *Curr Opin Chem Biol* 10:188
97. Roque ACA, Lowe CR (2006) *Biotechnol Adv* 24:17
98. Melissis S, Labrou NE, Clonis YD (2006) J Chromatogr A 1122:63
99. Melissis S, Labrou NE, Clonis YD (2007) Biotechnol J 2:121
100. Liu FF, Wang T, Dong XY, Sun Y (2007) J Chromatogr A 1146:41
101. Kitchen DB, Decornez H, Furr JR, Bajorath J (2004) Nat Rev Drug Discov 3:935
102. Yu HQ, Dong XY, Sun Y (2004) Biochem Eng J 18:169
103. Yu HQ, Dong XY, Sun Y (2004) Chromatographia 60:379
104. Kuntz ID, Blaney JM, Oatley SJ, Langridge R, Ferrin TE (1982) J Mol Biol 161:269
105. Kagawa M, Fujimoto Z, Momma M, Takase K (2003) J Bacteriol 185:6981
106. Rarey M, Kramer B, Lengauer T, Klebe G (1996) J Mol Biol 261:470
107. Liu FF, Dong XY, Wang T, Sun Y (2007) J Chromatogr A 1175:249
108. Lamba D, Bauer M, Huber R, Fischer S, Rudolph R, Kohnert U, Bode W (1996) J Mol Biol 258:117
109. Hamann CH, Hamnett A, Vielstich W (2007) Electrochemistry. Wiley, Weinheim
110. Nischang I, Tallarek U (2007) Electrophoresis 28:611
111. Paces M, Kosek J, Marek M, Tallarek U, Seidel-Morgenstern A (2003) Electrophoresis 24:380
112. Tan GM, Shi QH, Sun Y (2005) Electrophoresis 26:3084
113. Cole KD, Cabezas Jr H (1995) Appl Biochem Biotechnol 54:159
114. Leinweber FC, Tallarek U (2004) Langmuir 20:11637
115. Rudge SR, Basak SK, Ladisch MR (1993) AIChE J 39:797
116. Basak SK, Lad isch MR (1995) AIChE J 41:2499
117. van Kreveld ME, van den Hoed N (1973) J Chromatogr 83:111
118. Wan QH (1997) J Chromatogr A 782:181
119. Stol R, Kok WT, Poppe H (1999) J Chromatogr A 853:45
120. Venema E, Kraak JC, Poppe H, Tijssen R (1999) J Chromatogr A 837:3
121. Hoppe H, Stol R, Kok WT (2002) J Chromatogr A 965:75
122. Keim C, Ladisch MR (2000) Bioeng Biotechnol 70:71
123. Liu Z, Yin G, Feng SH, Wang DH, Ding FX, Yuan NJ (2001) J Chromatogr A 921:93
124. Cole KD (1997) Biotechnol Prog 13:289
125. Tan GM, Shi QH, Sun Y (2005) J Chromatogr *A* 1098:131
126. Yuan W, Zhao GF, Dong XY, Sun Y (2006) J Sep Sci 29:2383
127. Tong XD, Sun Y (2002) J Chromatogr A 977:173
128. Zhou X, Shi QH, Bai S, Sun Y (2004) Chin J Chem Eng 12:310
129. Tan GM, Dong XY, Sun Y (2006) J Sep Sci 29:684
130. Jia GD, Dong XY, Sun Y (2008) Sep Purif Technol 59:277
131. Chase HA (1994) Trends Biotechnol 12:296
132. Hjorth R (1997) Trends Biotechnol 15:2305
133. Hansson M, Stahl S, Hjorth R, Uhlen M, Mokes T (1994) Biotechnology 12:285
134. Jahanshahi M, Sun Y, Santos E, Pacek AW, Nienow AW, Lyddiatt A (2002) Biotechnol Bioeng 80:201

135. Hubbuch J, Thömmes J, Kula MR (2005) Adv Biochem Eng Biotechnol 92:101
136. Tong XD, Dong XY, Sun Y (2002) Biochem Eng J 12:117
137. Tong XD, Sun Y (2002) J Chromatogr A 943:63
138. Sun Y, Pacek AW, Nienow AW, Lyddiatt A (2001) Biotechnol Bioprocess Eng 6:419
139. Lyddiatt A (2002) Cur Opin Biotechnol 13:95
140. Zhou X, Shi QH, Bai S, Sun Y (2004) *Biochem Eng J* 18:81
141. Tong XD, Sun Y (2003) *Biotechnol Prog* 19:1721
142. Ding Y, Sun Y (2005) Chem Eng Sc*i* 60:917
143. Bruce LJ, Chase HA (2001) *Chem Eng Sci* 56:3149
144. Xue B, Tong XD, Sun Y (2003) AIChE J 49:2510
145. Tong XD, Xue B, Sun Y (2003) Biochem Eng J 16:265
146. Kaczmarski K, Bellot JC (2004) Biotechnol Prog 20:786
147. Kaczmarski K, Bellot JC (2005) J Chromatogr A 1069:91
148. Yang Z, Sun Y (2005) J Chromatogr A 1077:143

Adv Biochem Engin/Biotechnol (2009) 113: 255-274
DOI: 10.1007/10_2008_16

Published online: 21 May 2009

Recent Developments in Biodesulfurization of Fossil Fuels

Ping Xu, Jinhui Feng, Bo Yu, Fuli Li, and Cuiqing Ma

Abstract The emission of sulfur oxides can have adverse effects on the environment. Biodesulfurization of fossil fuels is attracting more and more attention because such a bioprocess is environmentally friendly. Some techniques of desulfurization have been used or studied to meet the stricter limitation on sulfur content in China. Recent advances have demonstrated the mechanism and developments for biodesulfurization of gasoline, diesel and crude oils by free cells or immobilized cells. Genetic technology was also used to improve sulfur removal efficiencies. In this review, we summarize recent progress mainly in China on petroleum biodesulfurization.

Keywords Biocatalyst, Biodesulfurization, Fossil fuels, Mechanism

---

P. Xu (✉)
Key Laboratory of Microbial Metabolism, Ministry of Education, School of Life Sciences and Biotechnology, Shanghai Jiao Tong University, Shanghai 200240, People's Republic of China and State Key Laboratory of Microbial Technology, Shandong University, Jinan 250100, People's Republic of China
e-mail: pingxu@sjtu.edu.cn

J. Feng (✉), F. Li, and C. Ma
State Key Laboratory of Microbial Technology, Shandong University
Jinan, 250100, People's Republic of China
e-mail: fengjinhui@mail.sdu.edu.cn

B. Yu
Institute of Microbiology, Chinese Academy of Sciences
Beijing, 100080, People's Republic of China

Contents

Abbreviations

2-HBP	2-Hydroxybiphenyl
4,6-DM-DBT	4,6-Dimethyl DBT
BDS	Biodesulfurization
BTH	Benzothiophene
CA	Carbazole
DBT	Dibenzothiophene
DMSO	Dimethylsulfoxide
DPS	Diphenylsulfide
EOR	Enhanced oil recovery
FCC	Fluid catalytic cracking
FGD	Flue gas desulfurization
GC-AED	Gas chromatographic atomic emission detection
GC-FID	Gas chromatographic flame ionization detection
HBPSi	2′-Hydroxybiphenyl 2-sulfinic acid
HDS	Hydrodesulfurization
MAM	Modification of A medium
PASHs	Polycyclic aromatic sulfur heterocycles
SR	Straight-run
TH	Thiophene
VHb	v*itreoscilla* hemoglobin

1 Introduction

China will displace the US as the largest energy consumer by 2010 based on current trends. Combustion of sulfur-containing compounds in fossil fuels leads to the emission of sulfur oxides, which can cause adverse effects on health, environment and economy. Today, China is the largest emitter of SO_2, and about 25.9 million tons of SO_2 were emitted in 2006 (http://www.zhb.gov.cn/ztbd/sjhjr/2007hjr/tpbd56/200706/P020070625532626111313.pdf). The implementation of the revised laws of clean air in the USA and China has brought stricter limitation on sulfur content in petroleum products. The maximum allowable sulfur for diesel in USA is currently 15 ppm, with 10 ppm sulfur by 2010 [1]. It is expected that SO_2 emission in China will be controlled to be within 23 million tons in 2010, while the rapid growth rate of economy is expected to continue. Although the growth rate of SO_2 emission has decreased in 2006, as shown in Fig. 1, it is still a big challenge for China to control SO_2 emission over the next few years.

Sulfur is the third most abundant element in crude oil, in which the content can vary from 0.05 to 10%. In addition to elemental sulfur, sulfate, sulfite, thiosulfate, and sulfide, more than 200 sulfur-containing organic compounds have been identified in crude oils. These include sulfides, thiols, thiophenes, substituted benzo- and dibenzothiophenes, benzonaphthothiophene and many considerably more complex molecules [2]. The condensed thiophenes are the most common form in which sulfur is present [3]. Dibenzothiophene (DBT) and substituted DBTs are the major sulfur-containing aromatic compounds in fuels, accounting for up to 70% of the sulfur content (Fig. 2) [4].

Some techniques of desulfurization have been used or studied to meet the stricter limitation on sulfur content in China. Hydrodesulfurization (HDS), oxidative desulfurization [5, 6], and biodesulfurization (BDS), have all been studied. In this review we summarize the recent progress – mostly in China – in petroleum desulfurization by microbial technologies.

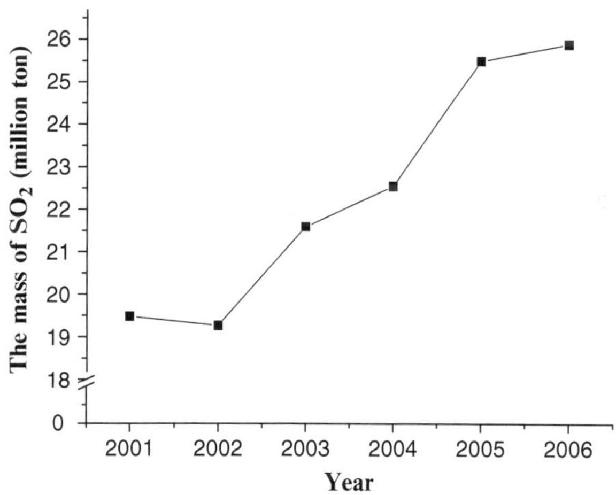

Fig. 1 The emission of sulfur oxides in China. The data was received from the website of the State Environmental Protection Administration of China (http://www.zhb.gov.cn)

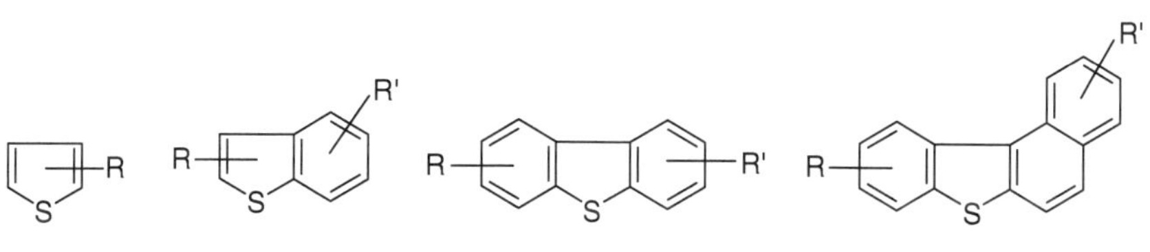

Fig. 2 The sulfur heteroaromatic compounds found in petroleum

2 Model Sulfur Compounds and Their Pathways of Biodesulfurization

At the present time, petroleum refining is mainly based on the use of physicochemical processes such as distillation and chemical catalysis that operate under extreme conditions, i.e., high temperatures and pressures. Such processes are energetically costly and cause high contamination. "Biorefining" operates at ambient temperature and pressure and is endowed with a high selectively, resulting in decreased energetic costs, low emissions and no generation of undesirable by-products; it is thought to be an interesting alternative or complement for development of new petroleum refining processes [7]. The derivatives of DBT and benzothiophene (BTH), as well as other polycyclic aromatic sulfur heterocycles (PASHs), are the most abundant heterocyclic compounds in petroleum, some of which have been reported to be difficult to remove by conventional HDS processes. Alkyl DBTs and alkyl BTHs are highly recalcitrant to chemical catalysts, especially when alkylated at the carbon 4 and 6 positions [3]. Much effort has been put into the investigation of biological desulfurization systems using DBT or alkylated DBTs as model compounds, and the metabolic pathway of desulfurization was proposed as the so-called "4S" pathway (Fig. 3) [1, 8–10].

In this pathway, DBT was first converted to DBT sulfoxide, then DBT sulfone, and then 2′-hydroxybiphenyl 2-sulfinic acid (HBPSi), and finally 2-hydroxybiphenyl (2-HBP), releasing sulfate into the medium. This process is called the "4S" pathway due to the four sulfur-containing compounds present. Isotopic labeling experiments have shown that the oxygen atom of the hydroxyl group of 2-HBP originates from molecular oxygen, implicating a role for an oxygenase or oxygenases in the pathway [11]. BDS processes using this strain will be ideal. In a BDS process the end product, 2-HBP and its derivatives, would partition back into the oil, thus preserving the fuel value. This technique has been the focus of BDS for the last decade. In summarizing the research of Gray et al. [12] and Oldfield et al. [11], the complete metabolism of DBT by *R. erythropolis* IGTS8 and the enzymes involved in the process are clarified

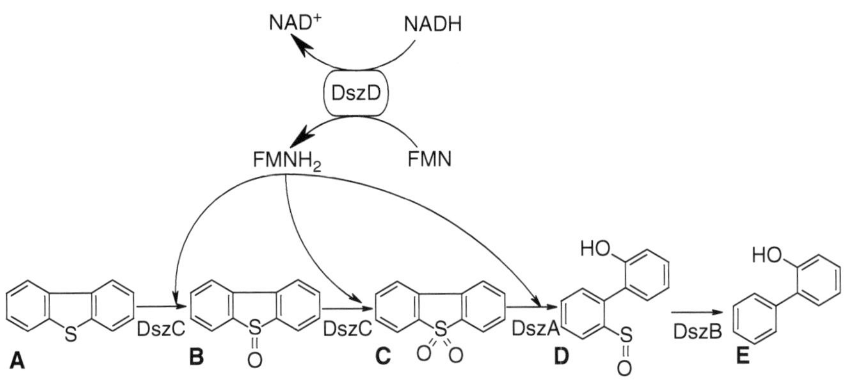

Fig. 3a–e Sulfur specific pathway for DBT desulfurization. DszC, DBT monooxygenase; DszA, DBT sulfone monooxygenase; DszB, HBPSi desulfinase; DszD, flavin reductase. **a** DBT. **b** DBT sulfoxide. **c** DBT sulfone. **d** 2-Hydroxybiphenyl-2-sulfinate. **e** 2-Hydroxybiphenyl

as follows: DszC catalyses the stepwise *S*-oxidation of DBT, first to DBT sulfoxide (DBTO) and then to DBT sulfone ($DBTO_2$); DszA catalyses the conversion of $DBTO_2$ to HBPSi and DszB catalyses the desulfurization of HBPSi to give HBP and sulfate. Studies with cell-free extracts show that DszA and DszC, but not DszB, require $FMNH_2$ for activity which appears to be supplied by the fourth enzyme DszD. $^{18}O_2$-labeling studies show that DszC is a monooxygenase, and DszA is an apparently unique enzyme which catalyses the reductive hydroxylation of $DBTO_2$ leading to the cleavage of the thiophene ring, and DszB is an aromatic sulfuric acid hydrolase. The structure of the *dsz* operon is shown in Fig. 4. The genes *dsz*A, *dsz*B, and *dsz*C are arranged in sequence [13]. The forth gene, *dsz*D, was not on the *dsz* operon.

2.1 Microorganisms

Many microorganisms such as *Mycobacterium* [14–17], *Rhodococcus* [18–21], *Microbacterium* [22], *Gordonia* [23], and *Pseudomonas* [24, 25] were found to degrade DBT at different sites in China. It is interesting to find that all of the microorganisms have the same or similar pathway to degrade DBT to 2-HBP, even though these microorganisms do not belong to the same genus. Horizontal gene transfer was assumed because the *dsz*ABC genes are transcribed as an operon located on a large plasmid. *Rhodococcus* predominates in the desulfurization strains. *R. erythropolis* XP [19, 20], *R. erythropolis* DS-3 [21, 26], *R. erythropolis* 1awq [18, 27], *R. erythropolis* LSSE8-1 [28], and *R. erythropolis* FSD-2 [29] were isolated and studied. All of these microorganisms have the ability to degrade DBT to 2-HBP and show different degrees of excellence.

Mycobacterium sp. ZD-19 can utilize a wide range of organic sulfur compounds as sole sulfur source. Thiophene (TH), BTH, diphenylsulfide (DPS), DBT, and 4,6-dimethyl DBT (4,6-DM-DBT) can all be degraded by *Mycobacterium* ZD-19 [16]. TH and BTH can be removed completely by strain ZD-19 within 42 h, and 100% of DBT and 4,6-DM-DBT was degraded within 56 h. About 80% of the DPS was degraded by strain ZD-19 after 90 h. *Mycobacterium goodii* X7B can degrade DBT, BTH and their alkyl derivatives below 45 °C [14, 17]. HDS-treated diesel oil, gasoline, and crude oil can all be desulfurized by *M. goodii* X7B and *R. erythropolis* XP [14, 15, 17, 19, 20]. *Gordonia* sp. ZD-7 grows well in aqueous/oil system and the biodesulfurization could be repeatedly performed for more than 193 h [23].

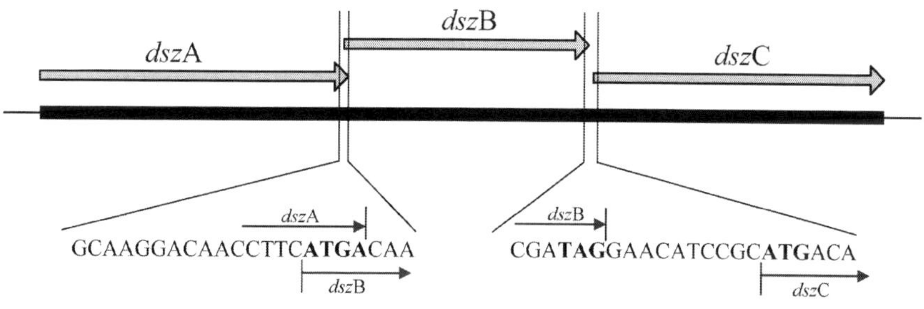

Fig. 4 The *dsz* operon of *R. erythropolis* IGTS8 [13]. The initiation codon and the termination codon are in *bold* type

2.2 Degradation Pathway

The degradation pathways of DBT by *Mycobacterium goodii* X7B and *Microbacterium* sp. ZD-M2 have been described in detail [17, 22]. As for *M. goodii* X7B, the intermediate metabolite of DBT degradation, $DBTO_2$ and 2-HBP were both detected by mass spectroscopy. Furthermore, in the broth of *M. goodii* X7B, 2-methoxybiphenyl (2-MBP) was identified. When 4,6-DM-DBT served as the sulfur source, the corresponding methoxy biphenyl was also detected [17]. Therefore, an extended "4S" pathway was revealed. The desulfurization pathway of DBT was as follows: DBT is degraded to 2-HBP in the initial steps as shown in Fig. 3, and then 2-HBP is converted to 2-MBP. It is interesting to note that *Microbacterium* sp. ZD-M2 can further produce biphenyl in addition to 2-MBP from 2-HBP or from the intermediate metabolite HBPSi [22]. 2-HBP has been used as a fungicide to control postharvest diseases of fruits [30]. Inhibition of desulfurization activity and cell growth occurred at high concentration of 2-HBP [31]. Therefore, the production of 2-MBP and biphenyl has the advantage of partially eliminating the enzyme inhibitory effect of 2-HBP.

In contrast to DBT-desulfurizing bacteria, little is known about bacteria that can desulfurize BTH. BTH predominates in gasoline. Oil contamination may impact the organisms that live in contaminated ecosystems because some of theses compounds, such as benzothiophene derivatives, have been reported to be mutagenic and carcinogenic [3]. Therefore the degradation pathway of BTH was also studied. *M. goodii* X7B can grow in modification of A medium (MAM) with BTH as the sole source of sulfur. The broth was extracted with ethyl acetate and then analyzed. The mass spectral data reviewed that *o*-hydroxystyrene and benzo(*c*)(1,2) oxathiin *S,S*-dioxide should be the intermediate metabolite [14]. A degradation pathway was proposed as shown in Fig. 5. The purified enzymes involved in BTH degradation by strain X7B would provide a detailed explanation for the degradation of both BTH and DBT by a single bacterial strain.

R. erythropolis XP was capable of growing on 3-methyl-benzothiophene (3-M-BTH) as sole sulfur source. Resting cells of *R. erythropolis* XP were also used

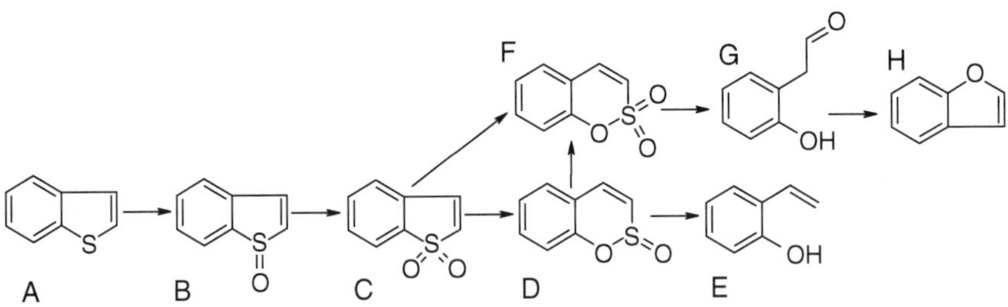

Fig. 5a–h The desulfurization pathway of BTH. **a** BTH. **b** BTH sulfoxide. **c** BTH sulfone. **d** Benzo(*e*)(1,2)oxathiin *S*-oxide. **e** *o*-Hydroxystyrene. **f** Benzo(*c*)(1,2)oxathiin *S,S*-dioxide. **g** 2-(2'-Hydroxyphenyl) ethan-1-al. **h** Benzofuran [14]

for degradation of 3-M-BTH. 3-M-BTH (0.5 mM) can be degraded completely in the aqueous phase. High concentrations of 3-M-BTH in *n*-octane were also degraded by resting cells of XP strain, and about 44% of 3-M-BTH was degraded in 22 h [19]. The degradation of BTH and its derivative by strains *M. goodii* X7B and *R. erythropolis* XP implied their capability for gasoline biodesulfurization.

3 Enhancement of Biodesulfurization by Recombinant Bacteria

The use of specific bacteria to remove sulfur pollutants from petroleum will reduce the amount of sulfur oxides released and therefore lessen the serious environmental pollution. However, genetic modifications to improve sulfur removal efficiencies are needed before a practical biorefining process for the removal of sulfur compounds from petroleum can be developed. Genetically engineered microorganisms are required to remove the sulfur compounds with higher activities in practical systems. Cultures with improved substrate ranges are also needed to address better the complicated mixture of chemicals present in petroleum. Some progress was also made in China.

3.1 Rearranging the dsz Gene Cluster

The expressional characteristics of the *dsz* operon of *R. erythropolis* DS-3 were investigated. The translation levels of desulfurization enzymes decreased according to their positions in the operon, and the ratio of mRNA quantity of *dsz*A, *dsz*B, and *dsz*C was determined to be 11:3.3:1. However, the expression level of DszC was far higher than that of DszB, because there is an overlapped region of the termination codon for *dsz*A and the initiation codon for *dsz*B. Therefore, the expression level of *dsz*B was determined both by its location on the *dsz* operon and by the baffled structure before the initiation codon of *dsz*B. When the expression level of *dsz*B was increased by removing the baffled structure, the DBT desulfurization rate was five times faster than that for the native *dsz* operon [26].

The catalytic capabilities of the Dsz enzymes were approximately 25:1:5 (DszA:DszB:DszC). Hence, the *dsz* operon was rearranged according to the catalytic capabilities of the enzymes. The expression levels of *dsz*B and *dsz*C were improved by rearranging the order of the *dsz* genes to generate operon *dsz*BCA, which contained *dsz*B, *dsz*C, and *dsz*A in tandem. The desulfurization rate of the recombinant strain containing the rearranged *dsz* operon was 12 times faster than that for the native *dsz* operon. The maximum desulfurization rate was only about 26 μmol DBT/g DCW/h for the strain contained the native *dsz* operon. After removing the baffled structure before the initiation codon of *dsz*B, the rate was 120 μmol DBT/g DCW/h. The recombinant strain containing the rearranged *dsz* operon showed the highest desulfurization rate, about 320 μmol DBT/g DCW/h.

Therefore, the enhanced expression levels of DszC and DszB increase the desulfurization rate of the recombinant strain [32].

Feng et al. found that the function of surfactant Tween 80 in the desulfurization was to decrease the product concentration associated with the cells, reducing product inhibition [27]. The *dsz* genes of *R. erythropolis* DS-3 were also integrated into the chromosome of *Bacillus subtilis* ATCC 21332, which can secrete biosurfactant, yielding the recombinant strain *B. subtilis* M29, which has higher desulfurization efficiency than *R. erythropolis* DS-3 and showed no product inhibition. The selective desulfurization efficiency of *R. erythropolis* DS-3 at 36 h was 13.1 mg DBT L^{-1} h^{-1}, while the efficiency of *B. subtilis* M29 was 16.2 mg DBT L^{-1} h^{-1} [21]. It should be noted that the biosurfactant secreted from *B. subtilis* M29 significantly varied the interfacial tension of the supernatant. The biosurfactant therefore has an important function in the degradation of DBT.

3.2 Introducing the Vgb Gene

A significant oxygen pressure is necessary during bioconversion because the Michaelis constant (*Km*) of the oxygenase for oxygen is relatively high. *Vitreoscilla* hemoglobin (VHb) can be used to improve the supply, transfer, and store of oxygen in vivo. The *vgb* gene (encoding VHb) was introduced into a specific desulfurization bacterium, *R. erythropolis* LSSE8-1. The DBT desulfurization ratios of recombinant strain and LSSE8-1 were 37.5% and 20.5%, respectively [28].

3.3 Construction of a Solvent-Tolerant Desulfurizing Bacterium

Most organic solvents are toxic to microorganisms even at low concentrations (e.g., 0.1 vol.%). Fuel oil is similar to organic solvents regarding toxicity to microorganisms. Microorganisms with a high tolerance to organic solvents are useful and important in many biotechnological fields such as biodesulfurization. Considering the difficulty of isolating strains that have both solvent tolerance and the desired catalytic activity from the environment, it might be preferable to combine solvent-tolerance with unique catalytic characteristics using the methods of genetic engineering. Tao et al. constructed a solvent-tolerant desulfurizing bacterium, *Pseudomonas putida* A4, by introducing the biodesulfurization gene cluster *dsz*ABCD from *R. erythropolis* XP into the solvent-tolerant strain *P. putida* Idaho [33]. The genes *dsz*A, *dsz*B, *dsz*C, and *dsz*D were amplified from the genome of desulfurizing bacterium *R. erythropolis* XP, separately, and inserted into the broad host plasmid pMMB66EH, resulting recombinant plasmid pMMABCD, as shown in Fig. 6. This plasmid was constructed and introduced into *P. putida* Idaho by the triparental mating method. The introduction and the screening method are shown in Fig. 7. The result strain was named *P. putida* A4.

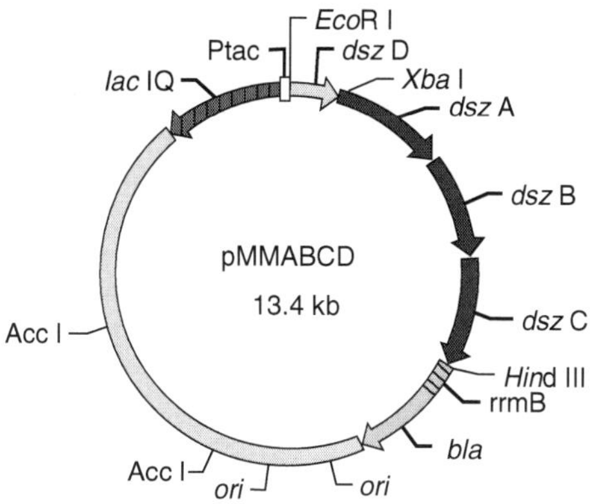

Fig. 6 Map of recombinant plasmid pMMABCD. The *arrows* indicate the direction of transcription of the genes. Only parts of restriction sites are shown (*Eco*RI, *Xba*I, *Hin*dIII, *Acc*I). *bla* ampicillin resistance gene; *lacIQ* the Lac repressor gene; *Ptac* promoter; *ori* sequence encoding the origin of replication for duplex DNA

P. putida A4 could degrade dibenzothiophene in the presence of 10 vol.% *p*-xylene, heptanol, styrene, ethylbenzene, cyclohexane, *o*-dichlorobenzene, diphenylether, isooctane, and *n*-dodecane. In the initial 6 h, 86% of the DBT was degraded by *P. putida* A4 in the presence of 10 vol.% *p*-xylene, while no DBT decrease was observed in the reactions performed by *R. erythropolis* XP [33].

3.4 Combination of Sulfur and Nitrogen Removal in One Biocatalyst

Carbazole (CA) and its derivatives are also common components of fossil fuels and their products, and have been detected in atmospheric samples, soil and groundwater at sites contaminated with petroleum, wood-preserving wastes and coal-processing residues. Importantly, the presence of nitrogen compounds can lead to significant poisoning of refining catalysts, resulting in a decrease in yield. So the removal of aromatic-nitrogen contaminants from petroleum is important.

CA is a heterocyclic compound shared the similar structure with DBT and is the most abundant nitrogen containing compound in many petroleum samples. The whole operon for the degradation of CA is shown in Fig. 8. The first step of CA degradation is catalyzed by CA dioxygenase. Yu et al. introduced the CA dioxygenase gene into the DBT degrader *R. erythropolis* XP, and the recombinant was designated SN8. The resting cells of SN8 could degrade 98% of DBT (92 mg L^{-1}) and 100% of CA (100 mg L^{-1}) within 6 h. The CA dioxygenase of SN8 can convert DBT to dibenzothiophene dihydrodiol and monohydroxy-dibenzothiophene, but the amount of both compounds was too little to have a significant effect

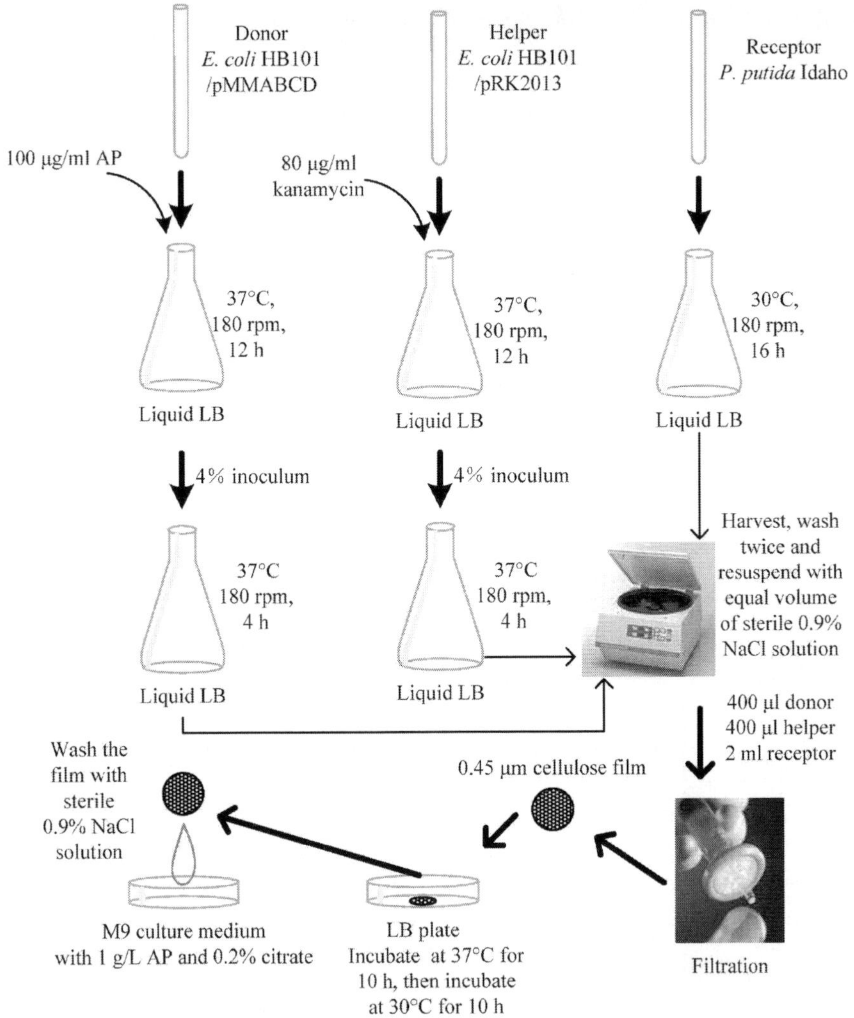

Fig. 7 Schematic map of tri-parental mate. The donor, helper, and receptor strains are *E. coli* HB101 containing plasmid pMMABCD, *E. coli* HB101 containing pRK2013, and *P. putida* Idaho, respectively. They were cultured in liquid LB and harvested and washed separately, with a subsequential mixture and filtration. Then the film was detached from the filter and put on a LB plate for an incubation at 37 °C for 10 h, then 30 °C for another 10 h. Finally the film was washed with sterile NaCl solution and spread on selection plate

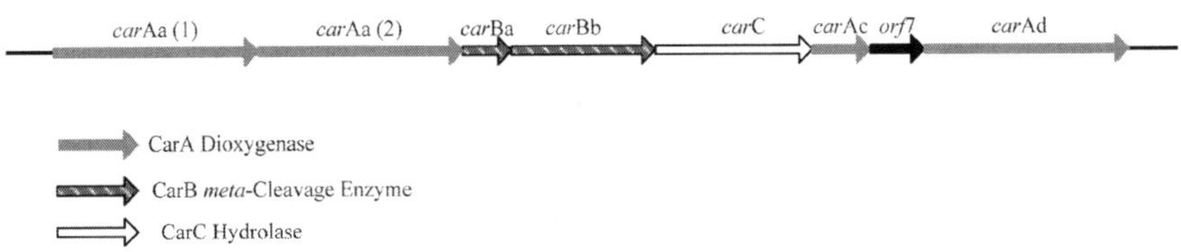

Fig. 8 The *car* operon from *Pseudomonas resinovorans* CA10. Functional analysis revealed that CarAa served as terminal oxygenase, CarAc as ferredoxin, and CarAd as ferredoxin reductase components [34]

on DBT degradation. When FS4800 crude oil was treated by SN8, most of alkylated derivatives, including methyl-, dimethyl-, and trimethyl-CA were removed [35]. Therefore, the recombinant SN8 can be used to remove recalcitrant nitrogen and sulfur compounds from petroleum.

The degradation of CA was also studied [34, 36, 37] and the pathway is shown in Fig. 9. The whole *car*ABC operon which encoded the enzymes involved in the CA degradation pathway was also introduced into *R. erythropolis* XP, and the recombinant was designated XPDN. DBT can be transformed to benzoic acid, CA to anthranilic acid, and dibenzofuran to salicylic acid simultaneously by the recombinant strain XPDN. The recombinant strain XPDN has the ability to remove 100% of the DBT (92 mg L^{-1}), 82.9% of the CA (200 mg L^{-1}), and 69.8% (200 mg L^{-1}) of dibenzofuran within 33 h. Therefore, the strain XPDN was suitable for the bioremediation to convert poisonous S, N, and O heterocycles in addition to petroleum biotreatment applications [38].

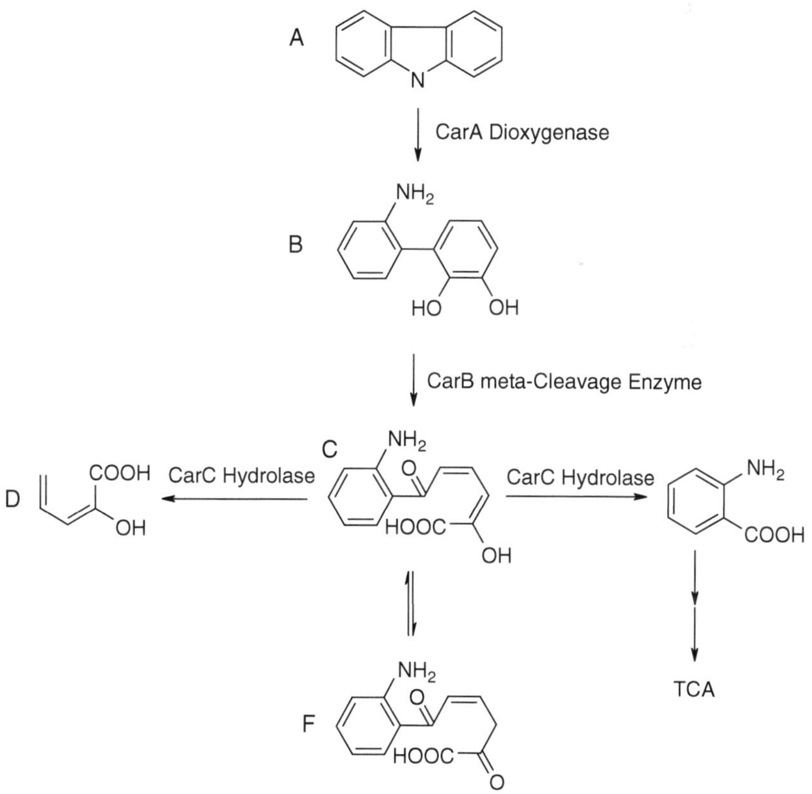

Fig. 9a–f The degradation pathway of CA. **a** CA. **b** 2′-Aminobiphenyl-2,3-diol. **c** 2-Hydrixy-6-oxo-6-(2′amino-phenyl)hexa-2,4-dienoic acid. **d** 2-Hydroxy-4-pentenoate. **e** Anthranilic acid. **f** 2,6-Dioxo-6-(2′-aminophenyl)-hexa-4-enoic-acid [34, 36, 37]

3.5 Getting Valuable Intermediates

In the "4S" pathway, the intermediate HBPSi should be noted because this compound can be used as substrate in the synthesis of surfactant (Fig. 10), which can decrease the interfacial tension, and can be used in enhanced oil recovery (EOR).

Studies are under way to knock out the *dsz*B gene, which encodes 2-hydroxybiphenyl-2-sulfinate sulfinolyase so that HBPSi can be accumulated in our group. Yu et al. have eliminated the *dsz*B gene using the recombinant PCR method in *E. coli* strains. The recombinant containing *dsz*A and *dsz*C could efficiently transform DBT and its alkyl derivatives into HBPSi and its corresponding alkylated derivatives [39]. It should be noted that HBPSi could cyclize under acidic conditions [40], as shown in Fig. 11. Therefore, a sultine would be recovered in the extraction of the culture supernatant incubation with DBT.

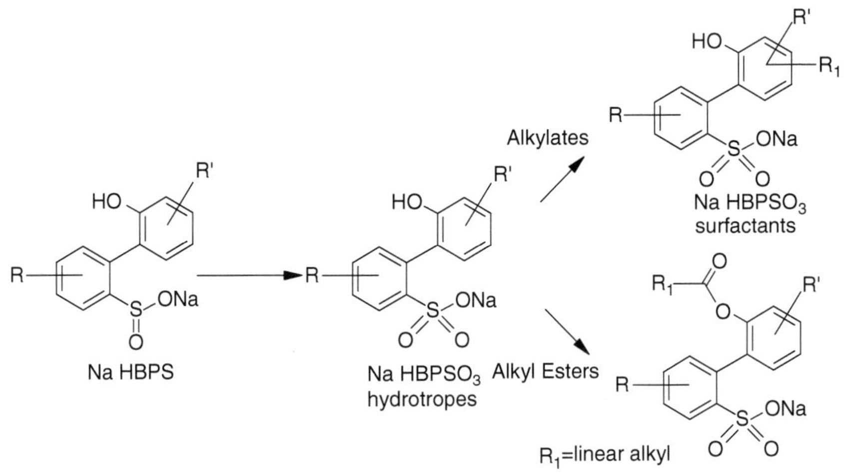

Fig. 10 Conversion of the BDS intermediate Cx-HBPS to value-added surfactants

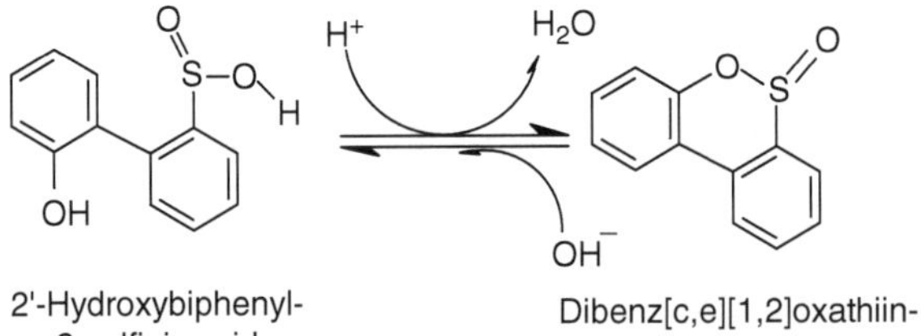

Fig. 11 Acid catalyzed cyclization of 2′-dydroxybiphenyl-2-sulfinic acid [40]

4 Biodesulfurization of Petroleum by Free-Cell Microorganisms

The feasibility of petroleum biodesulfurization was also investigated. Gas chromatographic flame ionization detection (GC-FID) and gas chromatographic atomic emission detection (GC-AED) analysis were used to evaluate the effects of biodesulfurization of petroleum, including gasoline, diesel and crude oils. GC-FID was generally used to study the desulfurization of model oils, while GC-AED was used specifically for the study of changes of sulfur contained compounds in oils. The typical GC-AED chromatogram of diesel oil is shown in Fig. 12.

The results on biodesulfurization of different kinds of oils acquired in China are summarized in Table 1.

M. goodii X7B was used in the biodesulfurization of diesel oil. When HDS-treated diesel oil was desulfurized by strain X7B, the total sulfur content was decreased from 535 to 72 ppm, corresponding to a reduction of 86%. The resting cells of *R. erythropolis* FSD-2 were used to desulfurize two kinds of diesel oils. After the biotreatment, 84% of the sulfur content of distillation diesel oil, from 666 to 105 ppm, and 94% of that of HDS-treated diesel oil, from 198 to 11 ppm, was removed in two consecutive reactions [29]. This result is interesting because the ultra-low sulfur content diesel oil required by current environmental regulations was obtained. The effect of surfactant on biodesulfurization of diesel oils was also investigated. The desulfurization rate was increased by 35% when 0.5%

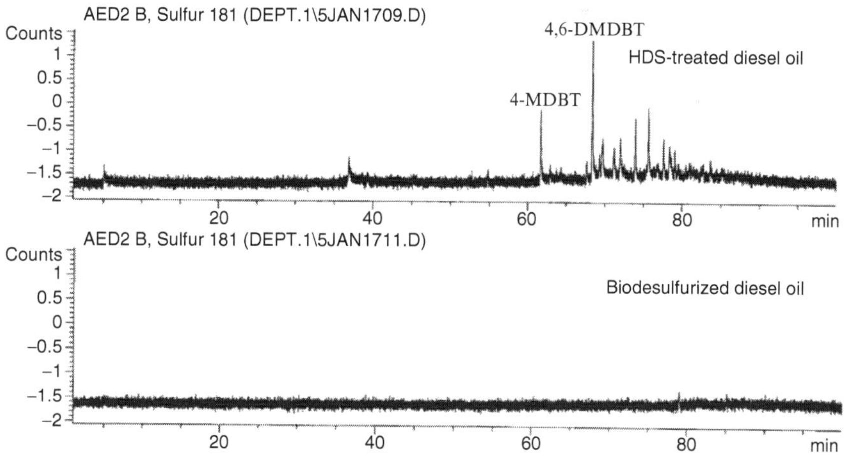

Fig. 12 GC-AED chromatography of diesel oil. The HDS-treated diesel oil was kindly provided by Fushun Research Institute of Sino Petroleum and Chemical Corporation, SINOPEC. The diesel oil was treated by *R. erythropolis* 1awq. The biocatalyst was suspended in 0.1 M phosphate buffer (pH 7.0) to reach a concentration of 8 g (of dry cells) L^{-1} and supplemented with hydrodesulfurized diesel oils (oil-to-water ratio of 1:9). The culture was given 2% glucose as carbon and energy sources and shaken at 30 °C for 48 h

Table 1 The biodesulfurization of fossil fuels

Microbial culture	Fossil fuels	Initial sulfur content (ppm)	Biodesulfurization treated sulfur content (ppm)	Effectiveness of biodesulfurization (%)	References
Mycobacterium goodii X7B	HDS-treated diesel oil	535	72	86	[17]
	Crude oil	3,600	1478	59	[15]
	SR gasoline	227	71[a]	69	[14]
	SR gasoline	275	54[a,b]	80	[14]
Rhodococcus erythropolis XP	Crude oil	3,210	1695	47.2	[20]
	Crude oil	1,237	466	62.3	[20]
	HDS-treated diesel oil	259	14	94.5	[20]
	FCC gasoline	495.3	338	29	[19]
	FCC gasoline	1,200	850	32	[19]
	SR gasoline	50.2	7.5	85	[19]
Rhodococcus erythropolis FSD-2	Distillation diesel oil	666	105	84	[29]
	HDS-treated diesel oil	198	11[b]	94	[29]
Rhodococcus erythropolis 1awq	HDS-treated diesel oil	406	176	56.6	[27]
	HDS-treated diesel oil	168	79	52.9	[27]
	HDS-treated diesel oil	406	89[c]	78.1	[27]
	HDS-treated diesel oil	168	59[c]	65	[27]

[a]The oil was treated by the immobilized cells
[b]The oil was treated by the two consecutive bioprocess
[c]The oil was treated by the culture with 0.5% Tween 80

Tween 80 was added. When HDS-treated diesel oils FHD406 and KHD168 were biodesulfurized in the presence of Tween 80, the effectiveness of desulfurization were 78.1% and 65%, respectively, compared to 56.6% and 52.9% without Tween 80 [27]. As for the strain *R. erythropolis* XP, the sulfur content of diesel oil was reduced by 94.5%, from 259 to 14 ppm [20]. These results were more noteworthy because the diesel oil had previously been treated by HDS so that only organosulfur compounds that were relatively recalcitrant to HDS remained. *R. erythropolis* XP was also adopted to desulfurize fluid catalytic cracking (FCC) and straight-run (SR) gasoline oils. About 29% of the total sulfur of Jilian FCC gasoline (from 495 to 338 ppm) and 32% of that of Qilu FCC gasoline (from 1,200 to 850 ppm) were removed, respectively. About 85% of the sulfur content of SR gasoline (from 50.2 to 7.5 ppm) was removed [19]. It is the first publication on the biodesulfurization of gasoline by free whole cells.

Biodesulfurization of crude oil was also investigated by strain *R. erythropolis* XP. After treated by resting cells, 62.3% of the total sulfur content in Fushun crude oil and 47.2% of that in Sudanese crude oil were removed. The initial total sulfur contents were 3,210 and 1,237 ppm, respectively. When Liaoning crude oil was treated by *M. goodii* X7B, 59% of the total sulfur content was removed, from 3,600 to 1,478 ppm. Biotreatments of petroleum, especially biodesulfurization of derived fractions of petroleum such as diesel oil, light gas oil and gasoline, all face the same problem that the ratio of water to oil is high which will significantly increase the reaction volume and make practical operation less advisable. In addition, techniques, including water flooding, were used to increase oil recovery in the oilfields. It is reported that up to 87% of the Chinese oilfields are flooded [41], and therefore a high content of water exists in the recovered crude oil. In old Chinese oilfields, there is more than 85% water in the recovered crude oil [42]. It would be more practical to incubate microorganisms having the capability to desulfurize crude oil in this water/oil system before the water is removed.

The methods for the preparation of a biodesulfurization biocatalyst were also investigated to reduce the biodesulfurization cost. Sulfur sources including DBT, dimethylsulfoxide (DMSO), sodium sulfate and mixed sulfur sources were used to study the cost of biocatalyst production. DMSO was selected as the sulfur source, because the desulfurization activity of the bacteria incubated in medium with DMSO as the sole sulfur source was shown to be similar to that with DBT as the sole sulfur source, and the price of DMSO was lower than that of DBT [18].

5 Biodesulfurization of Petroleum or Model Oils by Immobilized Microorganisms

The use of immobilized microorganisms rather than free cells in biodesulfurization can be advantageous to enhance the stability of the biocatalyst and the efficiency of catalysis, and to facilitate its recovery and reuse. Some researchers investigate the use of immobilized cells in biodesulfurization. The immobilized technologies include magnetic immobilization, adsorption, and assembling γ-Al_2O_3 nanosorbent, which have all been used in biodesulfurization.

Alginate was used for cell adsorption frequently. The process of immobilization could also minimize the toxicity of gasoline to the biocatalyst. Therefore, the immobilization of *M. goodii* X7B was used to treat different gasoline. When two kinds of Dushanzi straight-run gasoline were treated with immobilized cells of strain X7B, the total sulfur content decreased from 227 to 71 ppm, and from 275 to 54 ppm in two consecutive reactions [14]. Considering that benzothiophenes predominate in gasoline, and the toxicity of gasoline to bacterial cells, the biodesulfurization of gasoline is a significant achievement.

The immobilization of *P. delafieldii* R-8 in magnetic polyvinyl alcohol (PVA) beads has been studied. Compared with the immobilized cells in non-magnetic PVA beads, the magnetic immobilized cells' desulfurizing activity increased slightly from 8.7 to 9 mmol sulfur kg^{-1} (dry cell) h^{-1}. However, the magnetic immobilized cells maintained a high desulfurization activity after seven times repeated use, while only five times were repeated usages for non-magnetic immobilized cells [24]. The cells coated with magnetic Fe_3O_4 nanoparticles (10–15 nm) were also studied. The Fe_3O_4 nanoparticles were strongly absorbed by the surfaces of the cells and coated the cells. The coated cells had the same desulfurizing activity as free cells, and can be reused more than five times [43]. Furthermore, the magnetic immobilized cells have its intrinsic excellence in separation from the biodesulfurization reactor.

The immobilization of *P. delafieldii* R-8 in calcium alginate beads was also studied, and it was found that the specific desulfurization rate of 1.5 mm diameter beads was 1.4-fold higher than that of 4.0 mm. The desulfurization rate was also increased by the addition of 0.5% Span 80, about 1.8-fold compared with that of the untreated beads [44]. These results suggested that the rate of mass transfer of substrate to gel matrix could be the limited step of biodesulfurization. Therefore, assembling γ-Al_2O_3 nanosorbent onto the surfaces of cells was developed as a novel biodesulfurization technology leading to selective adsorption DBT from the organic phase. The rate of adsorption was far higher than that of biodesulfurization, and thus the transfer rate of DBT was improved in this system. The desulfurization rate of the cells assembled with nanosorbents was about twice as high as that of the free cells [45]. The technology of assembling γ-Al_2O_3 nanosorbent on the surfaces of cells could also be used in other biotransformation reactions performed in aqueous or biphasic systems when the mass transfer of substrate was the limited step.

A process of biodesulfurization was proposed based on the achievements of strains and the recombinant bacteria, the method of immobilized cells, and the technologies of fuels desulfurization. The processes of desulfurization, denitrogenation, and heavy metal removal could be performed by the same bacteria. The schematic diagram of the process is shown in Fig. 13.

6 Biodesulfurization of Flue Gas

The NO_x and SO_2 of flue gas from power plants can pollute the environment. Metal chelate was successfully used in the removal of SO_2 and NO_x. Fe(II)EDTA can promote the solubility of NO by the formation of Fe(II)EDTA-NO. The bound NO reacts with sulfite/bisulfite ions which are formed by the dissolution of SO_2, and then N_2O and nitrogen-sulfur compounds were produced [46]. Although Fe(II) EDTA has a high efficiency in flue gas desulfurization (FGD), it is also a secondary cause of pollution. Thus, a chelate compound of Fe(II) with citrate (Fe(II)Cit) was selected as the substitute. At all events, Fe(II)EDTA or Fe(II)Cit should be regenerated in time, and therefore the reaction can be continued. However, there are some drawbacks in the system, such as the oxidization of Fe(II) by oxygen to form Fe(III), and the low regeneration rate of Fe(II). In order to keep the concentration of Fe(II), bacteria were added to the system. The denitrifying bacteria can reduce bound NO to nitrogen under anaerobic conditions. In addition, some microorganisms can use O_2, NO_3^-, Fe(III) and S^0 as terminal electron acceptors, and therefore the addition of microorganisms can help to keep a low level of dissolved oxygen in the solution, and can reduce Fe(III) to Fe(II) directly. The sulfite/sulfate can also be

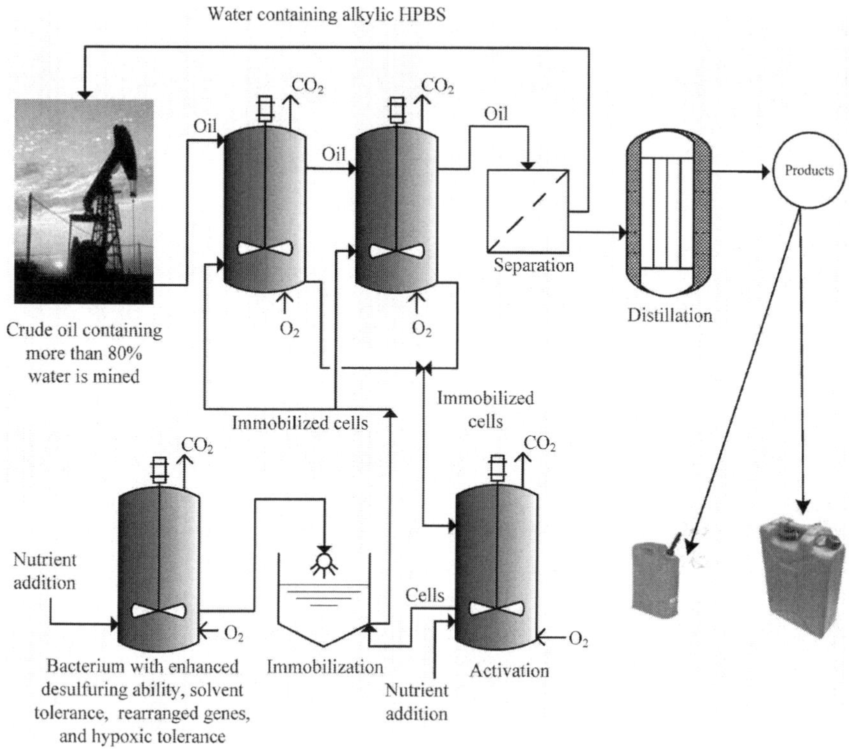

Fig. 13 A schematic diagram of petroleum biodesulfurization

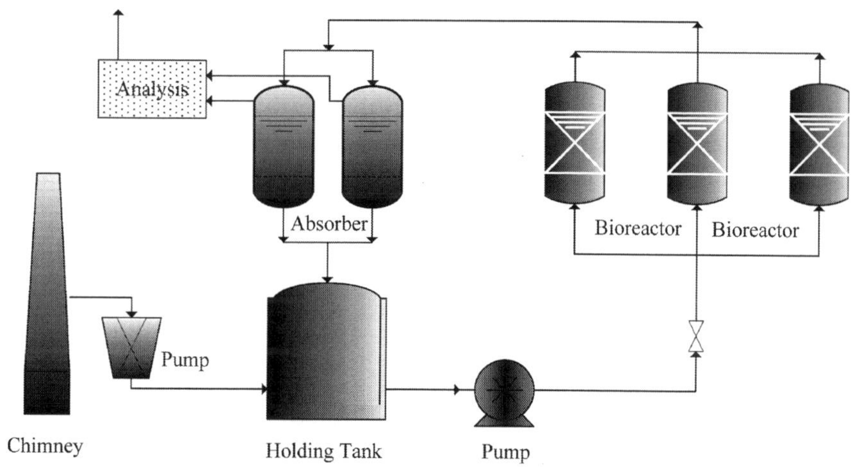

Fig. 14 A schematic diagram for the removal of NO and SO_2

reduced to H_2S by the microorganisms, which then reacts with excess SO_2. These processes can be expressed as follows:

$$\text{Fe(II)-NO+electdron donor} \xrightarrow{\text{microorganisms}} N_2\text{+Fe(II)}$$

$$\text{Fe(III)+electdron donor} \xrightarrow{\text{microorganisms}} \text{Fe(II)}$$

$$SO_3^{2-}/So_4^{2-}\text{+electron donor} \xrightarrow{\text{microorganisms}} H_2S$$

$$SO_2\text{+2}H_2S \rightarrow \text{3S+2}H_2O$$

When only the Fe(II)Cit solution was used, the NO removal efficiency was decreased quickly for the hindrance of Fe(II)Cit regeneration, and the removal efficiency was decreased to about 8% after 5 h of absorption. In contrast, when the microorganisms were added, the NO removal efficiency was kept at about 48% [47]. When Fe(II)EDTA was used in the system, almost 99% of SO_2 was removed during the operation, and the NO removal efficiency was not influenced significantly even changing the SO_2 concentration [46]. Therefore, the approaches combined with the FGD process have the potential for simultaneous removal of NO and SO_2. The schematic diagram of the set-up for the removal of NO and SO_2 of flue gas is shown in Fig. 14.

7 Concluding Remarks and Perspectives

Our understanding of how microorganisms metabolize sulfur heterocyclic compounds in petroleum has increased rapidly. All the studies outlined above are significant steps to explore the biotechnological potential for developing an efficient biodesulfurization process. However, these technologies have not yet been available for large-scale applications. We still need a far better understanding of all aspects of this pathway and its biotechnological process to turn it into a commercial process. Any progress that provides the possibility to remove sulfur in crude oil at higher temperature, with higher rate, or longer stability of desulfurization activity is considered to be a significant step toward industry level biodesulfurization. Microorganisms with a wider substrate range and higher substrate affinity in biphasic reaction containing toxic solvents or higher biodesulfurization activities could be engineered if the biocatalysts were to be used for petroleum treatment or the production of bio-derived compounds. The genomes of *Rhodococcus* sp. strain RHA1 and *Mycobacterium smegmatis* mc^{2}155 have been sequenced and published. The availability of the genome sequences will make it possible to regulate the metabolism of biodesulfurization genome-wide.

Acknowledgements We gratefully acknowledge the support for our research from the National Natural Science Foundation of China (grant numbers 30700024, 20590368) and the Chinese National Programs for High Technology Research and Development (Grant No. 2006AA06Z322).

References

1. Kilbane JJ, II (2006) Curr Opin Biotechnol 17:305
2. Monticello DJ, Finnerty WR (1985) Annu Rev Microbiol 39:371
3. Kropp KG, Fedorak PM (1998) Can J Microbiol 44:605
4. Kertesz MA, Wietek C (2001) Appl Microbiol Biotechnol 57:460
5. Li C, Jiang Z, Gao J, Yang Y, Wang S, Tian F, Sun F, Sun X, Ying P, Han C (2004) Chemistry 10:2277
6. Lu H, Gao J, Jiang Z, Yang Y, Song B, Li C (2007) Chem Commun (Camb)150
7. Borgne SL, Quintero R (2003) Fuel Process Technol 81:155
8. Gray KA, Mrachko GT, Squires CH (2003) Curr Opin Microbiol 6:229
9. Xu P, Yu B, Li FL, Cai XF, Ma CQ (2006) Trends Microbiol 14:398
10. Yan H, Kishimoto M, Omasa T, Katakura Y, Suga K, Okumura K, Yoshikawa O (2000) J Biosci Bioeng 89:361
11. Oldfield C, Pogrebinsky O, Simmonds J, Olson ES, Kulpa CF (1997) Microbiology 143:2961
12. Gray KA, Pogrebinsky OS, Mrachko GT, Xi L, Monticello DJ, Squires CH (1996) Nat Biotechnol 14:1705
13. Piddington CS, Kovacevich BR, Rambosek J (1995) Appl Environ Microbiol 61:468
14. Li FL, Xu P, Feng JH, Meng L, Zheng Y, Luo LL, Ma CQ (2005) Appl Environ Microbiol 71:276
15. Li FL, Zhang ZZ, Feng JH, Cai XF, Xu P (2007) J Biotechnol 127:222
16. Chen H, Zhang WJ, Chen JM, Cai YB, Li W (2008) Bioresour Technol 99:3630
17. Li FL, Xu P, Ma CQ, Luo LL, Wang XS (2003) FEMS Microbiol Lett 223:301
18. Ma CQ, Feng JH, Zeng YY, Cai XF, Sun BP, Zhang ZB, Blankespoor HD, Xu P (2006) Chemosphere 65:165
19. Yu B, Ma CQ, Zhou WJ, Wang Y, Cai XF, Tao F, Zhang Q, Tong MY, Qu JY, Xu P (2006) FEMS Microbiol Lett 258:284
20. Yu B, Xu P, Shi Q, Ma CQ (2006) Appl Environ Microbiol 72:54
21. Ma T, Li GQ, Li J, Liang FL, Liu RL (2006) Biotechnol Lett 28:1095
22. Li W, Zhang Y, Wang MD, Shi Y (2005) FEMS Microbiol Lett 247:45
23. Li W, Wang MD, Chen H, Chen JM, Shi Y (2006) Biotechnol Lett 28:1175
24. Shan GB, Xing JM, Luo MF, Liu HZ, Chen JY (2003) Biotechnol Lett 25:1977
25. Luo MF, Xing JM, Gou ZX, Li S, Liu HZ, Chen JY (2003) Biochem Eng J 13:1
26. Li GQ, Ma T, Li SS, Li H, Liang FL, Liu RL (2007) Biosci Biotechnol Biochem 71:849
27. Feng J, Zeng Y, Ma C, Cai X, Zhang Q, Tong M, Yu B, Xu P (2006) Appl Environ Microbiol 72:7390
28. Xiong XC, Xing JM, Li X, Bai XJ, Li WL, Li YG, Liu HZ (2007) Appl Environ Microbiol 73:2394
29. Zhang Q, Tong MY, Li YS, Gao HJ, Fang XC (2007) Biotechnol Lett 29:123
30. Suske WA, Held M, Schmid A, Fleischmann T, Wubbolts MG, Kohler HP (1997) J Biol Chem 272:24257
31. Chen H, Zhang WJ, Cai YB, Zhang Y, Li W (2008) Bioresour Technol 99:6928
32. Li GQ, Li SS, Zhang ML, Wang J, Zhu L, Liang FL, Liu RL, Ma T (2008) Appl Environ Microbiol 74:971
33. Tao F, Yu B, Xu P, Ma CQ (2006) Appl Environ Microbiol 72:4604
34. Riddle RR, Gibbs PR, Willson RC, Benedik MJ (2003) J Ind Microbiol Biotechnol 30:6
35. Yu B, Xu P, Zhu SS, Cai XF, Wang Y, Li L, Li FL, Liu XY, Ma CQ (2006) Appl Environ Microbiol 72:2235
36. Sato SI, Nam JW, Kasuga K, Nojiri H, Yamane H, Omori T (1997) J Bacteriol 179:4850
37. Sato SI, Ouchiyama N, Kimura T, Nojiri H, Yamane H, Omori T (1997) J Bacteriol 179:4841
38. Yu B, Ma CQ, Zhou WJ, Zhu SS, Wang Y, Qu JY, Li FL, Xu P (2006) Appl Environ Microbiol 72:7373
39. Yu B, Huang J, Cai XF, Liu XY, Ma CQ, Li FL, Xu P (2005) Chinese Chem Lett 16:935
40. Bressler DC, Norman JA, Fedorak PM (1997) Biodegradation 8:297

41. Tan TD (1995) China Offshore Oil Gas (Geol) (Chinese) 9:421
42. Zhou ZY, Zhang K (2004) Pet Explor Dev (Chinese) 31:84
43. Shan GB, Xing JM, Zhang HY, Liu HZ (2005) Appl Environ Microbiol 71:4497
44. Li YG, Xing JM, Xiong XC, Li WL, Gao HS, Liu HZ (2008) J Ind Microbiol Biotechnol 35:145
45. Guobin S, Huaiying Z, Weiquan C, Jianmin X, Huizhou L (2005) Biophys J 89:L58
46. Li W, Wu CZ, Shi Y (2006) J Chem Technol Biotechnol 81:306
47. Xu X, Chang SG (2007) Chemosphere 67:1628

Adv Biochem Engin/Biotechnol (2009) 113: 275-303
DOI: 10.1007/10_2008_29

Published online: 17 April 2009

Characterization, Modeling and Application of Aerobic Granular Sludge for Wastewater Treatment

Xian-Wei Liu, Han-Qing Yu, Bing-Jie Ni, and Guo-Ping Sheng

Abstract Recently extensive studies have been carried out to cultivate aerobic granular sludge worldwide, including in China. Aerobic granules, compared with conventional activated sludge flocs, are well known for their regular, dense, and strong microbial structure, good settling ability, high biomass retention, and great ability to withstand shock loadings. Studies have shown that the aerobic granules could be applied for the treatment of low- or high-strength wastewaters, simultaneous removal of organic carbon, nitrogen and phosphorus, and decomposition of toxic wastewaters. Thus, this new form of activate sludge, like anaerobic granular sludge, could be employed for the treatment of municipal and industrial wastewaters in near future. This chapter attempts to provide an up-to-date review on the definition, cultivation, characterization, modeling and application of aerobic granular sludge for biological wastewater treatment. This review outlines some important discoveries with regard to the factors affecting the formation of aerobic granular sludge, their physicochemical characteristics, as well as their microbial structure and diversity. It also summarizes the modeling of aerobic granule formation. Finally, this chapter highlights the applications of aerobic granulation technology in the biological wastewater treatment. It is concluded that the knowledge regarding aerobic granular sludge is far from complete. Although previous studies in this field have undoubtedly improved our understanding on aerobic granular sludge, it is clear that much remains to be learned about the process and that many unanswered questions still remain. One of the challenges appears to be the integration of the existing and growing scientific knowledge base with the observations and applications in practice, which this paper hopes to partially achieve.

X.W. Liu, H.-Q. Yu (✉), B.J. Ni, and G.-P. Sheng
Department of Chemistry, University of Science and Technology of China,
Hefei, 230026, China,
e-mail: hqyu@ustc.edu.cn

Keywords Aerobic granule, Biodegradation, Model parameters, Wastewater treatment

Contents

1 Introduction

In the last decade, intensive research has demonstrated that aerobic granular sludge technology is a novel and promising development in the field of biological wastewater treatment [1–7]. This process is based on sequencing batch reactors (SBRs), with a cycle configuration chosen such that a strict selection for fast settling aerobic granules and a frequent repetition of distinct feast and famine conditions occur. This leads to the growth of stable and dense granules [4, 5]. Compared to the conventional activated sludge systems, an aerobic-granular-sludge-based system has several advantages. An outstanding feature is the excellent settling ability (high settling velocity) which is a prerequisite to handle high liquid flows. Moreover, granular sludge provides a high and stable rate of metabolism, resilience to shocks and toxins due to the protection by a matrix of extracellular polymeric substances (EPS), long biomass residence times, biomass immobilization inside the granules, and therefore, the possibility for bioaugmentation [6–10]. In this sense, aerobic granular sludge technology will play an important role as an innovative technology alternative for activated sludge process in industrial and municipal wastewater treatment in the near future.

To facilitate and promote its practical application in wastewater treatment, researchers worldwide have excessively investigated the fundamentals of aerobic granulation. Several theories about aerobic granule formation have been proposed. Analysis of the great body of literature published in the last decade shows that the properties of aerobic granules formed in SBRs are influenced by many factors, including substrate composition, organic loading, hydrodynamic shear force, feast-famine regime, feeding strategy, dissolved oxygen (DO), reactor configuration, solids retention time, cycle time, settling time and volume exchange ratio. The knowledge of granule formation and critical process conditions were reviewed by Liu and Tay [11] and by de Kreuk et al. [12, 13].

The objective of this chapter is to discuss recent research advances in aerobic granular sludge. This review focuses on the factors affecting the formation of aerobic granular sludge, physicochemical characteristics, microbial structure and diversity of aerobic granules, modeling of aerobic granule, and practical applications of this technology. The current state of the art will be summarized and the future research will be addressed.

2 Definition of Aerobic Granular Sludge

Aerobic granules are dense spherical self-immobilized aggregates of micro-organisms with a strong compact structure and excellent settling ability. They have a well-defined appearance and are visible as separate entities larger than 0.1 mm in diameter after settling. However, a general definition of aerobic granular sludge had not been made until the first aerobic granular workshop was held in 2004 [12]. After a long discussion in this workshop, aerobic granule was defined as:

Granules making up aerobic granular activated sludge are to be understood as aggregates of microbial origin, which do not coagulate under reduced hydrodynamic shear, and which settle significantly faster than activated sludge flocs.

In the second aerobic granular workshop in 2006, a further explanation of the definition was discussed, regarding aggregates of microbial origin, no coagulation under reduced hydrodynamic shear, settling faster than activated sludge flocs, minimum size, method of harvesting [13]. When an aggregate fulfils all characteristics as described above, it can be called aerobic granular sludge. This simplifies the interpretation of experimental results and clarifies when to speak about aerobic granular sludge, activated sludge or biofilms.

3 Factors Affecting the Formation of Aerobic Granular Sludge

3.1 Substrate Composition and Organic Loading Rate

Aerobic granules have been cultivated with a wide variety of wastewater, as shown in Table 1. Generally aerobic granulation is independent of type of substrate. However, the morphology and microstructure of aerobic granules are highly dependent on the composition of the wastewater they were grown on. As an example, the structural instabilities of aerobic granules occurred due to filamentous outgrowth in the treatment of dairy wastewater [19]. The outgrowth of filamentous organisms was related to the fact that, in the dairy wastewater, easily biodegradable substances were slowly released at a low concentration due to slow hydrolysis of the initial polymeric substrates.

It is recognized that carbohydrates in wastewater and a low DO level lead to flocculation, due to the favored growth of filamentous microorganisms [23, 24]. Glucose-grown aerobic granules exhibited a filamentous structure, while acetate-grown aerobic granules had a nonfilamentous and very compact bacterial structure in which a rod-like species predominated [4]. At a high potential growth rate on certain substrates, it is more difficult to cultivate aerobic granules. It is easier to obtain a compact granule structure on methanol than on acetate because of the different growth rates of the microorganisms on the two substrates. The only observed exception was the growth of granular sludge on glucose. Usually, microorganisms have a high growth rate on glucose, but in biofilm and granule systems, populations are found to have a low growth rate on glucose and therefore dense and smooth structures are formed.

Table 1 Types of substrate used for cultivation of aerobic granules

Substrate	Organic loading rate (kg COD m^{-3} day)	Cultivation time (days)	Filamentous outgrowth	Size (mm)	Ref.
Molasses	0.42–1.20	70	Yes	2.35	[1]
Ethanol	2.5–7.5	63	Yes	3.2 ± 0.9	[2]
Glucose	6.0	21	Yes	2.4	[4]
Acetate	2.5	37	No	2.5	[5]
Soybean-processing wastewater	6.0	60	No	1.22 ± 0.85	[6]
Sucrose	3.75	60	No	1.0	[14]
Phenol	3.6	33	No	0.5–0.6	[15]
p-Nitrophenol/glucose	2.5	79	No	0.386 ± 0.016	[16]
2,4-Dichlorophenol/ glucose	2.8	39	No	1.0–2.0	[17]
Chloroanilines/ glucose/acetate	1.0–3.6	100	No	0.45–2.0	[18]
Dairy wastewater	2.4	105	Yes	2.0–3.0	[19]
Brewery wastewater	3.0	63	No	2.0–7.0	[20]
Domestic wastewater	1.0–1.6	36	Yes	0.5–2.0	[21]
Abattoir wastewater	2.6	76	No	1.7	[22]

[12] Aerobic nitrifying granules were also cultivated with an inorganic carbon source [25, 26]. The nitrifying granules showed excellent nitrification ability.

Accumulated evidence suggests that aerobic granules can form across a very wide range of organic loading rate (OLR) from 0.4 to 15 kg COD m^{-3} day [1, 11, 27]. This indicates that the formation of aerobic granule in an SBR is substrate-concentration-independent. However, it was reported that kinetic behavior and morphology of aerobic granules were also related to applied substrate loading [27, 28]. The mean size of aerobic granules increased from 1.6 to 1.9 mm with an increase in OLR from 3 to 9 kg COD m^{-3} day [11]. Furthermore, OLR could affect the stability of aerobic granules. Zheng et al. [28] reported that bacteria-dominated aerobic granules with a mean diameter of 1 mm could be cultivated in an SBR at a high OLR of 6 kg COD m^{-3} day in 30 days. However, under such high loading conditions, the bacteria-dominated granules were unstable and readily transited into large-sized filamentous ones. The instability of aerobic granules may be attributed to the mass transfer limitation and possible presence of anaerobes in the large-sized aerobic granules.

3.2 Seed Sludge

In most studies regarding aerobic granular sludge, SBRs were seeded with conventional activated sludge flocs. The important factors governing the quality of seed sludge for aerobic granulation appear to include the macroscopic characteristics, settling ability, surface properties, and microbial activity [11]. Little information on the role of seed sludge in aerobic granulation is available. Hu et al. [29] cultivated aerobic granules in SBRs by seeding anaerobic granular sludge using acetate-based synthetic wastewater. In their experiments, the inoculated anaerobic granules experienced a process of disintegration – recombination – growing up. The disintegrated anaerobic sludge might play a role of nucleus for the aerobic granulation. The bacterial community residing in seed sludge is important for aerobic granulation process, as the hydrophilic bacteria would be less likely to attach to sludge flocs compared with the hydrophobic counterpart, which constitutes the majority of free bacteria in the effluent from full-scale treatment plants. A greater number of hydrophobic bacteria in the seed sludge would result in a more rapid aerobic granulation with excellent settling ability.

3.3 Reactor Configuration

Reactor configuration influences the flow pattern of liquid and microbial aggregates in reactors [2, 30]. Until now, all aerobic granules were cultivated in pneumatically agitated reactors, including bubble column and airlift reactors with sequencing batch operation. An important operation factor in aerobic granulation and reactor operation is superficial air velocity. It exerts an influence on aerobic granules through oxygen supply and hydrodynamic shear stress. The effect of hydrodynamic shear stress on aerobic granules will be discussed in detail in Sect. 3.4. In a column-type upflow reactor a higher ratio of reactor height to diameter (H/D) can ensure a longer circular flow trajectory, which in turn provides a more effective hydraulic attrition to aerobic granules. However, with a higher H/D, the aerobic granules at the top face a low shear stress, and will grow out more readily with filametous and/or finger-type structure. In this case, no granules will be formed [12].

Shear stress provided by aeration rate depends on reactor configuration as well as reactor scale. Beun et al. [31] reported that much denser granules with a smaller diameter were obtained in an airlift reactor at the same substrate loading rate compared to a bubble column. To reduce aeration rate to a reasonable value, Liu et al. [32] designed a novel airlift loop reactor with divided draft tubes for aerobic granulation.

3.4 Operational Parameters of Reactor

3.4.1 Aeration Intensity

The positive role of a high hydrodynamic shear force provided a great aeration rate in the stable operation of biofilm and aerobic granule systems has been widely recognized [30]. Chen et al. [33] found that at shear forces of 2.4 and 3.2 cm s^{-1}, granules could maintain a robust and stable structure. Granules developed in low shear forces of 0.8 and 1.6 cm s^{-1} deteriorated to large-sized filamentous ones with irregular shape and loose structure, and resulted in poor performance and operation instability. Granules cultivated under high shear forces of 2.4 and 3.2 cm s^{-1} stabilized with clear morphology, dense and compact structure, and good performance in 120-day operation. In most studies on aerobic granulation, the upflow air superficial velocity in reactors was much higher than 1.2 cm s^{-1}. However, a high aeration rate means a high energy consumption.

Adav et al. [34] reported that the production of extracellular polysaccharides was closely associated with the shear force and the stability of aerobic granules. The extracellular polysaccharides content increased with the increasing shear force estimated in terms of superficial upflow air velocity. Thus, a high shear force stimulated bacteria to secrete more extracellular polysaccharides [35].

3.4.2 Cycle Time

The cyclic operation of an SBR consists of influent filling, aeration, settling and effluent removal. The settling time and exchange ratio of liquid volumes at the end of each cycle presents the main screening step to remove nongranular biomass from the reactor. A shorter cycle time results in a shorter hydraulic retention time (HRT), which provides a stronger selective pressure. Sludge loss was observed through hydraulic washout at a short cycle time because bacterial growth was unable to compensate [36]. Liu et al. [37] reported the influence of cycle time on the kinetic behavior of aerobic granules. The observed specific biomass growth rate of aerobic granules decreased from 0.266 to 0.031 L per day, while the observed biomass growth yield of granular sludge decreased from 0.316 to 0.063 g VSS g^{-1} COD when the cycle time was increased from 1.5 to 8 h.

The settling time acts as a major hydraulic selection pressure on microbial community in SBRs. A short settling time preferentially selects for the growth of rapid settling bacteria and the sludge with a poor settling ability is washed out. It is recognized that the selection pressure imposed by short settling time should be more important in fully aerobic granule systems, but in anaerobic–aerobic alternative systems with phosphate accumulating organisms (PAOs), the settling time seemed to be less important because of the inherent tendency of PAOs to aggregates [12]; meanwhile Meyer et al. [38] cultivated aerobic granules containing glycogen accumulating organisms at a settling time of 25 min.

3.4.3 Feeding Strategy

The unique feature of an SBR over a continuous activated sludge reactor is its cycle operation, which in turn results in a periodical starvation phase during the operation. It is proposed that such a periodical starvation would be somehow important to the aerobic granulation [4]. Although starvation was proposed not to be a prerequisite for aerobic granulation [39], increases in hydrophobicity on carbon-starvation have been reported [40].

McSwain et al. [41] enhanced aerobic granulation by intermittent feeding. In fact, pulse feeding to the SBR contributed to compact aerobic granules. Li et al. [42] observed that the aerobic granulation process was initiated by starvation and cooperated by shear force and anaerobic metabolism. A shorter starvation time resulted in a faster granulation [43].

3.4.4 Dissolved Oxygen, Temperature and pH

Aerobic granules were successfully cultivated at a DO concentration above 2 mg L^{-1} [34, 44]. However, Peng et al. [45] observed that small granules (diameter of 0.3–0.5 mm) were agglomerated into big flocs during settling in an SBR at a DO level of 1 mg O_2 L^{-1}. Mosquera-Corral et al. [46] reported that reducing the oxygen saturation to 40% caused deterioration, a decreased density and finally breaking of the granules. Based on the literature available, DO concentration is not a dominating factor for aerobic granulation.

The rates of biological processes depend on temperature. Most research on aerobic granular sludge was carried out at room temperatures (20–25 °C). In an investigation into the effect of temperature changes on the conversion processes and the stability of aerobic granular sludge, de Kreuk et al. [47] found that temperature change could affect the performance of an aerobic granular sludge reactor to a large extent. The start-up of a reactor at low temperatures led to the presence of organic COD in the aeration phase, deterioration of granule stability and even biomass washout. Once a reactor was started up at a higher temperature it was possible to operate a stable aerobic granular sludge system at a lower temperature. Thus, they concluded that start-up should take place preferentially during warm summer periods, and that decreased temperatures during winter periods should not be a problem for granule stability and pollutant removal in a granular sludge system [47].

In microbial growth pH is an important environmental factor. However, information regarding the effect of pH on species selection and aerobic granulation is still limited. Yang et al. [48] evaluated the effect of feeding alkalinity and pH on the formation of aerobic sludge granules. In an SBR with a low alkalinity of 28.7 mg $CaCO_3$ L^{-1} in the influent and a reactor pH of 3.0, rapid formation of fungi-dominating granules was achieved in 1 week. In another SBR with a high alkalinity of 301 mg $CaCO_3$ L^{-1} and a reactor pH of 8.1, formation of bacteria-dominating granules was achieved after 4 weeks of operation. These results suggest that microbial communities and structural features of aerobic granules could be formed through controlling the feeding alkalinity and reactor pH.

4 Physicochemical Characteristics of Aerobic Granular Sludge

4.1 Morphology

The definition of aerobic granules doesn't refer to the morphology, which varies substantially, depending on substrate types and cultivation conditions [12]. However, many studies confirm that the morphology of aerobic granules is completely different from that of sludge flocs [1–6]. Aerobic granules have a well-defined appearance and are visible as separate entities larger than 0.1 mm in diameter after settling, as shown in Fig. 1 [48]. The granule size is an important parameter in the characterization of aerobic granulation. The average diameter of aerobic granules varies in a range of 0.2–5 mm. For the quantitative differentiation of the structure characteristic and morphology of aerobic granules, Su and Yu [6] used fractal geometry to describe their characteristics. The matured aerobic granules had a fractal nature with a fractal dimension of 1.87 ± 0.34.

4.2 Settling Properties

An evident advantage of aerobic granules over conventional activated sludge flocs is their excellent settling properties [1–4], which determine the efficiency of

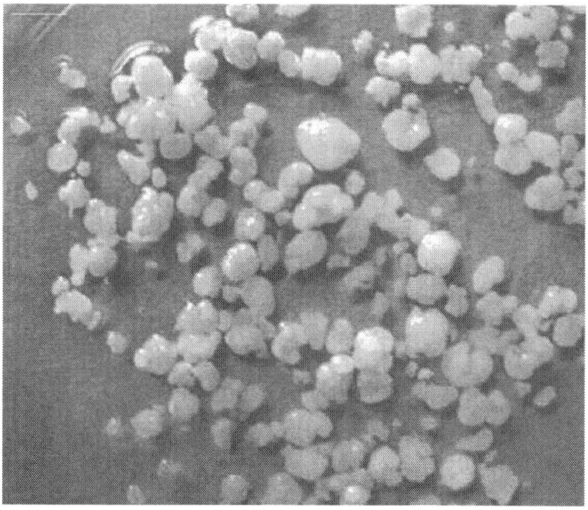

Fig. 1 Image of aerobic granules cultivated with glucose as carbon source from [48]. *Scale bar* = 10 mm

solid–liquid separation and are essential to the stable operation of wastewater treatment systems. The sludge volume index (SVI) of aerobic granules can be lower than 50 mL g^{-1}, which is much lower than that of conventional sludge flocs. The settling velocity of aerobic granules is associated with granule size and structure, and is as high as 30–100 m h^{-1}, while that of flocs is usually lower than 9 m h^{-1} [14, 20]. Liu et al. [49] developed a new model to describe the settling velocity of aerobic granules. In this model the settling velocity of aerobic granules is the function of SVI, mean size of granules and granule concentration. The established model could satisfactorily match the experimental results obtained in the course of aerobic granulation under different conditions. When the granule size is sufficiently small, the model is changed into the well-known Vesilind equation in the form of

$$V_S = V_O e^{-kX}, \tag{1}$$

where V_S is the settling velocity, V_O is the initial settling velocity, and k is the empirical settling parameter.

4.3 Strength

The strength of aerobic granules is essential for their stability and for solid/liquid separation of effluent from reactors. The granules with a high physical strength could withstand compression and high shear (abrasion). Zheng [50] characterized the compressive strength of aerobic granules with a diameter range from 1 to 4 mm using a novel device. They found that the aerobic granules with a diameter of 2 mm exhibited the best compressive properties and could withstand the compressive strength of 24 N cm^{-2}. The abrasion experimental results demonstrated that an increase in size enhanced the erosion of primary particles at a certain shear rate and this had a significantly negative effect on the solid/liquid separation process.

4.4 Extracellular Polymetric Substances (EPS)

EPS produced by bacteria are a rich matrix of polymers, including polysaccharides, proteins, glycoproteins, nucleic acids, phospholipids, and humic acids [51]. EPS are crucial in the formation of a gel-like network that keeps bacteria together in biofilms, cause the adherence of microorganisms to surfaces, and protect them against noxious environmental conditions. Because EPS are a major component of cell flocs and biofilms, they are hypothesized to play a central role in all types of biofilm formation, including flocculation and granulation [51].

4.4.1 Role of EPS

Aerobic granule stability governs the feasibility of long term granulation process [52]. Tay et al. [53] reported that an increased aeration rate in an SBR resulted in an increased polysaccharide content and that granules became disintegrated when polysaccharides were lost from the granules. Aerobic granules could not be formed in reactors with a reduced aeration rate [25, 54]. It is concluded that hydrodynamic shear force increases the production of cellular polysaccharides, which aid in the formation and stability of aerobic granules. An overview of the effects of EPS on the formation and stability of granules has been given by Liu et al. [52].

Wang et al. [55] and McSwain et al. [56] reported that EPS had a significant contribution to the granule stability. Wang et al. [55] determined that nonsoluble β-polysaccharide formed of the outer shell of aerobic granules and provided granule strength. Conversely, McSwain et al. [56] and Zhang et al. [57] argued that a non-cellular protein core in aerobic granules provided stability. Thus, controversial information exists regarding the roles of different components of EPS in the structural stability of aerobic granules. Recently, Adav et al. [58] demonstrated that the selective enzymatic hydrolysis of proteins, lipids, and α-polysaccharides had a minimal effect on the three-dimensional structural integrity of granules. Conversely, selective hydrolysis of β-polysaccharides fragmented the granules. The β-polysaccharides were expected to form the backbone of a network-like outer layer with embedded proteins, lipids, α-polysaccharides, and cells to support the mechanical stability of granules.

Table 2 Strains for probing EPS in aerobic granules

Strains	N_S^a	Ref.
Syto 63, FITC, Con A	3	[56]
Con A-FITC	1	[55]
BacLight Live/Dead staining kit	2	[55]
FITC, calcofluor white, Syto 63,Syto Blue, Con A	5	[62]
FITC, calcofluor white, Syto 63,Syto Blue, Con A	5	[63]

N_S^a: Number of strain applied on the same aerobic granule

4.4.2 Distribution of EPS

Extensive work has been carried out to investigate the characteristics and distribution of EPS in flocs and biofilm, and various approaches have been used to determine the EPS content [59–61]. McSwain et al. [56] compared the EPS contents of aerobic granules with/without homogenization. It was observed that homogenization had a significant effect on the amount of EPS extracted from aerobic granules. Thus, they suggested that homogenization should be used when samples with vastly different shapes and sizes were compared. For the exploration of EPS spatial distribution in aerobic granules, McSwain et al. [56] used confocal laser scanning microscopy (CLSM) in situ to visualize the distribution of cells, polysaccharides, and proteins in granules (Table 2). Wang et al. [55] reported that the EPS content in the inner layer of aerobic granules was about four times greater than that in outer layer. Chen et al. [63, 64] described the distribution of EPS (proteins, α- and β-polysaccharides, and lipids) and cells (total and dead) in aerobic granules using a novel sixfold staining scheme and CLSM. In this work, lipids, proteins, and α- and β-polysaccharides in phenol-degrading granules were selectively hydrolyzed using specific enzymes.

4.5 *Cell Surface Hydrophobicity*

Cell surface hydrophobicity plays a key role in the self-immobilization and attachment of cells to a surface [65, 66]. In a thermodynamic sense, an increase in cell hydrophobicity simultaneously causes a decrease in the excess Gibbs energy of the surface, which promotes the self-aggregation of bacteria from liquid phase to form a new solid phase [66, 67], i.e., microbial granules. The hydrophobic binding force is of a prime importance for the cell-to-cell approach and interaction, and the cell hydrophobicity can act a driving force for the initiation of cell-to-cell aggregation, which is the first step towards aerobic granulation, and further keep the aggregated bacteria tightly together [68].

4.6 Rheology

Rheology is a powerful tool for characterizing the non-Newtonian properties of sludge suspensions, as it can quantify flow behaviors in real processes on a scientific basis [69]. Rheological parameters are very important in sludge management, not only as design parameters for transporting, storing, landfilling and spreading operations, but also as monitoring parameters in many biological treatment processes [70]. Su and Yu [6] characterized the non-Newtonian rheological behavior of aerobic granules and demonstrated that the granules containing liquor were shear thinning, and their rheological characteristics could be described using the Herschel–Buckley equation as follows:

$$\tau = \tau_{Y} + K\gamma^{m}, \tag{2}$$

where τ is shear stress, τ_Y is yield stress, γ is shear rate, and K and m are constants. The suspended solids concentration, pH, temperature, diameter, settling velocity, specific gravity, and sludge volume index all had an effect on the apparent viscosity of the mixed liquor of aerobic granules.

4.7 Permeability

Unlike microbial flocs and suspended cells, the dense aerobic granules may encounter problems associated with the limited diffusion of substrates into and metabolites out of the granules. The pore size distribution and permeability of aerobic granules are closely related to substrate and metabolites transport [71]. Therefore, their permeability feature has a significant influence on substrate and product transportation in granules [72]. The pore size distribution and available porous volume to which substrate can penetrate are the main factors governing the granule activity. Tay et al. [73] used CLSM to evaluate the porosity profiles. To evaluate quantitatively the porosity and permeability of aerobic granules, Zheng and Yu [71] used size-exclusion chromatography, in which polyethylene glycols and distilled water were, respectively, used as solute and mobile phase. The porosity of the aerobic granules varied from 68 to 93%. The EPS of the granules might clog the pores and might be responsible for the reduced porosity.

The porosity and permeability also influence the settling velocity of aerobic granules and the settling velocity of a particle depends greatly on its drag coefficient. Thus, Mu et al. [74] provided a reliable approach to evaluate the drag coefficient of the porous and permeable microbial granules. It was realized by taking the porosity and permeability of the granules into consideration. The drag coefficient of the microbial granules was found to be less than that of smooth rigid spheres and biofilm-covered particles. In addition, this study demonstrates that the drag coefficient of microbial granules depended heavily on their permeability and porosity. A fractal-cluster model was found to be able to predict the distribution of the primary particles in the microbial granules:

$$\Omega = \frac{2\xi^3 - \xi^2 \tanh \xi}{2\xi^3 + 2\xi - 3\tanh \xi} \quad (3)$$

where Ω is the reduction in the drag force, the dimensionless permeability of the aggregate; ζ, is calculated from the following equation:

$$\xi = \frac{d}{2\sqrt{\kappa}} \quad (4)$$

From Eq. (3) it can be seen that the parameter Ω is solely a function of the permeability factor, ζ.

4.8 Mass Transfer

Aerobic granules have a dense surface layer and compact interior core that can generate significant mass transfer resistance to oxygen and nutrient intake [75]. In biodegradation processes, substrates and DO are initially transferred from the bulk liquid to the external surface of aerobic granules before diffusing or being carried by an advective flow into the interior for biodegradation [62]. Oxygen diffusivity is commonly applied as an input parameter for modeling the transport processes in granules. Morgenroth et al. [1] assumed that the granule center is anaerobic, although the surrounding liquid has a relatively high DO content. Liu et al. [76] noted a decline in specific substrate removal rate from large granules, revealing the inhibition induced by mass transfer resistance. Oxygen limitation, if it occurred, produced anaerobic core in aerobic granules. Anaerobic bacteria were found to exist in the anaerobic core for large-sized aerobic granules [77].

Recently, the microelectrode method has been adopted to evaluate the microtransport processes in aerobic granules [78]. Chiu et al. [79] estimated the oxygen diffusivity in acetate- and phenol-grown aerobic granules by probing the DO level at the granule center using microelectrodes with a sudden change in the DO of the bulk liquid. The flow velocity across the granule was set such that the external mass transfer resistance was negligible. They also measured the steady-state and transient DO with step changes in surrounding DO levels at various depths in granules using microelectrodes [62]. A marked decrease in DO was also observed over this surface layer. No aerobic oxidation could occur beneath the active layer, indicating the oxygen transfer limitation. Liu et al. [75] successfully fabricated a novel microelectrode using photolithography to determine the DO distribution in aerobic

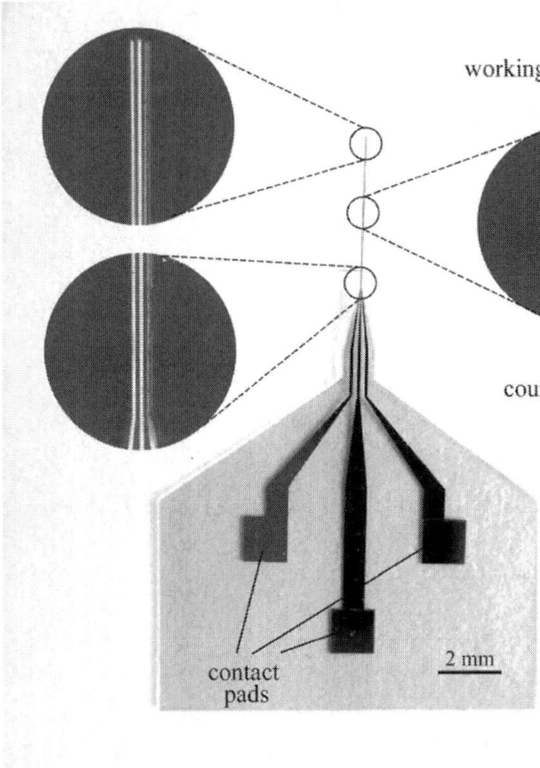

Fig. 2 Photograph of the innovative microelectrode and three magnified microscopic images of different sections of the needle [75]

nitrifying granules, as shown in Fig. 2. It was found that the main active part of the nitrifying granule was the upper 150-μm layer. Consequently, Liu et al. [76] manufactured a solid state microelectrode through electrochemical codeposition of Pt–Fe nanoparticles on a gold microelectrode fabricated using photolithography for in situ nitrite determination in aerobic nitrifying granules. At the outside of the granular surface, the nitrite concentration dropped rapidly because of its diffusion from the granular surface to the bulk solution.

5 Microbial Community Structure of Aerobic Granular Sludge

Recent development of modern molecular biology techniques is giving new insight into the structure and function of microbial communities in natural and engineered systems, including biological wastewater treatment reactors [80].

Microbial community structure analysis of aerobic granules, primarily based on 16S rRNA gene sequencing [26, 81], is becoming the most powerful tool to investigate the microbial populations present in aerobic granular reactors. The microbial diversity of aerobic granules is closely related to the composition of culture media, in which they are developed and structure of aerobic granules. Heterotrophic, nitrifying, denitrifying, P-accumulating bacteria, glycogen-accumulating bacteria, and fungi have been identified in aerobic granules developed under different conditions [10, 81–85]. Jiang et al. [84] isolated ten isolates from matured phenol-fed granule of which six have taxonomic affiliations with *β-Proteobacteria*, three with *Actinobacteria* and one with *γ-Proteobacteria*. Gram and Neisser stains and FISH analyses showed that most of the filamentous bacteria in aerobic granules cultivated in brewery wastewater belong to the genus *Thiothrix* or to *Sphaerotilus natans* [10]. Anaerobiosis and dead cells have been documented at the centers of aerobic granules [86]. The presence of anaerobic bacteria in aerobic granules is likely to result in the production of organic acids and gases within the granules.

The relative abundance and population dynamics of functional groups in aerobic granules have also been studied using a range of techniques: 16S rRNA-targeted oligonucleotide hybridization [26], polymerase chain reaction (PCR), and denaturing gradient gel electrophoresis (DGGE) analysis of 16S rRNA genes [84]. Community structure and function of aerobic granules is one of the most intriguing topics for researchers and several important descriptions have been recently reported to address the fundamental questions: which microbes are present, where are they located and what are they really doing there? The spatial distribution of microorganisms within aerobic granules has been visualized in thin-sections of granules using 16S rRNA-targeted fluorescence in situ hybridization (FISH) analysis with conventional fluorescence microscopy or CLSM, as shown in Table 3 [10, 25, 81–86]. The FISH–CLSM technique identified that obligate aerobic ammonium-oxidizing bacterium *Nitrosomonas* spp. was mainly at a depth of 70–100 μm from the granule surface, while anaerobic bacterium *Bacteroides* spp. was detected at a depth of 800–900 μm from the granule surface [87]. Lemaire et al. [85] reported that in granules >500 mm in diameter, *Accumulibacter* spp. was dominant in the outermost 200 mm region of the granule while *Competibacter* spp. dominated in the granule central zone, as shown in Fig. 3. The stratification of these two populations between the outer aerobic and inner anoxic part of the granule was highly significant ($P < 0.003$). They concluded that the GAO *Competibacter* spp., and not the PAO *Accumulibacter* spp., was responsible for denitrification in this SBR. This is undesirable for SNDPR as savings in carbon demand cannot be fulfilled with phosphorus removal and denitrification being achieved by different groups of bacteria. Thus, these techniques are bridging the 'gap' between engineers (reactor operation) and microbiologists (culture-based study) and lead to an interactive communication between them for optimizing microbial population and improving reactor performance [88, 89].

Table 3 16S rRNA targeted probes used for FISH detection of microorganisms in aerobic granules

Probe	Sequence 5′–3′	Specificity	Ref
EUK 516	ACCAGACTTGCCCTCC	Fungi	[10]
EUB338	GCTGCCTCCCGTAGGAGT	Eubacteria	[25]
NSO190	CGATCCCCTGCTTTTCTCC	Ammonia-oxidizing bacteria of the β subclass of the *Proteobacteria*	[25]
NIT3	CCTGTGCTCCATGCTCCG	*Nitrobacter* sp.	[25]
GAOQ 431	TCCCCGCCTAAAGGGCTT	Some Competibacter	[81]
GAOQ 989	TTCCCCGGATGTCAAGGC	Some Competibacter	[81]
PAO 462	CCGTCATCTACWCAGGGTATTAAC	Candidatus Accumulibacter phosphatis	[82]
PAO 651	CCCTCTGCCAAACTCCAG	Candidatus Accumulibacter phosphatis	[82]
NEU 653	CCCCTCTGCTGCACTCTATTCCATCCCCCTCTGCCG	*Nitrosomonas*	[83]
ARCH 915	GTGCTCCCCCGCCAATTCCT	Archaeal	[84]
PAO 846	GTTAGCTACGGCACTAAAAGG	Candidatus Accumulibacter phosphatis	[85]
Bacto 1080	GCACTTAAGCCGACACCT	*Bacteroides*, *Prevotella*, and *Porphyromonas* subgroups	[86]

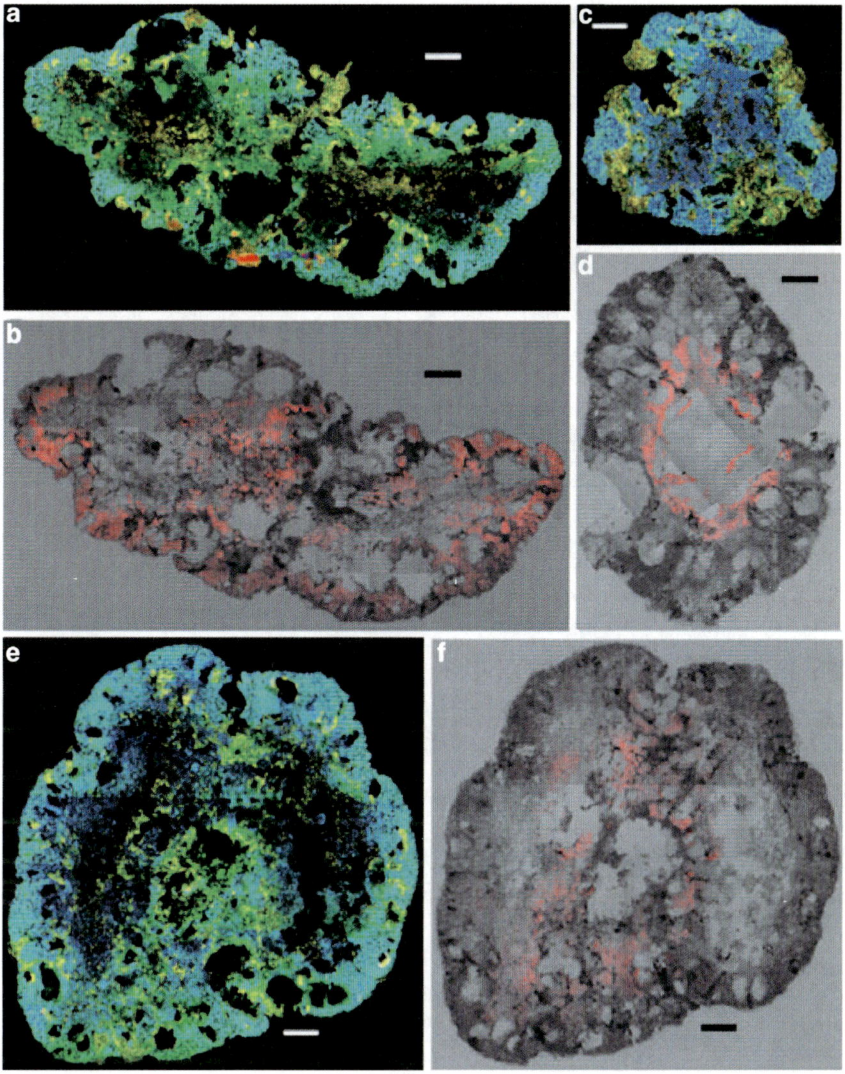

Fig. 3 Reconstructed CLSM images of FISH micrographs (**a**,**c**,**e**) and Nile Blue stain micrographs (**b**,**d**,**f**) of entire granule sections. In **a**,**c**,**e** *Accumulibacter* spp. cells are *cyan* (overlay of *blue* PAOmix and *green* EUBmix), *Competibacter* spp. Cells are *yellow* (overlay of *red* GAOmix and *green* EUBmix) and other bacteria are *green* (*green* EUBmix). In **b**,**d**,**f** overlays were of transmitted light images (*black and white*) and Nile Blue-stained PHAs in *red*. Subsequent granule sections (7 mm apart) of two different granules are presented in **a**,**b** and **e**,**f**. **c**,**d** are not images of the same granules. The granules were sampled at the end of the anaerobic (**a**–**c**) and at the end of the aerobic (**d**–**f**) periods. *Scale bar* = 100 mm [85]

6 Modeling of Aerobic Granular Sludge

Mathematical modeling has proven very useful to study complex processes, such as the aerobic granular sludge systems [3, 90]. Biological processes in aerobic granules are determined by concentration gradients of oxygen and diverse substrates. The substrate and DO concentration profiles are the result of many factors, e.g., diffusion coefficient, conversions rate, granule size, biomass spatial distribution, and density. All of these factors tightly influence each other; thus the effect of separate factors cannot be studied experimentally. Moreover, due to the long sludge age in granule systems (usually up to 70 days) [91], lengthy experiments should be carried out before a steady state is reached. Therefore, a good computational model for the granular sludge process provides significant insight into the most important factors that affect the nutrient removal rates and the distribution of different microbial populations in granules. Furthermore, models could also be used for process optimization and the scale-up and design of a full-scale aerobic granule reactor [90].

6.1 *Modeling Formation of Aerobic Granules*

Based on Monod equation, Yang et al. [92] developed a kinetic model that can describe the growth of aerobic granules developed at different conditions. Recently, Xavier et al. [93] introduced a multiscale model, which described the granular sludge reactor in details, from the metabolism of microbial groups, through the spatial structure of granules, to the dynamics of the reactor. With this model a preferential distribution of species along radial distances was shown, which were more heterogeneous than in strict layers. This heterogeneous structure and growth in microcolonies underlined the difficulty of representative microelectrode measurements. Mainly the outer layers of granules will be eroded, which contain less inert material. Therefore, aerobic granules are expected to contain more inert materials resulting from biomass decay than activated sludge.

6.2 *Modeling the Microbial Species in Aerobic Granules*

The architecture of aerobic granules is relevant to their microbial ecology, as they are macroscopic microbial consortia [83]. They consist of two main different microbial groups, i.e., autotrophic and heterotrophic bacteria [94, 95]. Both autotrophs and heterotrophs coexist and interact in aerobic granules. For the in-depth understanding of the growth of autotrophs and heterotrophs and their activity in aerobic granules, Ni et al. [96] developed a model to describe the simultaneous autotrophic and heterotrophic growth in granule-based SBR. A schematic drawing of the

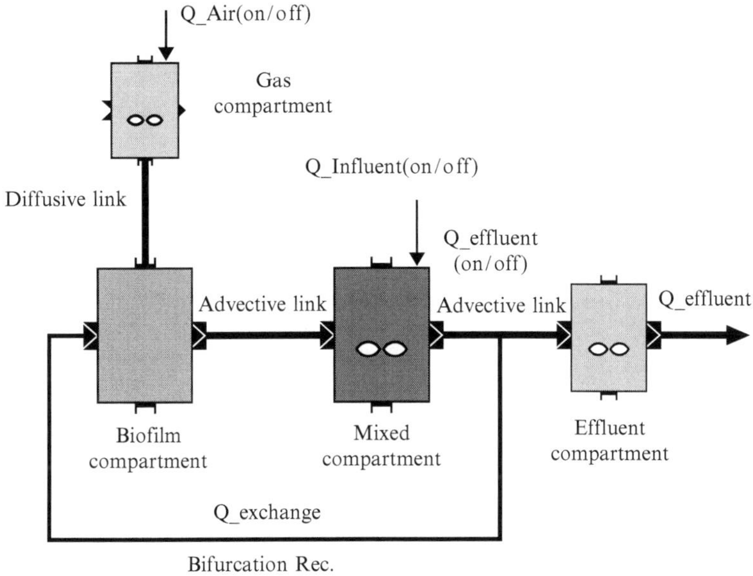

Fig. 4 Schematic diagram of the granule-based SBR as implemented in the AQUASIM: a biofilm compartment is connected to a completely mixed compartment with an advective link and a high recirculation flow rate, another completely mixed compartment is connected to the first one with an advective link to simulate the effluent draw, and aeration is simulated by a gas compartment connected with the biofilm one using a diffusive link. Biofilm compartment contains all soluble and particulate components of the biological model [96]

aerobic-granule-based SBR configuration simulated with AQUASIM is illustrated in Fig. 4. The model is able to simulate the reactor performance and get insight in autotrophic and heterotrophic growth in the granules. With the established model, the fractions of the active biomass (autotrophs and heterotrophs) and inert biomass are predicted to be 55.6 and 44.4% of the total mixed liquid volatile suspended solid, respectively, at a solids retention time (SRT) of 20 days. Biomass content increases with the increasing SRT, but active biomass ratio decreases. Autotrophs have no significant effect on the total biomass content, although they play an important role in nitrogen removal.

6.3 Modeling Nutrient Removal

De Kreuk modeled the substrate degradation kinetics of an aerobic granule SBR [90]. The effect of process parameters on the nutrient removal rates was evaluated in the model. The model is based on previously developed models.

The SBR and granular sludge descriptions are principally as in the model of Beun et al. [3], which described heterotrophic organisms storing acetate in a feast-famine regime in combination with autotrophic organisms for nitrogen removal. The existing aerobic granular sludge model was extended with conversion processes involving PAO as described in the models of continuous activated sludge systems by Hao et al. [97]. The simulation results demonstrated that the ratio between aerobic and anoxic volume in the granule strongly determines the N-removal efficiency as it was shown by model simulations with varying oxygen concentration, temperature, and granule size. The optimum granule diameter for maximum N- and P-removal in the standard case operating conditions (DO 2 mg L^{-1}, 20 °C) was found between 1.2 and 1.4 mm and the optimum COD loading rate was 1.9 kg COD m^{-3} day. When all ammonia is oxidized, oxygen diffuses to the core of the granule inhibiting the denitrification process. In order to optimize the process, anoxic phases can be implemented in the SBR-cycle configuration, leading to a more efficient overall N-removal. Phosphate removal efficiency mainly depends on the sludge age; if the SRT exceeds 30 days, not enough biomass is removed from the system to keep effluent phosphate concentrations low.

6.4 *Modeling the Storage Process in Aerobic Granules*

Aerobic granules in SBR are subjected to alternative feast and famine conditions, and are able to take up carbon substrate in wastewater rapidly and to store it as intracellular storage products when the substrate is in excess. This phenomenon could not be described by the widely used activated sludge model No.3 (ASM3). Taking adsorption process, microbial maintenance and substrate diffusion into account, Ni and Yu [98] established a new model to investigate the simultaneous growth and storage processes in aerobic granules cultivated with soybean-processing wastewater. Simulation results show that the approach is appropriate for elucidating the fates of major model components. Comparison between ASM3 and the model established in this work demonstrates that the latter is better to describe the substrate removal mechanisms and simultaneous growth and storage processes in aerobic granules.

For the future elucidation of the storage process in aerobic granules, Ni and Yu [99, 100] also developed a model to describe the storage and growth activities of denitrifiers in aerobic granules under anoxic conditions recently. In this model, mass transfer, hydrolysis, simultaneous anoxic storage and growth, anoxic maintenance, and endogenous decay are all taken into account. The modeling results explicitly show that the external substrate is immediately utilized for storage and growth at feast phase. More external substrates are diverted to storage process than the primary biomass production process. The model simulation indicates that the nitrate utilization rate (NUR) of granules based denitrification process includes four linear phases of nitrate reduction. Furthermore, the methodology for determining the most important parameter in this model, that is, anoxic reduction factor, is established.

6.5 Modeling Aerobic Granule-Based SBR

The aerobic-granule-based SBR is a complex biological system with numerous internal interactions among process variables and sludge characteristics. In addition to the biological reactions, mass transfer, hydrodynamic of reactors and the characteristics of granules have been proven to be influential to the overall performance of the aerobic granule-based SBR. Su and Yu [94, 95] established a generalized model to simulate an aerobic granule-based SBR with considerations of biological processes, reactor hydrodynamics, mass transfer, and diffusion. The ASM1 was modified and applied for describing the biological reactions. This model established could successfully simulate and predict the performance of an aerobic-granule-based SBR.

7 Application of Aerobic Granulation Technology in Biological Wastewater Treatment

7.1 Organic Wastewater Treatment

Granulation of activated sludge can lead to a high biomass retention in reactors because of its compact and dense structure. Biomass concentrations as high as 6.0–12.0 g L^{-1} have been obtained in SBRs operated with a volumetric exchange ratio of 50% [101]. Thus, aerobic granular sludge is advantageous for the wastewater treatment over the conventional activated sludge. A considerable amount of work has been conducted for the treatment of synthetic wastewater using aerobic granules, as shown in Table 1. The feasibility of applying aerobic granulation technology for the treatment of an industrial wastewater was demonstrated by Schwarzenbeck et al. [19], who examined the ability of aerobic granules to treat dairy wastewater. Aerobic granular sludge can be successfully cultivated in an SBR treating dairy wastewater. After complete granulation and the separation of biomass from the effluent, removal efficiencies of 90% total COD, 80% total nitrogen and 67% total phosphorus was achieved at a volumetric exchange ratio of 50% and a cycle duration of 8 h. Wang et al. [20] cultivated aerobic granules with brewery wastewater. After granulation, high and stable removal efficiencies of 88.7% COD and 88.9% NH_4^+-N were achieved at a volumetric exchange ratio of 50% and cycle duration of 6 h. The aerobic granules had smooth regular shapes, as shown in Fig. 5.

Fig. 5 Image of aerobic granules cultivated with real brewery wastewater from [20]

However, Schwarzenbeck et al. [19] and Arrojo et al. [83] found that a significant fraction of suspended solids was present in the reactor effluents. Thus, a post-treatment step is required.

7.2 Nutrient Removal

In the biological nutrient removal process, inorganic nitrogen in the form of ammonium is removed through aerobic, autotrophic nitrification followed by anoxic, heterotrophic denitrification. The process is normally split into two stages as the environmental conditions required for each process are very different. In the granular sludge system, the substrate is supplied in a short period of time. A high substrate concentration in the bulk liquid causes a penetration of substrate deeper into granules than oxygen. During the period in which the external substrate is available (feast period), oxygen is quickly consumed in the outer layers by growth, substrate storage and nitrification. Oxygen has only a limited penetration depth in this period. Acetate storage (as poly-β-hydroxybutyrate or PHB) and growth occur aerobically in the outermost layer or anoxically inside the granule (as indicated in Fig. 6) [46, 102]. During the period without external substrate (famine period), growth takes place on the internally stored PHB at a much lower growth rate [103]. The oxygen penetration depth during the famine period will be higher because of this decreased respiration rate of the heterotrophs, but will still be limited due to the increased nitrification relative to the feast phase [103]. The nitrate produced can be simultaneously denitrified inside the granules using the

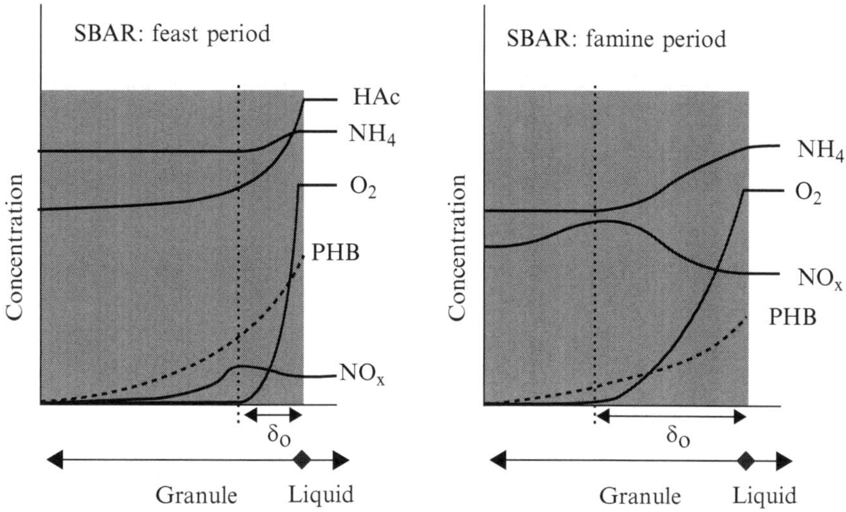

Fig. 6 Schematic profiles of substrate inside an SBR (feast and famine periods; external mass transfer neglected) from [46]

stored PHB as electron donor. Tsuneda et al. [25] investigated the granulation of nitrifying bacteria in an aerobic upflow fluidized bed reactor for the treatment of an inorganic wastewater containing 500 mg L^{-1} of NH_4^+–N. The ammonia removal rate reached 1.5 kg N m^{-3} day, suggesting that the nitrifying granules were able achieve a stable and efficient nitrification.

Phosphorus can be removed biologically through biological phosphorus removal processes, exploiting the ability of PAOs to take up P in excess of metabolic requirements and accumulate it intracellularly as polyphosphate. Lin et al. [104] cultivated phosphorus-accumulating granules at different substrate P/COD ratios in a range of 1/100 to 10/100 by weight in SBRs. The soluble COD and PO_4–P profiles showed that the granules had typical P-accumulating characteristics, with a concomitant uptake of soluble organic carbon and the release of phosphate in the anaerobic stage, followed by rapid phosphate uptake in the aerobic stage.

Recent work on aerobic granules suggests that granules could be beneficial for simultaneous nitrification, denitrification and phosphorus removal (SNDPR). In granules there is oxygen mass transfer limitation that is important in facilitating aerobic/anoxic zones required for SNDPR. Lemaire et al. [85] cultivated aerobic granules for SNDPR and observed that simultaneous phosphorus removal and nitrification occurred throughout the SBR operation. Good phosphorus removal and nitrification occurred throughout the SBR with a dominancy of *Accumulibacter* spp. (PAO) and *Competibacter* spp. (GAO). *Accumulibacter* spp. was dominant in the outermost 200 μm region of the granule while *Competibacter* spp. dominated in the granule central zone. Yilmas et al. [105] investigated the biological removal of nitrogen and phosphorus from nutrient-rich abattoir wastewater using granular sludge. It is demonstrated that the granules could be sustained and indeed further developed with the use of abattoir wastewater. The organic, nitrogen and phosphorus loading rates applied were 2.7 kg COD m^{-3} day, 0.43 kgN m^{-3} day and 0.06 kgP m^{-3} day, respectively. The removal efficiency of soluble COD, soluble nitrogen and soluble phosphorus were 85, 93 and 89%, respectively. However, the high suspended solids in the effluent limited the overall removal efficiency to 68, 86 and 74% for total COD, total nitrogen (TN) and total phosphorus (TP), respectively. *Accumulibacter* spp. was found to be responsible for most of the denitrification, further reducing the COD requirement for nitrogen and phosphorus removal. Mineral precipitation was evaluated and was found not to contribute significantly to the overall nutrient removal. Thus, compared to conventional sludge flocs, aerobic granules are relatively large, compact, dense microbial aggregates of different bacterial species with an approximately spherical external appearance. Their size and dense, compact structure are expected to contribute positively to the oxygen mass transfer limitation required for reliable SNDPR systems. In addition, the excellent settling ability of granular sludge allows for more biomass to be maintained in a relatively small reactor volume, enhancing the ability of the reactor to withstand high loading rates. This is of great interest for the treatment of nutrient-rich wastewaters.

7.3 Domestic Wastewater Treatment

For aerobic granules, Tay et al. [106] reported that it was difficult to form granules with an OLR lower than 2 kg COD m^{-3} day. It is expected that a low-strength wastewater will result in the difficulties of granulation or long start-up of aerobic granular system from activated sludge. Thus, the possibility of forming aerobic granules on domestic sewage in an SBR should be explored. de Kreuk and van Loosdrecht [21] studied aerobic granulation with presettled sewage as influent. After 20 days of operation at a high COD loading (about 2 g COD L^{-1} day with a 2 h cycling time), heterogeneous aerobic granules were formed, with an SVI after 10 min settling of 38 mL g^{-1} and an average diameter of 1.1 mm. Utilization of a high COD load was found to be critical to the formation of aerobic granules on low-strength domestic wastewater. Therefore, a short cycle time is preferred to form granules in an SBR when domestic wastewater is used as the influent.

7.4 Xenobiotic Contaminants Biodegradation

The aggregation of microorganisms into compact aerobic granules also has additional benefits, such as protection against predation and resistance to chemical toxicity

[84]. Cell self-immobilization creates a diffusional resistance and establishes a concentration gradient that shelters the microbial cells beneath the protective barrier by diluting the toxic chemical below some threshold value to allow continued microbial activity and substrate utilization [107]. Tay et al. [107] and Jiang et al. [108] investigated the feasibility of treating phenol-containing wastewater with aerobic granular sludge. The granules displayed an excellent ability to degrade phenol [108]. Adav et al. [109] reported that aerobic granules could degrade phenol at 1.18 g phenol/g VSS day. Jiang et al. [8, 84, 110, 111] isolated ten bacterial strains from aerobic phenol-degrading granules and identified their potential for degrading phenol. The PG-01 strain, a member of α-*proteobacteria*, is common in granules and is the predominant strain in phenol degradation. Another strain affiliated with β-*proteobacteria*, PG-08, has minimal phenol degradation capability, and a high propensity for self-aggregation. Adav et al. [109] isolated yeast strain, *Candida tropicalis* from their phenol degrading granules and reported as a functionally dominant strain in the phenol degrading granules. Hence, different strains on aerobic granules may have specific roles in granule structural integrity and phenol degradation.

The cultivated phenol-fed aerobic granules were also able to degrade pyridine. The granules completely degraded 250–1,500 mg L^{-1} pyridine at a constant rate with no time lag, and with 12 and 15 h time lag at 2,000 and 2,500 mg L^{-1} pyridine concentration, respectively [112]. Yi et al. [16] developed aerobic granules for cometobolic transformation of *p*-nitrophenol with glucose as a cosubstrate. It was found that glucose could promote aerobic granulation and facilitate the biodegradation of recalcitrant xenobiotics. Nancharaiah et al. [113] successfully cultivated aerobic granules for the biodegradation of nitrilotriacetic acid. Development of aerobic granules for the biological degradation of 2,4-dichlorophenol (2,4-DCP) in an SBR was reported by Wang et al. [17]. After operation of 39 days, stable granules with a diameter range of 1–2 mm and a clearly defined shape and appearance were obtained. After granulation, the effluent 2,4-DCP and COD concentrations were 4.8 and 41 mg L^{-1}, with high removal efficiencies of 94 and 95%, respectively. Specific 2,4-DCP biodegradation rates in the granules followed the Haldane model for substrate inhibition, and peaked at 39.6 mg 2,4-DCP g/VSS h at a 2,4-DCP concentration of 105 mg L^{-1}. Efficient degradation of 2,4-DCP, nitrilotriacetic acid and phenol by the aerobic granules suggests their potential application in the treatment of industrial wastewaters containing chlorophenols and other xenobiotic contaminants.

7.5 *Biological PHB Production*

Because of the dense, compact structure and high biomass retention, aerobic granular sludge system could be compact and has a very high volumetric loading rate. Wang and Yu [114] cultivated PHB-rich aerobic granules through adopting a two-step strategy for the treatment coupled with PHB production in an SBR. After granulation, the

PHB content in the aerobic granules exceeded 40%, which significantly improved the volumetric PHB production rate compared to activated sludge flocs.

8 Summary

Based on this review of the literature, the current state of the art with regard to aerobic granulation technology can be summarized as follows.

Aerobic granulation is a very promising technology from an engineering as well as economic point of view and offers an attractive alternative for the treatment of different types of wastewater. This granular system allows, in many cases, a more stable operation, and the treatment of larger loads, removal of multiple toxic pollutants, inferior volumes for the settling systems and production of better quality effluents than the conventional systems. The aspects related to aerobic granules, including the factors affecting their formation, physicochemical characteristics, microbial structure and diversity, modeling of granule-based systems, and practical applications of this technology have been presented.

9 Future Directions

Over the years, our knowledge on aerobic granulation has significantly improved, largely due to the high number of valuable studies from various disciplines. However, unanswered questions about this technology still remain in relation to some issues. Work in the following areas is necessary in order to gain a greater understanding of aerobic granulation technology.

Although aerobic granulation technology is a compact reactor technology, it still has limitations, such as a relative high SS concentration in the effluent. Thus, further research is needed to investigate the feasibility of being coupled to a post-treatment unit.

For further improvements in aerobic granulation biotechnology, we need to expand our knowledge of microbial communities and population dynamics related to their function in aerobic granules. Linking the microbial population with the operational factors that influence aerobic granules is likely to be the key issue to resolve. To acquire this knowledge, detailed molecular surveys of microbial communities in aerobic granules are required.

The factors affecting the formation of aerobic granules have been extensively explored. However, the mechanisms behind are still unclear. The formation mechanisms of aerobic granules that quorum-sensing effect of microorganisms might be involved in are to be elucidated.

Acknowledgements The authors wish to thank the Natural Science Foundation (NSFC) of China (Grants. 20577048, 50625825 and 50738006) for the partial support of this study.

References

1. Morgenroth E, Sherden T, Van Loosdrecht MCM, Heijnen JJ, Wilderer PA (1997) Water Res 31:3191.
2. Beun JJ, Hendriks A, Van Loosdrecht MCM, Morgenroth E, Wilderer PA, Heijnen JJ (1999) Water Res 33:2283
3. Beun JJ, Van Loosdrecht MCM, Heijnen JJ (2001) Biotechnol Bioeng 75:82
4. Tay JH, Liu QS, Liu Y (2001) J Appl Microbiol l91:68
5. Beun JJ, van Loosdrecht MCM, Heijnen JJ (2002) Water Res 36:702
6. Su KZ, Yu HQ (2005) Environ Sci Technol 39:2818
7. Adav SS, Chen MY, Lee DJ, Ren NQ (2007) Chemosphere 67:1566
8. Jiang HL, Tay JH, Maszenan AM, Tay STL (2006) Environ Sci Technol 40:6137
9. Tay STL, Zhuang WQ, Tay JH (2005) Environ Sci Technol 39:5774
10. Weber SD, Ludwig W, Schleifer KH, Fried J (2007) Appl Env Micr 73:6233
11. Liu Y, Tay JH (2004) Biotechnol Adv 22:533
12. De Kreuk MK, De Bruin LMM, Van Loosdrecht MCM (2005) Aerobic granular sludge, S. Bathe, MK De Kreuk, BS McSwain, N Schwarzenbeck, eds., IWA, London, 111.
13. De Kreuk MK, Kishida N, Van Loosdrecht MCM (2007) Water Sci Technol 55:75
14. Zheng YM, Yu HQ, Sheng GP (2005) Process Biochem 2005:645
15. Jiang HL, Tay JH, Tay STL (2002) Letters in Applied Microbiology 35:439
16. Yi S, Zhuang WQ, Wu B, Tay STL, Tay JH (2006) Envion Sci. Technol 40:239.
17. Wang SG, Liu XW, Zhang HY, Gong WX, Sun XF, Gao BY (2007) Chemosphere 69:769
18. Zhu L, Xu XY, Zheng Y (2005) Acta Scientiae Circumstantiae 25:1148 (in Chinese)
19. Schwarzenbeck N, Borges JM, Wilderer PA (2005) Appl Microbiol Biotechnol 66:711.
20. Wang SG, Liu XW, Gong WX, Gao, BY, Zhang DH, Yu HQ (2007) Bioresource Technol 98:2142
21. De Kreuk MK Van Loosdrecht MCM (2006) J Environ Eng 132:694
22. Cassidy DP, Belia E (2005) Water Res 39:4817
23. Gaval G, Pernell JJ (2003) Water Res 37:1991
24. Martins AMP, Heijnen JJ, Van Loosdrecht MCM (2003) Appl. Microbiol. Biotechnol. 62:586.
25. Tsuneda S, Nagano T, Hoshino T, Ejiri Y, Noda N, Hirata A (2003) Water Res 37:4965.
26. Tsuneda S, Ogiwara M, Ejiri Y, Hirata A (2006) Water Sci. Technol 53(3):147
27. Moy BYP, Tay JH, Toh SK, Liu Y, Tay STL (2002) Letters in Applied Microbiology 34:407.
28. Zheng YM, Yu HQ, Liu SJ, Liu XZ (2006) Chemosphere 63:1791.
29. Hu L, Wang J, Wen X, Qian Y (2005) Process Biochemistry 40:5.
30. Liu Y, Tay JH (2002) Water Res 36:1653.
31. Beun JJ, van Loosdrecht MCM, Heijnen JJ (2000) Water Sci Tech 41:41
32. Liu YQ, Tay JH (2007) Biochem Eng J 34:1
33. Chen Y, Jiang W, Liang DT, Tay JH (2007) Appl. Microbiol. Biotechnol. 76:1199
34. Adav SS, Lee DJ, Lai JY (2007) Appl. Microbiol. Biotechnol. 77:175
35. Ramasmy P, Zhang X (2005) Water Sci. Technol 52(7):217.
36. Pan S, Tay JH, He YX, Tay STL (2004) Letters in Applied Microbiology 38:158
37. Liu YQ, Tay JH (2007) Enzyme Microb. Technol 41:516
38. Meyer RL, Saunders AM, Zeng RJ, Keller J, Blackall LL (2003) Fems Ecology 45:253
39. Liu YQ, Wu WW, Tay JH, Wang JL (2007) Appl Microbiol Biotechnol 76:211
40. Sanin SL, Sanin FD, Bryers JD (2003) Process Biochem 38:909
41. McSwain BS, Irvine RL, Wilderer PA (2004) Water Sci Technol 49:19
42. Li ZH, Kuba T, Kusuda T (2006) Enzyme and Microbial Technology 38: 670
43. Liu YQ, Tay JH (2007) Appl Microbiol Biotechnol 75:205
44. Yang SF, Tay JH, Liu Y (2005) J Environ. Eng 131:86
45. Peng DC, Bernet N, Delgenes JP, Moletta R (1999) Water Res. 33:890

46. Mosquera-Corral A, De Kreuk MK, Heijnen JJ, Van Loosdrecht MCM (2005) Water Research 39:2676
47. De Kreuk MK, Pronk M, Van Loosdrecht MCM (2005) Water Research 39:4476
48. Yang SF, Li XY, Yu HQ (2007) Process Biochem. 43:8
49. Liu Y, Wang ZW, Liu YQ, Qin L, Tay JH (2005) Biotechnol. Prog. 21: 621
50. Zheng YM (2006) PhD Dissertation Hefei, China. University of Science & Technology of China
51. Wingender J, Neu T R, and Flemming HC 1999 Microbial extracellular polymeric substances: characterization, structure, and function. Springer, Berlin, Germany.
52. Liu YQ, Liu Y, Tay JH (2004) Appl Microbiol Biotechnol 65:143
53. Tay JH, Liu QS, Liu Y (2001) Letters in Applied Microbiology 33:222
54. De Kreuk MK, Van Loosdrecht MCM (2004) Water Sci. Technol. 49:9.
55. Wang ZW, Liu Y, Tay JH (2005) Appl Microbiol Biotechnol. 69:469
56. McSwain BS, Irvine RL, Hausner M, Wilderer PA (2005) Appl Environ Microbiol 71:1051
57. Zhang L, Feng X, Zhu N, Chen J (2007) Enzyme and Microbial Technology 41:551
58. Adav SS, Lee DJ, Tay JH (2007) Water Res. 42:1641
59. Frølund B, Palmgren R, Keiding K, Nielsen PH (1996) Water Res 30:1749.
60. Wilén BM, Jin B, Lant P (2003) Water Res 37:2127
61. Sheng GP, Yu HQ (2007) Appl. Microbiol. Biotechnol. 74:208
62. Chiu ZC, Chen MY, Lee DJ, Wang CH, Lai JY (2007) Water Res 41:884
63. Chen MY, Lee DJ, Tay JH (2007) Appl Microbiol Biotechnol 73:1463
64. Chen MY, Lee DJ, Tay JH, Show KY (2007) Appl Microbiol Biotechnol 75: 467
65. Azeredo J, Visser J, Oliveari R (1999) Colloids and Interfaces B Biointerfaces 14:141
66. Liu XM, Sheng GP, Yu HQ (2007) Environ Sci Technol 41:4620
67. Liu Y, Yang SF, Qin L, Tay JH (2004) Appl Microbiol Biotechnol 64:410
68. Liu Y, Yang SF, Tay JH, Liu QS, Qin L, Li Y (2004) Enzyme and Microbial Technology 34:37
69. Dentel SK (1997) Water Sci Technol 36(11):1
70. Mu Y, Yu HQ (2006) Water Res 40:3596
71. Zheng YM and Yu HQ (2007) Water Res. 41:39.
72. Alphenaar PA, Perez MC, Willem JH, van Berkel GL, Lettinga G (1992) Appl. Microbiol. Biotechnol. 36:795
73. Tay JH, Tay STL, IVanov V, Pan S, Jiang HL, Liu QS (2003) Lett. Appl. Microbial. 36:297
74. Mu Y, Ren TT, Yu HQ (2008) Environ. Sci. Technol 42:1718
75. Liu SY, Liu G, Tian YC, Chen YP, Yu HQ, Fang F (2007) Environ. Sci. Technol. 41:5447
76. Liu SY, Chen YP, Fang F, Liu SH, Ni BJ, Liu G, Tian YC, Xiong Y, Yu HQ (2008) Environ. Sci. Technol. 42: 4467
77. Liu YQ, Liu Y, Tay JH (2005) Lett. Appl. Microbiol. 40:312
78. Ivanov V, Tay JH, Tay STL, Jiang HL (2004) Water Sci. Technol. 50(12):147
79. Chiu ZC, Chen MY, Lee DJ, Tay STL, Tay JH, Show KY (2006) Biotechnol. Bioeng. 94:505
80. Hof-man J, Zheng D, Westermann P, Ahring BK, Raskin L (2003) Adv. Biochem. Eng. Biot. 81:151
81. Meyer RL, Saunders AM, Zeng RJ, Keller J, Blackall LL (2003) FEMs Ecology. 45:253
82. Kishida N, Kim J, Tsuneda S, Sudo R (2006) Water Res. 40:2303
83. Arrojo B, Mosquera-Corral A, Garrido JM, Mendez R (2004) Water Res. 38:3389
84. Jiang HL, Tay JH, Maszenana M, Tay STL (2004) Applied and Environmental Microbiology 70:6767
85. Lemaire R, Yuan Z, Blackall LL, Crocetti GR (2008) Environmental Microbiology 10:354
86. Tay STL, Ivanov V, Yi S, Zhuang WQ, Tay JH (2002) Microb. Ecol. 44:278
87. Adav SS, Lee DJ, Tay JH (2007) Environ Technol 28:1227
88. Sekiguchi Y, Kamagata Y, Harada H (2001) Current Opinion in Biotechnology 12:277
89. Yuan Z, Blackall LL (2002) Water Res. 36:482
90. De Kreuk MK, Picioreanu C, Hosseini M, Xavier JB, Van Loosdrecht MCM (2007) Biotechnol. Bioeng. 97:801
91. De Kreuk MK, Van Loosdrecht MCM (2004) Water. Sci. Technol. 49(11–12):9
92. Yang SF, Liu QS, Tay JH, Liu Y (2004) Letters in Applied Microbiology 38:106

93. Xavier JB, de Kreuk MK, Picioreanu C, van Loosdrecht MCM (2007) Environ. Sci. Technol. 41:6410
94. Su KZ, Yu HQ (2006) Environ. Sci. Technol. 40:4703
95. Su KZ, Yu HQ (2006) Environ. Sci. Technol. 40:4709
96. Ni BJ, Yu HQ, Sun YJ (2008) Water Res 42:1325
97. Hao X, van Loosdrecht MCM, Meijer SCF, Heijnen JJ, Qian Y. (2001) J Environ Eng 127:112
98. Ni BJ, Yu HQ (2008) Biotechnol. Bioeng. 100: 664
99. Ni BJ, Yu HQ (2008) Biotechnol. Bioeng. 99:314
100. Ni BJ, Yu HQ, Xie WM (2008) Biotechnol. Bioeng. 99:324
101. Tay JH, Liu QS, Liu Y (2002) Water Sci Technol 46:13
102. Qin L, Liu Y, Tay JH (2005) Water Res 39:1503
103. Third KA, Burnett N, Cord-Ruwisch R (2003) Biotechnol Bioeng 83:706
104. Lin YM, Liu Y, Tay JH (2003) Appl Microbiol Biotechnol 62:430
105. Yilmaz G, Lemaire R, Keller J, Yuan Z (2008) Biotechnol. Bioeng. 100:529
106. Tay JH, Pan S, He YX, Tay STL (2004) J. Environ. Eng. 130:1094
107. Tay JH, Jiang HL, Tay STL (2004) J Environ Eng 130:1415
108. Jiang HL, Tay, JH, Tay STL (2002) Lett. Appl Micro 35:439
109. Adav SS, Chen MY, Lee DJ, Ren NQ (2007) Biotechnol Bioeng 96:844
110. Jiang HL, Tay STL, Maszenan AM, Tay JH (2006) FEMS Microbiol Ecol 57:182
111. Jiang HL, Maszenan AM, Tay JH (2007) Appl Microbiol Biotechnol 75:1191
112. Adav SS, Lee DJ, Ren NQ (2007) Water Res 41:2903
113. Nancharaiah YV, Schwarzenbeck N, Mohan TVK, Narasimhan SV, Wilderer PA (2006) Water Res 40:1539
114. Wang J, Yu HQ (2006) Water Sci Tech: Water Supply 6(6):81

Index